Molecular mechanisms of drug action

Molecular mechanisms of drug action

by
Christopher J. Coulson
Glaxo Group Research, UK

Taylor & Francis
London · New York · Philadelphia
1988

UK   Taylor & Francis Ltd, 4 John St., London WC1N 2ET

USA   Taylor & Francis Inc., 242 Cherry St., Philadelphia,
PA 19106–1906

© Taylor & Francis Ltd 1988

British Library Cataloguing in Publication Data

Coulson, Christopher J.
Molecular mechanisms of drug action.
1. Pharmacology
I. Title
615'.7   RM300

ISBN 0–85066–376–8
ISBN 0–85066–396–2 Pbk

Library of Congress Cataloging-in-Publication Data

Coulson, Christopher J.
   Molecular mechanisms of drug action.
   Includes bibliographies and index.
   1. Molecular pharmacology.   I. Title.   [DNLM:
1. Drugs—metabolism.   2. Molecular Biology.
QV 38 C855m]
RM301.65.C66   1987   615'.7   87–18023
ISBN 0–85066–376–8
ISBN 0–85066–396–2 (pbk.)

Cover design by Russell Beach
Typeset by Katerprint Typesetting Services, Oxford
Printed in Great Britain by Taylor & Francis (Printers) Ltd,
Basingstoke, Hants.

Contents

To Annette

Preface

As my experience is that of an industrial biochemist who has worked in the pharmaceutical industry for over 20 years, this book is written from a practical standpoint. It includes the mechanism of drug action, encompassing drugs for infectious disease, as well as for 'endogenous conditions' such as cancer, arthritis, heart disease, schizophrenia etc. It is written from the point of view of targets — whether enzymes in pathways, receptors or ion channels in membranes.

The alternative procedure, where disease is made the central focus, has been used in a number of other publications and so I have approached the subject in a different way which was better able to link the practical aspects of drug targetting with the biochemical and pharmacological background. My aim was to bring together our present state of knowledge with respect to drug mechanisms (information that is otherwise scattered throughout the literature) in a readable and accessible form, which could assist both those teaching the subject, and students who wish to study it. I felt that there was a gap between academia and industry which such a book might help to close.

The introductory chapter is concerned with the basic principles that cover enzyme inhibition and receptor binding by drugs. The rest of the book is divided into two sections; the next seven chapters deal with drugs that modulate biochemical pathways, both of synthesis and breakdown, while the last four are concerned with organizational structures of the cell. Chapters 2 to 5 are concerned with the biosynthesis of DNA, protein, carbohydrate and cell-walls. Much of the chemotherapy of cancer, viruses and bacteria is to be found within these chapters. Next are two chapters on lipid biosynthesis; one covering the pathways of sterol and steroid interconversions and the second on the various pathways that lead from arachidonic acid. Although not a pathway, I have next included a discussion on inhibitors of zinc metalloenzymes as they form a coherent group.

In the second section the agonists and antagonists at neurotransmitter receptors are discussed first, followed by those agents that interfere directly with membranes. In these two chapters I have deliberately covered membrane enzymes side by side with ion channels and receptors. Microtubule ligands form the subject of the 11th chapter while hormone modulators are discussed in Chapter 12. In the appendix I discuss the basis for the measurement of binding constants for ligand/macromolecule binding.

Some readers may feel that there are notable lacunae in the coverage of pharmaceuticals. This has occurred because I have tried to cover those drugs that show recognized principles in their mode of action; some have not been included because their mode of action is not known and the work done on them cannot be related in a coherent fashion in a textbook of this sort. No attempt has been made to cover all drugs. Others are included, even though their mode of action is not fully understood, if their development illustrates a useful point. The use of cromoglycate in asthma is such a case. Other drugs of interest have been discussed if they have been designed particularly for a given condition and may be approaching the market — although not yet fully launched.

I have not tried to cover drug delivery systems, metabolism, side-effects etc. except where these are germane to the mechanism. The quantity of material available in these areas is so great as to have increased the length of the book to unmanageable proportions if it had been included. Furthermore, the amount of coverage does not relate to market value or quantity of drug sold, but rather to scientific interest, so that a widely used drug whose mode of action is clearly defined may not warrant a long coverage, while an interesting but less used drug may have a greater coverage.

The book is intended for third, and possibly second, year students studying subjects or modules in microbiology, pharmacology, biochemistry, pharmacy and medicine. It may also be useful for medicinal scientists in the drug industry who are changing fields and need a quick entrée into a fresh area. Comprehensive reviews have been listed, if available, to help the reader to follow up points of interest and to go more deeply into the subject. These are largely in accessible journals and the occasional book.

I am deeply indebted to Charles Ashford, John Foreman, Ian Kitchen, Hugh White, Alan Wiseman, Helen Wiseman and David Williams for reading all or part of the manuscript and who made many useful suggestions for improvement and corrections.

Chapter 1

General principles

1.1 Background

It would have been impossible to write a book of any length detailing mechanisms of drug action more than 20 years ago, because very little was known at that time. Penicillin, for example, had been on the market for 20 years, but we were still far from detailing its precise mode of action. The philosophical attitude, prevalent in those days and still common now, was that drugs were just used because they worked: enquiry into their mode of action was deemed unnecessary. The fact that, at that time, our detailed understanding of biological processes was very limited, may not be unconnected. Many of us entering the drug industry in the 1960s hoped, however, to be able to design drugs more effectively on the basis of molecular structure, whether it be of the enzyme or the receptor.

There have been a number of major advances in the intervening years. It was, after all, only 16 years ago that the connection between aspirin and prostaglandin biosynthesis was realized by Vane (1971), which led in its turn to a much greater understanding of the biosynthesis of prostanoids. In fact drugs have frequently provided researchers in the medical sciences with tools to unravel a 'knotty' scientific and/or medical problem.

1.2 Do drugs have a specific mode of action?

Perhaps one question that should be considered is whether drugs do have a singular mechanism of action that can be reduced to molecular terminology. The concept of a drug as a 'magic bullet' that can penetrate through a myriad of biochemical systems, acting at only one specific site to have the desired effect, is perhaps reductionist to the point of absurdity. Nevertheless, in practical terms the history of drug development suggests that, in the initial analysis, it is possible to view the process in this fashion. The basis for this assertion lies in the linking of enzyme inhibition, receptor binding or ion channel blocking with expected effects on substrate levels or events inside and outside the cell. These must lead to organ effects or death of parasitic micro-organisms, and improvement of the patient's condition. These factors must all be consistent with the proposed mode of action.

Nevertheless drugs do act at other sites — molecular, cellular and/or organ — besides the intended one more often than not. As a consequence, they show undesirable side-effects. Indeed, the process of developing a drug is intended to reduce undesirable activities to an acceptable level. Unless the aspect of the structure of the drug molecule that gives rise to the activity also gives rise to the side-effect, it is usually possible to reduce the undesirable activity by making minor changes to the structure.

In the last analysis the crucial factor is the therapeutic ratio. This is the ratio between the dose required to treat the condition and the dose which gives rise to unacceptable side-effects. There are occasions when the side-effect is apparently random (rare but often lethal) as for example the development of aplastic anaemia associated with some drugs, notably chloramphenicol (see Chapter 3). Difficult decisions then have to be made weighing up the possible risk to the patient against the benefit.

Another factor that enters the calculation here is the condition for which the drug is prescribed. A drug for a condition that is frequently lethal (such as cancer) can be tolerated with a lower therapeutic ratio than one intended for reducing blood pressure. For the viral diseases there are very few drugs; permission might therefore be given to use a drug with a low safety margin if there was no other drug available for a certain condition.

We must not forget that in the end a patient has to take this drug. The side-effects, therefore, need to be minimal and not of such a nature as not to be distressing in any way. At present, our need is for effective, non-toxic drugs to treat a variety of conditions in a preventive or palliative manner.

1.3 Basic processes

Drugs can be used to help the body reject an invading pathogenic organism (whether parasite, fungus, bacterium or virus) or to modify some aspects of the metabolism of the body that is functioning abnormally. In the former case, a drug is normally used which is toxic to the pathogenic organism but not to the host, and stimulation of the body's normal processes for combating invaders may play a part. In the latter case, the approach is usually more subtle with a modulation of a process as the requirement.

In all cases, the drug can have biological activity only by interacting with the molecules of the target organ or organism. These molecules are usually proteins — enzymes that catalyse reactions essential for the functioning of the organism or proteins called receptors which transmit signals by interacting with messenger molecules such as hormones and neurotransmitters. The drug has its effect by binding to either an active site or a secondary site that influences the active site on the enzyme or receptor and thereby prevents access by the normal substrate or ligand (inhibitor or antagonist) or it provokes a signal where none was wanted (substrate or agonist).

The interaction of the drug with the protein molecule can be measured in terms of the strength of inhibition of an enzyme reaction or the strength of drug binding to a receptor. The former may be quantified as I_{50} or IC_{50}, the concentration of inhibitor required to reduce the rate of a reaction or the binding of a ligand by half. This, however, varies with the amount of substrate or ligand available to the enzyme or receptor thereby making comparisons between data obtained under different conditions almost impossible.

With more data, a binding constant for ligand attaching to receptor or inhibitor constant for enzyme reaction may be determined. These quantities are usually expressed in terms of the dissociation constant of the ligand–receptor complex (K_d) or enzyme–inhibitor complex (K_i). For the ligand–receptor interaction this constant is equal to the ligand concentration at which 50% of the receptors are occupied by ligand and it is assumed, although this is not always correct, that the measured response has fallen to half its maximum value. For typical drug–receptor interactions, the dissociation constants are of the order of 10^{-7} to 10^{-10} M.

1.4 Drug binding to enzymes

Inhibitors can bind to enzymes in several ways. They may bind either reversibly or irreversibly, and with either the active site or with another part of the enzyme. In the latter case the binding can cause a conformational change in the enzyme, thus producing distortion of the active site. Different effects of inhibitor concentration upon the kinetics of the enzyme in these various cases allow evidence to be obtained on the type of binding involved in any particular situation.

Reversible inhibition can occur in various ways including competitive, non-competitive and uncompetitive binding. If the inhibitor binds to the enzyme alone, the inhibition is said to be competitive. In this case, the inhibitor constant (K_i) is defined as the dissociation constant of the enzyme/inhibitor complex (see the Appendix for a more detailed discussion of K_i). In principle, increasing the substrate concentration can eventually restore the reaction rate to what it was in the absence of inhibitor. Frequently, the substrate competes with the inhibitor for the substrate binding site, but there are cases in which competitive inhibition is observed although substrate and inhibitor bind to distinct sites.

In non-competitive inhibition the inhibitor can bind to the enzyme/substrate complex as well as to the enzyme itself, at a site separate from the substrate binding site. In this case, increasing the substrate concentration may reduce but cannot eliminate the inhibition. Uncompetitive inhibition is found when the inhibitor will only bind to the enzyme/substrate complex.

The binding constants noted above are, in principle, independent of ligand or substrate concentration and so results obtained in different laboratories can be compared with some confidence. They do not, however, necessarily bear any relation to the I_{50} values mentioned above. (See Appendix 1 for a discussion.)

The links between drug and target are usually of a reversible nature and so the drug will eventually be released from the complex. The complex is usually formed instantaneously, although this can take many hours if the binding constant is of the order of 10^{-9} M or smaller. In some cases, however, the drug and target will react covalently, and this is often characterized by time-dependent inhibition, and by lack of enzyme recovery after dialysis. If the ligand is bound reversibly, it dissociates during dialysis and enzyme activity is recovered. If irreversible inhibition has occurred, the activity is not recoverable; instead synthesis of fresh protein will be needed to restore the original level of the target. Some reversible inhibitors whose binding is extremely tight may cycle on and off the enzyme so slowly that for all practical purposes the enzyme is inactivated by their presence.

Tight binding inhibitors therefore need a considerable time for preincubation for the inhibition to reach its maximum effect. They do not satisfy the normal kinetic criteria of enzyme reactions in that the equilibrium is not set up rapidly and the concentrations of inhibitor and substrate are not very much higher than that of the enzyme. In fact, the inhibitor and enzyme concentrations are often very similar and in some cases stoichiometric or one-to-one. Under these conditions the Lineweaver-Burk plots (double reciprocal plots of reaction velocity against substrate concentration) will have both curved and linear portions while the progress curves will be aptly named because, in effect, a portion of the enzyme is being put gradually out of action (Morrison, 1969, 1982). An example is methotrexate binding to dihydrofolate reductase with a K_d of 10^{-11} M (Chapter 2).

Irreversible inhibition is particularly useful if the step in a pathway chosen as the target for drug action is not the rate-limiting one. Reversible inhibitors are likely to be ineffective in these situations because substrate levels increase and, even with non-competitive inhibitors, the inhibition is reduced. Most drugs are administered on at least a daily basis. Irreversible inhibitors, however, probably do not need to be administered daily since it may take several days for enough enzyme to be re-synthesized to return to steady state levels. A dosage regime in which the drug is only given on alternate days is likely to be more useful than one requiring daily dosage. Otherwise the effect of the drug is likely to become magnified to the point of producing side-effects. A case in point is the use of monoamine oxidase inhibitors (see Chapter 9) which were given on a daily basis, and eventually lowered the level of the enzyme in the stomach wall to such a level that tyramine and dopamine ingested from food were not deaminated and caused hypertensive crises that were sometimes lethal. This was known as the 'cheese effect' because cheese contains a considerable amount of tyramine.

A special type of irreversible inhibition occurs when the drug is chemically unreactive but the enzyme converts it to a highly reactive species that inactivates the enzyme. This has been given the name of mechanism-based irreversible inhibition and the drugs are called suicide substrates or K_{cat} inhibitors because they require the catalytic activity of the enzyme. Examples of this include allopurinol inhibiting xanthine oxidase and 5-fluorodeoxyuridylate inhibiting thymidylate synthetase (both in Chapter 2), clavulanate inhibiting β-lactamase (Chapter 5), and the monoamine oxidase inhibitors (Chapter 9). For a fuller discussion see Walsh (1982).

Another class of enzyme inhibitors is described as transition state inhibitors. Every reaction proceeds through an intermediate state in which the structure attached to the enzyme is neither substrate nor product but an intermediate form. The transition state is the structure of highest energy on the pathway from substrate to product. As enzymes catalyse a specific reaction, compounds that interfere with the transition state are likely to be highly selective in their action. One example of a transition state inhibitor is 2-deoxycoformycin which inhibits adenosine deaminase (see Chapter 2) (Lienhard, 1973).

An inhibitor does not always bind to the active site of an enzyme. It may modulate the enzyme activity by binding to another part of the protein, and is called an allosteric effector — allosteric is derived from Greek words meaning 'other shape'. Allosteric effectors are thus distinguished from those competitive inhibitors which resemble the substrate in shape and are termed isosteric. Inhibition of this type is also known as negative cooperativity, whereas activators that bind allosterically give rise to positive cooperativity. An enzyme that behaves in this way is normally composed of more than one subunit. Ribonucleotide reductase is a case in point where the effectors are the purine and pyrimidine trinulcleotides (Chapter 2).

1.5 Drug binding to receptors

It is important to be clear about what constitutes a receptor. For the binding of a ligand to a receptor to be a genuine physiological phenomenon and not just non-specific binding, the ligand binding should:

(a) be saturable, in which case a plot of amount of drug bound against drug concentration will level off and reach a plateau;
(b) be high affinity (binding constants less than 10^{-6} M) and low capacity for the binding to be regarded as specific;
(c) obey the Law of Mass Action (see Appendix);
(d) be linked to a pharmacological response characteristic of the particular ligand.

If these criteria are not fulfilled, the data may indicate, fortuitously, the existence of a binding protein for a particular ligand but not a receptor with physiological significance. It is also advisable, if possible, to have a series of agonists and antagonists, preferably of different structural types, that are specific for a given receptor, and can thus define its role.

An agonist is an agent that interacts with the receptor to produce a clearly defined change in the target cell. An antagonist also binds to the receptor, but lacks the structural features essential to initiate any further changes and so can block the action of an agonist. In a cell-free system, therefore, both agonist and antagonist will bind, but it will be impossible to distinguish between them without data on activity on a cellular or tissue basis.

The measurement of the reversible binding of a radioactive ligand to a receptor preparation (radioligand binding) has greatly increased our understanding of receptors, and, provided it is understood that the binding occasionally may have to be discounted because it is an artefact of the preparation, the knowledge gained has been of great value in identifying, and in some cases, isolating the receptor. One of the greatest challenges in modern pharmacology is (a) to link the binding data with pharmacological effect and (b) to understand how in molecular terms the signal is transduced in a given cell to produce a given response (see Chapter 9).

1.6 Further considerations

The attachment of a drug to its target is only part of the process that leads to an effective drug. Antibacterial drugs have to be able to reach the bacterium in sufficient quantity to be able to kill it (bactericidal action). In some cases the drug merely prevents the bacterium from growing and does not kill it (bacteristatic action), but the host may still recover from the infection by the antibacterial action of the immune system. Similarly, some antifungal drugs merely stop fungal growth, but here the situation is more serious because fungal infections frequently occur in individuals with impaired immune systems, for example in cancer patients or in patients who are receiving immune suppressant drugs after an organ transplant. It is therefore most important to have fungicidal drugs.

The activity of antibacterial and antifungal drugs is usually measured by the minimum concentration at which the drug completely inhibits the growth of the microorganism. This concentration is known as the minimum inhibitory concentration (M.I.C. for short).

Furthermore, absorption, serum binding and metabolism must all be taken into account in developing an effective drug. The drug, if given by mouth, must be able to cross the stomach wall, reach the bloodstream, and be transported in sufficient quantity to the site of action. Serum proteins may play a part in this process — notably albumin, which is particularly effective at carrying drugs that exist in anionic form at pH 7·4, such as the non-steroidal anti-inflammatory drugs (see Chapter 7). Albumin binding can however, if too strong, prevent a compound that shows good activity *in vitro* from demonstrating efficacy *in vivo* (clorobiocin activity against gram-negative bacteria is a case in point — see Chapter 2). Other compounds may fail to show activity *in vivo* because they are metabolized to inactive metabolites.

1.7 Partition coefficient

A partition coefficient is a measure of how a solute distributes itself between two immiscible solvents. Often denoted P and expressed as the logarithm (log P) they are particularly useful in describing the potential of a drug for reaching the brain. The brain (and the spinal cord) is surrounded by a fatty sheath and a considerable degree of lipid partitioning or lipophilicity (i.e. preferential transfer into lipid, not just solubility) is required for the drug to cross into the brain. Molecular weight and charge are also of importance here. This concept of blood/brain transport is of crucial importance if a drug is expected to act within the brain, as with an antipsychotic agent, or to handle infections in the membrane that surrounds the brain and spinal cord (meninges) such as meningitis. Conversely, it is essential to avoid side-effects of a mental nature such as drowsiness or hallucinations when the drug is intended to act elsewhere in the body.

The partition coefficient is measured by allowing the compound to come to equilibrium, assisted by shaking, between an aqueous and an organic medium, usually *n*-octanol or hexane. The quantity of material in each phase at equilibrium is determined by a physical method such as ultraviolet absorption, and the ratio between the concentration in the lipid phase and that in the aqueous phase is known as the partition coefficient.

The partition coefficient as measured above views the compound as an uncharged species. The distribution coefficient (D) is a measure of how a drug distributes itself between aqueous and lipid phases at a given pH, usually close to neutrality, which may be more relevant conceptually to biologists since the blood plasma and intracellular pH are normally about pH 7·4. This is particularly relevant to molecules which are capable of ionization since the charged species is less likely to be able to enter a lipophilic membrane (see for example the local anaesthetics discussed in Chapter 10). The aqueous medium chosen for the measurement is often phosphate buffer at pH 7·4.

For ionizable drugs, the tendency to ionize is measured in terms of a dissociation coefficient K, which, like hydrogen ion concentration, is often expressed as its negative logarithm: $-\log K$ is called pK. If the pK of the drug is within one log unit of the pH at which the measurement is made, more than 5% of the compound will be in the charged form. The pK is identical to the pH at which equal numbers of the drug molecules are charged and uncharged. pK_a refers to the molecule ionizing as an acid (releasing a proton) while pK_b is the similar position for a base (accepting a proton). Usually only the non-ionized form of the drug will move into the lipid phase.

It should be noted that we are here concerned with passive transport across membranes. Active transport is a completely different phenomenon that requires the expenditure of energy, and also is saturable and inhibitable in a similar fashion to enzyme reactions. There is an active transport system into the brain for the aromatic amino acids, for example. There is often confusion in the literature about which is intended.

An interesting use of the partition coefficient alone to rationalize binding of ligands to macromolecules has been made in the case of various families of drugs and other compounds binding to serum albumin. This transport protein carries tryptophan, fatty acids and other naturally occurring materials around the body, and also a number of foreign molecules (xenobiotics) including drugs. The binding of members of a closely related group of compounds, such as penicillins, is directly proportional to log P; no other parameter appears to be necessary to explain the binding. This is because the same factors of hydrophobic versus hydrophilic interaction govern the albumin binding as determine the partition coefficient. Sulphonamide binding, on the other hand, may be defined by a term representing the pK_a of the sulphonamide side-chain as well as lipophilicity (see Jusko and Gretch, 1976 for a review).

Other attempts to rationalize the activity of families of related structures on biological systems (structure/activity analyses) have been made using various physical constants including log P, together with molecular size, electronic effects in aromatic rings and shape of substituents. Ligands may bind to macromolecules by ionic, hydrogen or covalent bonds or by hydrophobic or van der Waals interactions. The intention of this analysis is to make predictions for new, more active, structures and also to reduce toxicities. A detailed discussion on this subject is beyond the scope of this book but is available in the review by Hansch (1985).

1.8 Drug nomenclature

Drug nomenclature is often rather confusing so it may be helpful to explain the different names that may attach to a drug and how they were derived.

(a) The detailed chemical name (according to IUPAC or other systematic nomenclature).

(b) The name by which the drug is often known (the generic name) which is usually a shortening of the chemical name, and is usually the name approved by drug regulatory authorities. These can often vary from country to country.

(c) The manufacturer's patented trade-name by which the drug is sold, which is usually short and 'punchy' so that it is easily remembered by hard-pressed general practitioners. There may be several of these if different drug companies are marketing the same product, possibly in differing formulations.

(d) There is also a number given by a drug company to any compound, synthesized by the organic chemists or extracted from a fermentation culture medium, that is usually of the form of an initial letter or letters followed by a number which may extend to six digits. Usually when the decision is taken to progress the compound towards the market, it is given a generic name. Earlier publications may have the number only, and it may be necessary to correlate name and number from later publications.

An example of these names is provided by the following structure of a drug which is used for the treatment of infections caused by anaerobic micro-organisms (see Chapter 4).

(a) The detailed chemical name is 1-(2'-hydroxyethyl)-2-methyl-5-nitroimidazole.

(b) The generic name is metronidazole. We can see how this name was arrived at from the detailed chemical name with the relevant letters in capitals:

1-(2'-hydroxyethyl)-2METhyl-5-NitROimIDAZOLE (there has been a slight change in the order to make a more harmonious sound).

(c) The trade-name of metronidazole is Flagyl.

(d) The original company number was 8823 RP (short for Rhône-Poulenc).

Metronidazole

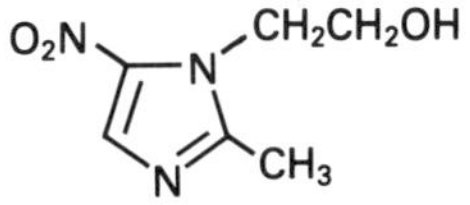

Generic names are not always entirely based on the chemical name but in some cases the use of the drug is taken into account. The anti-herpes virus drug originally numbered BW 248U, was subsequently named acyclovir, although it might have been given the generic name acycloguanosine since guanosine is the purine base in the structure (Chapter 2). Zovirax is the tradename.

Acyclovir

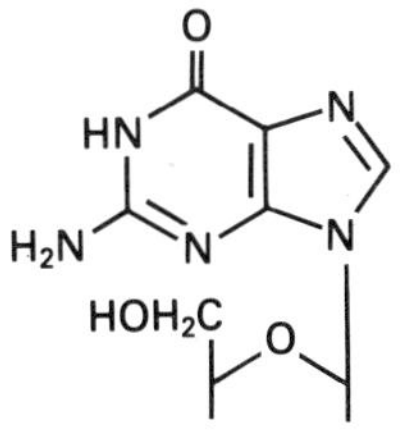

Throughout this book the generic name of the drug is used, except in one or two instances when the best known tradename and generic name are both quoted.

1.9 The future

It may be that we are standing on the threshold of an explosion of biological knowledge which will allow us to treat the major diseases which are outstanding today. Such a leap forward is obviously to be desired. One of the most important messages that this book is intended to convey is that most of the major diseases are still, despite the advances of recent years, neither fully understood, nor curable.

Thus most of the drugs for cancer are nearly as toxic to the host as to the tumour. The inflammation of arthritis may be suppressed but the underlying disease process still cannot be halted and indeed is poorly understood. In the case of mental illness we treat schizophrenics with drugs that largely sedate the patient and do not cure, although they may partially arrest the course of the disease. In spite of our hard-won success in dealing with bacterial disease we still have to be eternally vigilant (a) to cope with mutants that are resistant to many antibiotics and (b) to devise new drugs to kill fungi colonizing the areas rendered vacant by the destruction of bacteria. Nature does, indeed, abhor a vacuum!

References

Hansch, C. (1984–5). *Drug Metab. Rev.*, **15**, 1279–1294.
Jusko, W.J. and Gretch, M. (1976). *Drug Metab. Rev.*, **5**, 43–140.
Lienhard, G.E. (1973). *Science*, **180**, 149–154.
Morrison, J.F. (1969). *Biochim. Biophys. Acta*, **185**, 269–286.
Morrison, J.F. (1982). *Trends in Biochem. Sci.*, **7**, 102–105.
Vane, J.R. (1971). *Nature New Biology*, **231**, 232–236.
Walsh, C. (1982). *Tetrahedron*, **38**, 871–909.
Wolfenden, R. (1976). *Adv. Biophy. Bioeng.*, 271–306.

Chapter 2

Nucleic acid biosynthesis and catabolism

2.1 Introduction

As nucleic acids are a fundamental part of any organism and have to be synthesized early on in the growth cycle, it is not surprising that many of the drugs in clinical use for the chemotherapy of neoplastic disease and microbial infection have the enzymes of nucleic acid synthesis as their primary target. A number of enzymes involved in viral and bacterial replication are either unique to that organism, or are sufficiently different from the host enzyme which catalyses a similar reaction for that difference to be exploited.

Cancer cells are characterized by rapid, uncontrolled growth; there do not seem to be any major differences between tumour enzymes and those from normal cells. It is, therefore, the rate of synthesis of DNA that forms the target for chemotherapeutic agents. Consequently, toxicity problems do arise by virtue of inhibition of these enzymes in rapidly growing non-tumour tissue — hence the practice of delivering antitumour drugs in 'cocktails' of three or four drugs at a time, working on different pathways, so that the toxicity of any one drug does not become overwhelming. An outline scheme of the drugs and their targets discussed in this chapter is presented in Table 2.1.

Another difference between microbial infection and cancer lies in the efficacy of the immune system. Usually with bacterial and viral infection we rely on the immune system to assist in, and eventually take over, the healing process. In cancer, and also in some fungal infections, the immune system is weakened and, therefore, a complete kill of the tumour or fungal cells is required; drugs that merely arrest the growth of these cells are frequently not sufficient to cure the disease.

The type of interference with enzymes that we find in this area is not only the straightforward competitive inhibition of the rate-determining step in a pathway — although that may occur. Other approaches include the concept of false substrates whereby one enzyme accepts a foreign substrate and yields a product that can react with another enzyme; in the cancer field this is known as the anti-metabolite approach. Furthermore, the synthesis of a macro-molecular product allows the possibility of incorporation of false substrate

Table 2.1 Drugs discussed in Chapter 2

Target	Drug	Disease
2.2 Nucleotide biosynthesis		
Dihydropteroate synthase	Sulphonamides	Bacterial infection
Dihydrofolate reductase	Trimethoprim	Bacterial infection
	Pyrimethamine	Malaria
	Methotrexate	Cancer
Ribonucleotide reductase	Hydroxyurea	Cancer
Orotate phosphoribosyltrans-		
ferase (thymidylate synthetase)	5-Fluorouracil	Cancer
Cytosine deaminase		
(thymidylate synthetase)	5-Fluorocytosine	Candidosis
Inosine monophosphate		
dehydrogenase	Ribavirin	Influenza
2.3 DNA biosynthesis		
Thymidine kinase	{ Acyclovir	Herpesvirus infection
(viral DNA polymerase)	{ Vidarabine	Herpesvirus infection
Deoxycytidine kinase	Cytarabine	Cancer
(mammalian DNA polymerase)		
DNA gyrase	Ciprofloxacin	Bacterial infection
	Norfloxacin	Bacterial infection
	Novobiocin	Bacterial infection
DNA topoisomerase II*	Doxorubicin	Cancer
DNA polymerase (RNA primed),		
reverse transcriptase	Azidothymidine	Infection by human immunodeficiency virus (AIDS)
2.4 Nucleotide catabolism		
Adenosine deaminase	2-Deoxycoformycin	Cancer
Xanthine oxidase	Allopurinol	Gout
Hypoxanthine phosphoribosyl		
transferase	Allopurinol	Leishmaniasis

Where two enzymes are recorded as targets, the first enzyme uses the drug as a false substrate while the enzyme in brackets is the ultimate target.

* This assignation remains controversial.

into the polymer chain, resulting in either chain termination or the production of a faulty macromolecule.

We also find the occasional occurrence of a suicide substrate (allopurinol, see Section 2.4.2) and transition state analogues (2-deoxycoformycin, see Section 2.4.1) which have aroused considerable interest (see Chapter 1 for an outline of these concepts). The recent concentration on the widespread screening of secondary microbial metabolites (i.e. those compounds formed that are not on the normal pathways required for cell growth) and of the products of organic chemical synthesis, has greatly increased the scope of chemical structures that are under investigation.

2.1.1 Overall scheme

Nucleic acids in mammalian organisms are constructed from nucleotides which in turn are made up from a purine or pyrimidine base, a sugar (ribose in ribonucleic acid or deoxyribose in deoxyribonucleic acid) and a phosphate group attached to the 5′-position of the ribose ring (see Figure 2.1). The synthesis of these moieties may be *de novo* (see Figure 2.2) — i.e. constructed from small molecular building blocks — or via salvage pathways which effectively recycle a pre-formed base. Purines are synthesized by attaching an amino group to a ribose ring and then using this as the anchor from which to build up an imidazole ring to yield 5-aminoimidazole-4-carboxamide ribotide (AICAR).

One more step leads to the formation of inosine monophosphate (the first purine nucleotide) from which the pathways diverge to yield guanosine and adenosine monophosphates. The monophosphates have a second and third phosphate group added to give the triphosphates that are required for RNA biosynthesis. The dinucleotides may be reduced to deoxynucleotides, which after conversion to the trinucleotides, are used for DNA biosynthesis.

Folic acid analogues are essential cofactors for the transfer of some one-carbon units. The particular conversions of interest in the nucleotide biosynthetic pathways are the addition of a formyl (-CHO) group to glycinamide

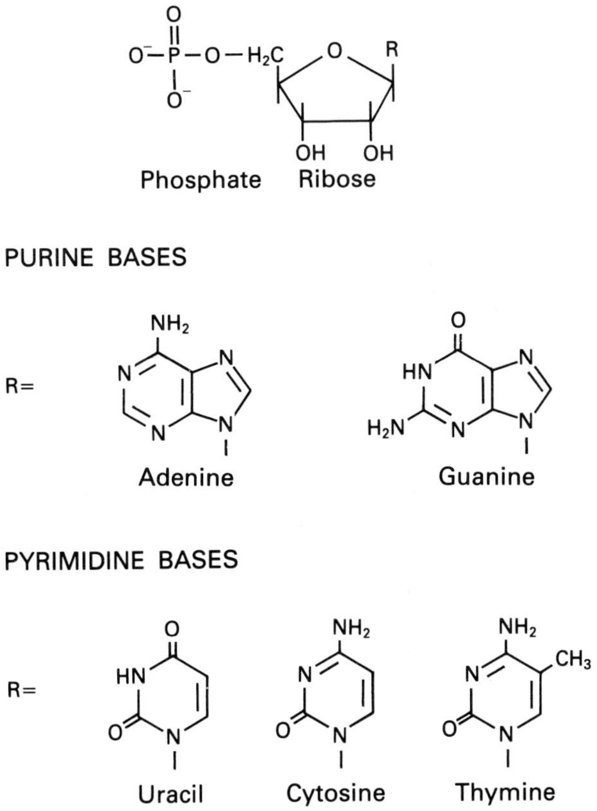

Figure 2.1 Nucleotide structure.

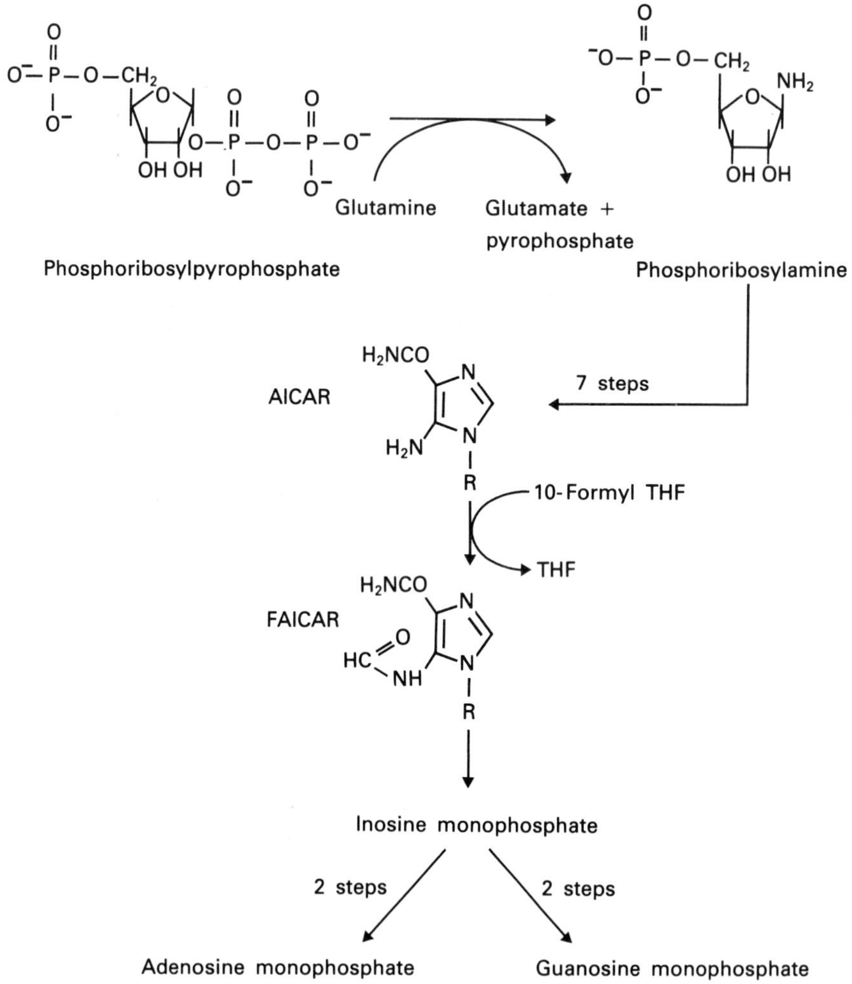

R is ribose 5′-phosphate; AICAR is 5-aminoimidazole-4-carboxamide ribotide; FAICAR is 5-formamidoimidazole-4-carboxamide ribotide.

Figure 2.2 Outline of purine nucleotide biosynthesis.

ribotide and subsequently to AICAR by 10-formyltetrahydrofolate in the purine pathway, and the methylation of deoxy-UMP to deoxy-TMP catalysed by thymidylate synthetase in pyrimidine biosynthesis. The first step in the biosynthesis of folate is the formation of dihydropteroate from *p*-aminobenzoate and 2-amino-4-hydroxy-6-methylpteridine (see Figure 2.4). This step is the target for the sulphonamide antibacterials (hence the build-up of AICAR in cells treated with sulphonamides — see below).

The heterocyclic ring of pyrimidines, on the other hand, is constructed before the ribose is attached (see Figure 2.3). Indeed, in mammals the first

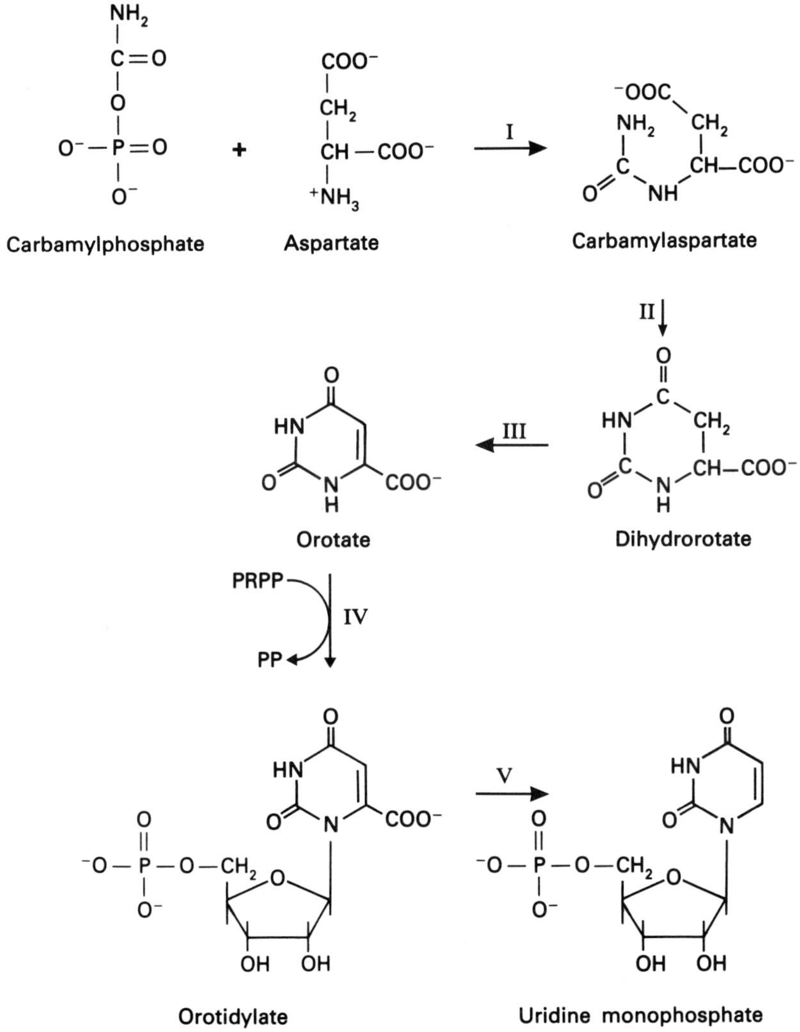

Figure 2.3 Pyrimidine nucleotide biosynthesis.

I: Aspartate transcarbamylase
II: Dihydro-orotase
III: Orotate dehydrogenase
IV: Orotidylate pyrophosphorylase
PRPP: Phosphoribosylpyrophosphate
PP: Pyrophosphate

three steps to yield dihydro-orotate are all catalysed by the same protein molecule, although the activities are on separate proteins in bacteria. Orotate is converted to the first pyrimidine nucleotide, orotidine 5-phosphate, which loses a carboxyl group to form uridine 5-monophosphate. The addition of two more phosphate groups forms uridine triphosphate from which cytidine triphosphate can be made. Thymidine monophosphate is formed by the methylation of uridine monophosphate catalysed by thymidylate synthetase. Again, as with the purine nucleotides, it is the ribo*di*nucleotides which are reduced to the deoxy form.

The action of DNA polymerases, using the four deoxyribotrinucleotides, produces DNA. RNA is synthesized from a DNA template in a process (transcription) requiring the four ribotrinucleotides, and messenger and transfer RNA are employed to generate protein on the ribosome (translation — see Chapter 3). Exceptions to this sequence are found with viruses that contain RNA as the genetic code. Two examples are the influenza virus, where the RNA acts as its own template, and the virus that produces the disease known as acquired immune deficiency syndrome (AIDS), where the RNA acts to form DNA, catalysed by a viral enzyme known as reverse transcriptase (see Section 2.3.5).

Purine nucleotides are first broken down to nucleosides by removal of the phosphate group and ribose groups to produce inosine. Subsequent changes include oxidative steps to hypoxanthine, xanthine and uric acid (Figure 2.8). Pyrimidine nucleotides are also converted to the free bases, uracil and 5-methyluracil), that eventually yield malonate (from cytosine) and 2-methyl malonate (from thymine).

2.2 Nucleotide biosyntheses — enzyme targets of drugs

2.2.1 Dihydropteroate synthetase

One family of drugs that were recognized early to have antibacterial properties was the sulphonamides. They have been in medical use since the 1940s — indeed Winston Churchill was given one of the earliest compounds, M & B 693, in North Africa in February 1943 when he had an attack of pneumonia, and it is likely that it played a major role in his recovery (Churchill, 1951). These drugs interfere with the purine biosynthetic pathway; 5-aminoimidazole-4-carboxamide ribotide (AICAR) accumulates in bacterial cells treated with sulphonamides. The mechanism of interference is indirect in that AICAR accumulates not because the next enzyme in the pathway is inhibited, but because there is insufficient folate cofactor to service the enzyme. Mammals, however, require preformed folate in the diet as an essential vitamin, and so inhibition of this step is not likely to lead to mammalian toxicity.

The biosynthesis of folate requires *p*-aminobenzoic acid as a starting material, and sulphonamides are sufficiently close in structure to act not only as inhibitors of dihydropteroate synthetase but also as false substrates (reviewed in Woods, 1962) (see Figure 2.4). Any agent that can penetrate a bacterium which needs a functioning enzyme is very likely to have activity against that bacterium.

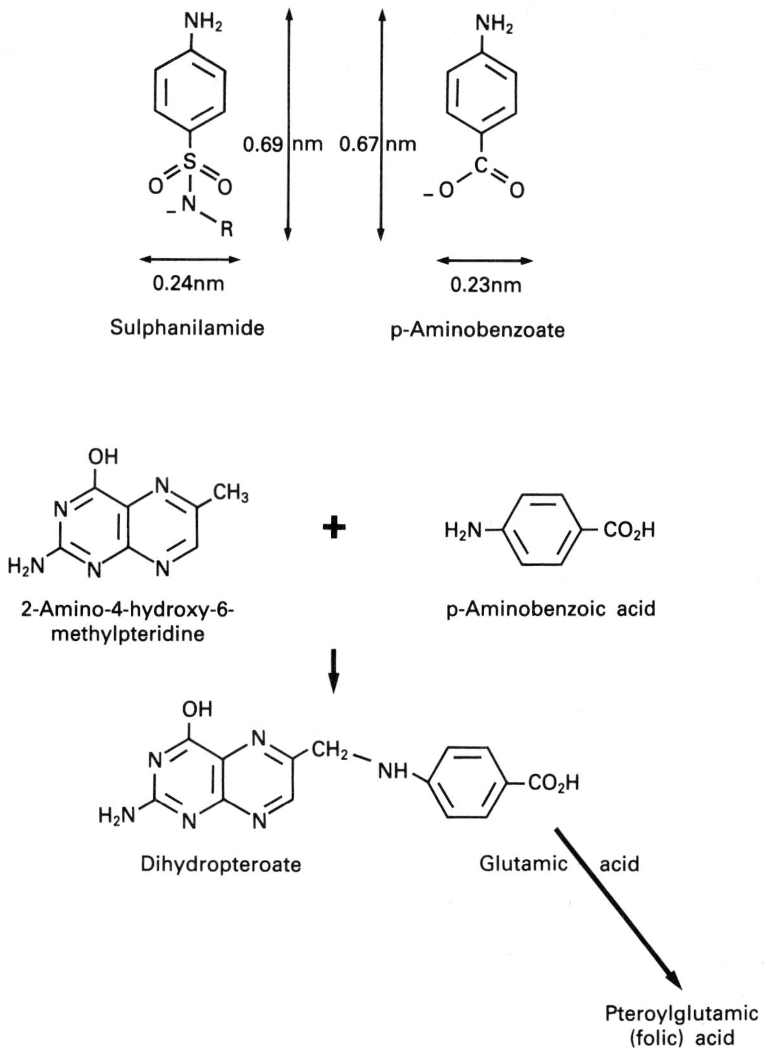

Sulphanilamide   p-Aminobenzoate

2-Amino-4-hydroxy-6-
methylpteridine

p-Aminobenzoic acid

Dihydropteroate   Glutamic   acid

Pteroylglutamic
(folic) acid

The conversion of 2 amino-4-hydroxy-6-methylpteridine to dihydropteroate and to folic acid is shown.
A typical sulphonamide is compared with *p*-aminobenzoate and is shown to be structurally similar.

Figure 2.4 Dihydropteroate synthetase reaction

Sulphonamides are not usually used alone to treat infections in man nowadays, although they may be used in conjunction with inhibitors of dihydrofolate reductase (see below), as there are many other more effective families of antibacterials now available. An exception to this is dapsone, which is sometimes used alone for the treatment of leprosy (Shepard *et al.*, 1976).

2.2.2 Dihydrofolate reductase (DHFR)

The reaction carried out by this enzyme is crucial to the recycling of the folate cofactors involved in methylation of the DNA bases:

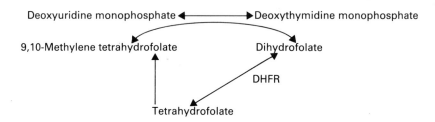

Dihydrofolate reductase has turned out to be an effective target for both anti-microbial and anticancer drugs, possibly because one folate molecule has to be reduced for every deoxythymidine monophosphate molecule synthesized, and in situations of rapid growth this requirement could be considerable. The mechanism of thymidylate synthetase requires that the one-carbon fragment is transferred as a methylene group to the 5-position of the uracil ring of deoxyUMP. This is then reduced to a methyl group by the pteridine ring of the coenzyme to yield dihydrofolate. Dihydrofolate reductase is required to reform tetrahydrofolate.

Unlike the case of dihydropteroate synthase above, mammals carry out the same interconversion, and so selectivity becomes of prime importance. In this instance trimethoprim, used as an antibacterial agent, and pyrimethamine, an effective agent for killing the protozoal parasites (plasmodia) that cause malaria, are far less inhibitory towards a mammalian liver dihydrofolate reductase than for the enzyme from their respective microbial targets. This is shown by the following figures (the effect of the drugs on the parasite, *Plasmodium berghei*, that infects mice is shown for comparison):

Concentration for 50% inhibition of DHFR ($\times 10^{-9}$ M)

	Rat liver	E. coli	P. berghei
Pyrimethamine	700	2500	0·5
Trimethoprim	260 000	5	70

(from Ferone *et al.*, 1969)

DHFR has been confirmed as the target for trimethoprim by the identification of an altered, and largely uninhibited, DHFR in resistant organisms (reviewed in Hitchings and Smith, 1980).

These agents are able to penetrate their target organisms by non-ionic diffusion. Methotrexate, a diaminopyrimidine which is used for choriocarcinoma and trophoblastic tumours (Hammond *et al.*, 1981), strongly inhibits DHFR in tumour cells with a K_i of 1×10^{-9} M. The drug forms a ternary complex with enzyme and NADPH, as shown by X-ray crystallography (Matthews *et al.*, 1978). The action of the drug *in vivo* relies on the greater need of the tumour cell for purine nucleotides compared with normal cells, and so gives rise to considerable toxicity in rapidly dividing cells such as in the bone marrow. Methotrexate is transported into mammalian cells by the active transport system for folate, to which it bears a marked resemblance. Micro-organisms do not have such a transport system and, although methotrexate is as effective against the microbial enzymes as against the mammalian, it has little anti-microbial activity (Ferone *et al.*, 1969).

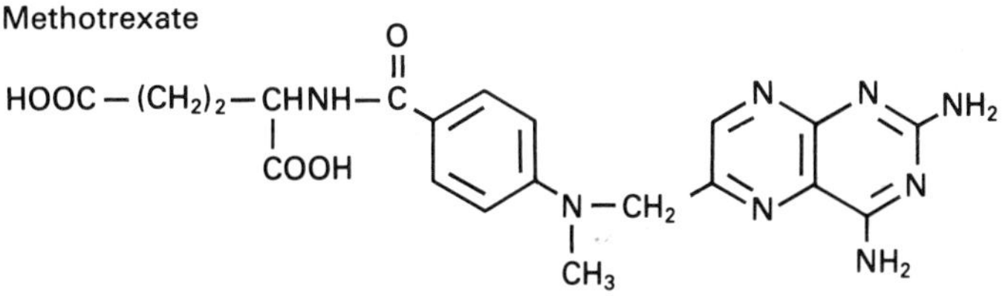

As the targets for sulphonamides and diaminopyrimidines lie on different pathways that intersect, it seemed appropriate to test combinations of these drugs to establish whether they could be used in conjunction more effectively i.e., whether they showed synergy. Synergy was found to occur both *in vitro* (Moody and Young, 1975) and *in vivo* (Ramachandran *et al.*, 1978) when trimethoprim was combined with sulphamethoxazole against bacterial infections. With pyrimethamine similar effects have been observed against human malaria, but in this case the value of such combinations is neither to reduce toxicity nor to obtain high potency, but rather to lower the dose of pyrimethamine such that there is less likelihood of resistance developing. Resistance is a serious problem with the malarial parasite at present.

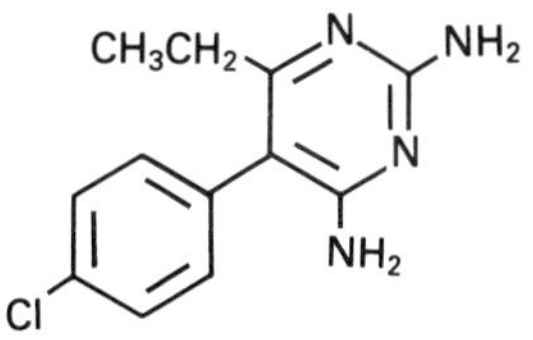

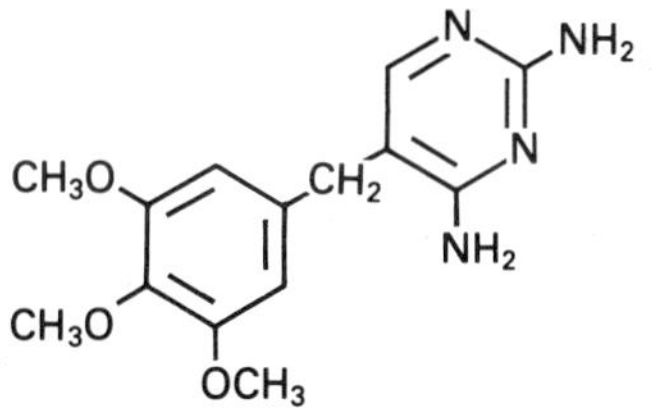

Combinations of trimethoprim with other sulphonamides such as sulphadiazine (Hurly, 1959) or dapsone (Lucas *et al.*, 1969) were shown to be very effective in man. Sulphadoxine appears to be the sulphonamide of choice for synergy with trimethoprim at present as it has a particularly long half-life for a sulphonamide — seven to nine days in plasma. This is clearly of great value for reducing the number of doses required for treatment.

Sulphadiazine

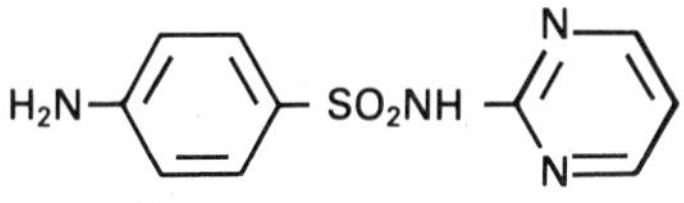

H$_2$N⟨⟩—SO$_2$NH ⟨N⟩

Sulphamethoxazole

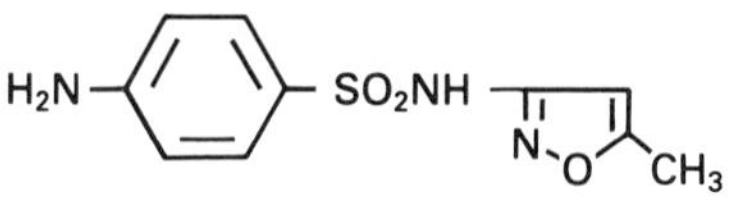

H$_2$N⟨⟩—SO$_2$NH ⟨ O-N CH$_3$⟩

Dapsone

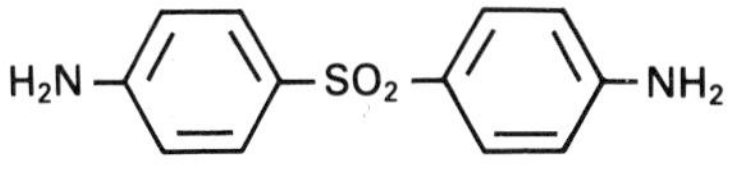

H$_2$N⟨⟩—SO$_2$—⟨⟩—NH$_2$

Sulphadoxine

H$_2$N⟨⟩—SO$_2$NH ⟨N N⟩ CH$_3$O   OCH$_3$

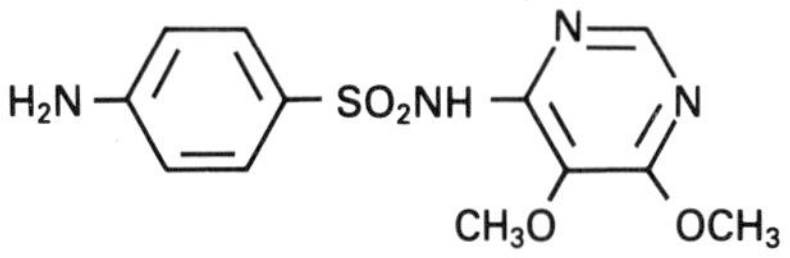

2.2.3 Ribonucleotide reductase

The reduction of the nucleoside diphosphates to their respective deoxy counterparts is the first committed step in deoxyribonucleotide biosynthesis, and is catalysed by ribonucleotide reductase. The activity of this enzyme has been found to follow closely the regulation of DNA biosynthesis, although this connection may not be any more than a correlation (Thelander and Reichard, 1979). The same enzyme reduces all four ribonucleotide diphosphates by directly replacing the 2-hydroxyl group on the ribose ring by a hydrogen atom. With the mammalian enzyme, magnesium and ATP are also required for activity. The enzyme consists of two subunits and a number of allosteric effectors have been noted (Erikkson *et al.*, 1979) (see Chapter 1 for a discussion on allostery). In order to carry out the reduction the enzyme requires the iron–sulphur protein thioredoxin together with thioredoxin reductase which uses NADPH to regenerate thioredoxin. Thioredoxin contains two free sulphydryl groups positioned in such a way that a disulphide bond can be formed between them:

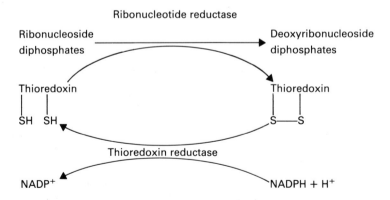

Since the activity of ribonucleotide reductase correlates well with the level of DNA biosynthesis it might be expected to be a logical target for chemotherapy (reviewed in Elford *et al.*, 1981), but it is perhaps surprising that so far only one drug on the market, hydroxyurea, owes its effectiveness to inhibition of this enzyme. This drug is used for the treatment of chronic granulocytic leukaemia and malignant melanoma (Kennedy and Yarbro, 1966; Ariel, 1970). The use of hydroxyurea suffers from the drawback that frequent large doses are required to maintain effective concentrations *in vivo*. This is understandable in view of the high drug concentration, 5×10^{-4} M, required to produce 50% inhibition of the enzyme (Elford *et al.*, 1979). Various attempts have been made to improve on hydroxyurea, but although 2,3,4-trihydroxybenzamide is more potent than hydroxyurea *in vitro*, and *in vivo* against the L1210 leukaemia (Elford *et al.*, 1979), no analogue has yet reached the market.

The precise mode of action of hydroxyurea is not yet fully understood. The enzyme inhibition is reversible, and non-competitive with respect to CDP and ATP when thioredoxin and a thioredoxin-generating system are employed (Moore, 1969). Dithioerythritol can be used to substitute for this regenerating system, and is able to reverse the inhibition partially. Ferrous iron can also partially protect the enzyme against inhibition by hydroxyurea, and this has given rise to the view that hydroxyurea inhibits by binding to the iron in ribonucleotide reductase. Nevertheless, the fact that iron can only partially protect has led to the suggestion that the dithioerythritol may either keep the iron in a reduced state and/or induce a more active conformation in the enzyme which protects the iron from hydroxyurea (Moore, 1969).

2.2.4 Thymidylate synthetase

The methylation of uridine monophosphate to yield thymidine monophosphate is a crucial step in the provision of pyrimidine bases for the synthesis of DNA, and so is another possible target for cancer chemotherapy. 5-Fluorouracil is believed to exert its anti-tumour effect, after conversion to ribo- and deoxyribo-nucleotide, partly on thymidylate synthetase and partly on RNA biosynthesis (reviewed in Heidelberger *et al.*, 1983).

5-Fluorouracil (5-FU) is converted to 5-fluorouridine monophosphate (5-FUMP), probably in one step catalysed by orotate phosphoribosyltransferase using phosphoribosyl-1-pyrophosphate as co-substrate. 5-FUMP is phosphorylated to the diphosphate (5-FUDP), reduced in the presence of ribonucleotide reductase to 5-FdUDP and then hydrolysed to 5-FdUMP. This latter compound forms a covalently bound ternary complex with the folate cofactor $N^{5,10}$-methylene-tetrahydrofolate and thymidylate synthetase, which closely resembles the transition state of the normal reaction. 5-FdUMP acts as a false substrate since the folate methylene group is transferred to the 5-position of the uracil ring, but further conversions cannot take place because of the presence of the fluorine atom at position 5 of the uracil ring. As a consequence, a covalent link remains between the methylene group and either the 5 or the 10 positions of the pteridine ring of the cofactor (see Figure

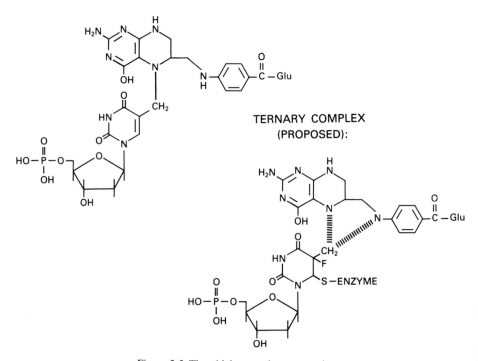

Figure 2.5 Thymidylate synthetase reaction.

2.5). The inhibitor, therefore, behaves as a suicide substrate, and the synthesis of thymidylate is shut off.

Recent studies, however, have tended to suggest that thymidylate synthetase may not be the only target. The incorporation of 5-fluorouracil into RNA to yield a faulty ribosomal RNA may also play a part in the drug's efficacy, and both mechanisms may be important to varying extents in different situations (Heidelberger *et al.*, 1983). 5-Fluorouracil is used with good effect in certain types of carcinoma of the breast and of the gastrointestinal tract (Bonadonna and Valogussa, 1981; Kisner *et al.*, 1981).

A closely related compound, 5-fluorocytosine, has found favour as a treatment for fungal infections, namely candidosis and cryptococcosis, particularly when used in conjunction with amphotericin B (Cohen, 1982). 5-Fluorocytosine is not recommended as a single medication in view of the rapid development of resistance. Up to 30% of the patients in one trial treated with 5-fluorocytosine alone developed resistance (Block *et al.*, 1973) — hence the concomitant use of amphotericin. 5-Fluorocytosine is converted to 5-fluorouracil by a fungal enzyme, cytosine deaminase; similar metabolic conversions to those for 5-fluorouracil in mammalian cells follow. 5-Fluorouracil is initially converted to 5-FUMP by UMP pyrophosphorylase (an enzyme that normally acts to salvage uracil in *Candida albicans*). This appears to be the point at which resistance is induced, since resistant *C. albicans* has little or no detectable enzyme (Whelan and Kerridge, 1984). Cytosine deaminase is believed to be largely absent from the host, although Diasio *et al.* (1978) found some evidence of 5-fluorouracil production in man.

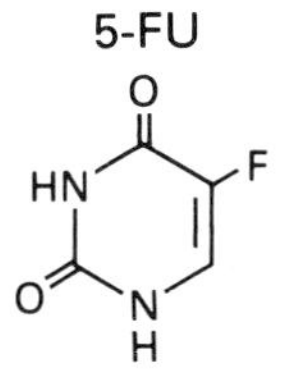

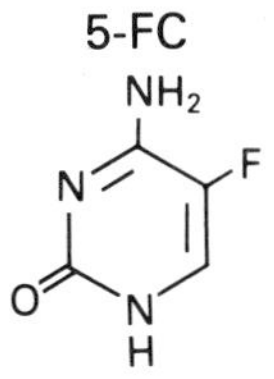

5-FU   5-FC

2.2.5 Inosine monophosphate dehydrogenase

Ribavirin (virazole) has shown variable clinical effectiveness against a variety of viruses including those with RNA as the genetic material, such as influenza, and those containing DNA, for example, herpes, measles and some adenoviruses (reviewed in Chang and Heel, 1981). There is some uncertainty about the effectiveness of the drug, but, at least in the case of influenza, the delivery of the drug by aerosol is more effective than by the oral route. The latter may deliver insufficient drug to the lungs for activity to be shown (McClung *et al.*, 1983). Ribavirin resembles adenosine sufficiently to be phosphorylated intracellularly by host adenosine kinase, and is subsequently converted to the di- and trinucleotides by other host enzymes. The drug structure can also be drawn to resemble guanosine (Gilbert and Knight, 1986).

Intracellular pools of GTP are reduced by up to 45% in influenza-infected cells treated with ribavirin (Wray *et al.*, 1985). This results from inhibition of the first committed enzyme in *de novo* synthesis of GMP; the NAD-linked inosine monophosphate dehydrogenase, by the monophosphate of ribavirin (Streeter *et al.*, 1973). This enzyme converts IMP to xanthosine monophosphate which is then transformed into GMP by GMP synthetase (see Figure 2.6). The inhibition is competitive with a K_i of 2.5×10^{-7} M against IMP dehydrogenase from Ehrlich ascites tumor cells (K_m for IMP is 1.8×10^{-5} M).

This inhibition accounts for part of the antiviral effect, but the fact that guanosine can only partly reverse the inhibition of proliferation of influenza virus in canine kidney cells suggests that other mechanisms may be involved. This is supported by the observation that the lowering of the GTP levels reaches a minimum at 2.5×10^{-5} M ribavirin but the antiviral effect continues

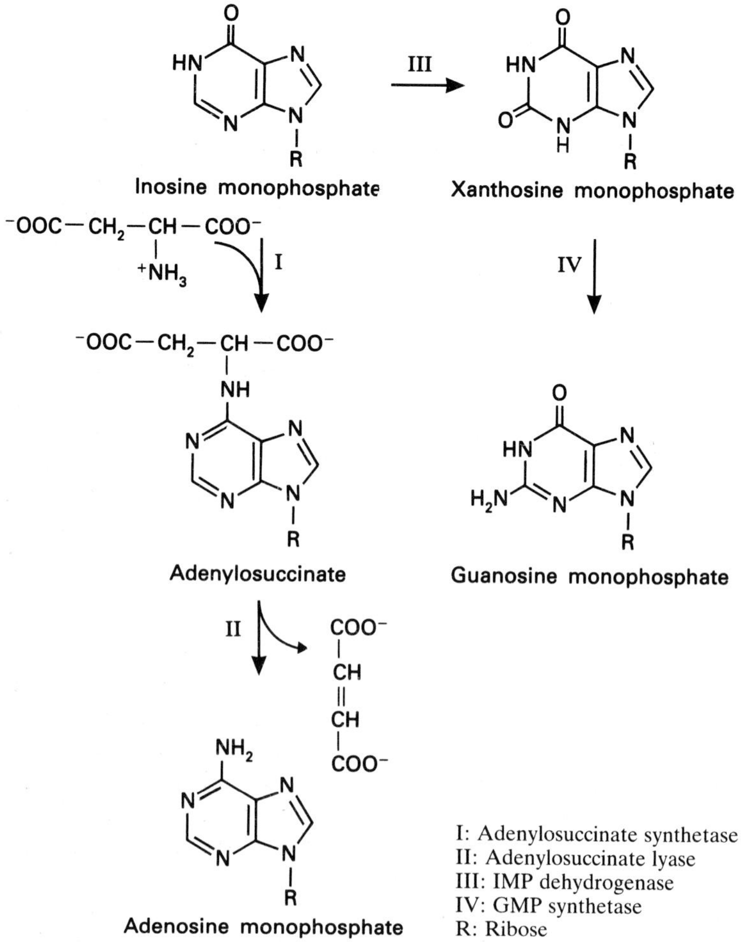

I: Adenylosuccinate synthetase
II: Adenylosuccinate lyase
III: IMP dehydrogenase
IV: GMP synthetase
R: Ribose

Figure 2.6 The biosynthesis of GMP.

to increase up to 1.4×10^{-4} M drug (Wray *et al.*, 1985). Other possible targets include ribavirin triphosphate inhibition of (a) the capping reaction of viral messenger RNA (GTP is added to the 5′ end of freshly synthesized viral mRNA to render it effective), and (b) the viral RNA polymerase complex (see review by Gilbert and Knight, 1986).

Ribavirin

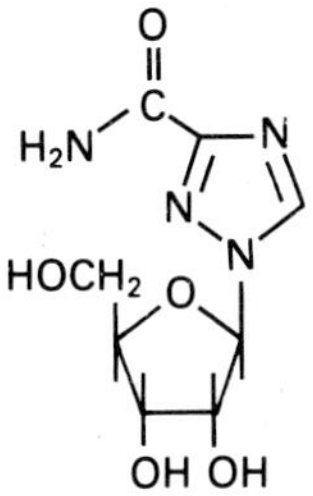

2.3 DNA biosynthesis

So far we have been considering drugs that interfere with the process of nucleotide biosynthesis. We now turn to the next step in the formation of nucleic acid; the magnesium-dependent process in which the four nucleoside triphosphates are utilized in the polymerization of DNA. In mammalian cells more than one enzyme is required to do this conversion; two enzymes α and β are responsible for the formation of chromosomal DNA while γ catalyses the synthesis of mitochondrial DNA.

The viral genome, on the other hand, codes for a number of enzymes which it needs to synthesize a fresh virus particle. Foremost among these is the RNA or DNA polymerase with which, after subversion of the host cell's protein synthesizing machinery, more genome is synthesized. Although all DNA polymerases require magnesium, the enzyme coded by herpesviruses 1 and 2 differ from their mammalian counterparts in a number of ways, including a five-fold activation by 6×10^{-2} M ammonium sulphate while the host cell enzymes are completely inhibited at this concentration (Mar and Huang, 1979).

A few drugs that interfere with DNA biosynthesis have been developed for the treatment of cancer and for viral disease. Indeed, the main drugs used to treat viral infection may be found in this section, which is interesting considering how many other activities of the virus could be targets, such as the binding of the virus to the outside of the target cell; the removal of the protein outer coating of the virus and the penetration of the nucleic acid into the cell; the induction of protein as well as nucleic acid biosynthesis by the cellular apparatus; assembly of protein and nucleic acid into infectious virus particles followed by their exit from the cell.

This process destroys the cell, whether bacterial or mammalian, and can lead to a fresh round of virus replication if other cells are nearby. A zone of lysis, known as a plaque, is then formed wherever replicating viral particles are located. If host cells are grown on a suitable nutrient in a plate, plaque formation can be used to quantify the amount of active virus present.

2.3.1 Herpesvirus DNA polymerase

9-[(2-Hydroxyethoxy)methylguanine (generic name, acyclovir), an analogue of a purine nucleoside, is the first of a new family of drugs for the treatment of infections caused by the family of DNA viruses, known as herpesvirus, which range in severity from the ubiquitous mouth ulcer, via serious cases of blindness, to life-threatening disease. Herpesvirus infectious to humans may be divided into two classes; one responsible mainly for mouth ulcers and eye damage while the other, herpesvirus 2, gives rise to serious genital infections.

Acyclovir owes its mode of action to its conversion to a trinucleotide that subsequently inhibits viral DNA polymerase with a tight binding constant of 1.6×10^{-9} M (De Clerq and Walker, 1984). Furthermore, the enzyme cata-lyses the insertion of the false purine base into viral DNA, bringing DNA synthesis to a premature end, since acyclovir does not have a 3′-hydroxyl group — obligatory for further chain elongation.

Another virally coded enzyme, thymidine kinase, is responsible for con-verting the drug to a nucleotide analogue. In practice the enzyme is unfortu-nately named because it is not specific for pyrimidines, and also shows thymidylate kinase activity. Nevertheless, acyclovir is a poor substrate for the enzyme from herpesvirus 1 with a K_m of 4×10^{-4} M and a V_{max} of 6.1×10^{-12} mol/min/μl of enzyme preparation. The respective figures for thymi-dine, in contrast, are 8.5×10^{-6} M and 218 (Ashton *et al.*, 1982). Acyclovir monophosphate is converted to the diphosphate by the action of viral thymi-dine kinase or by cellular guanosine monosphosphate kinase, and a number of kinases can add a third phosphate (Miller and Miller, 1982). Acyclovir is not toxic to uninfected cells because there it is not converted to a triphosphate analogue.

Acyclovir

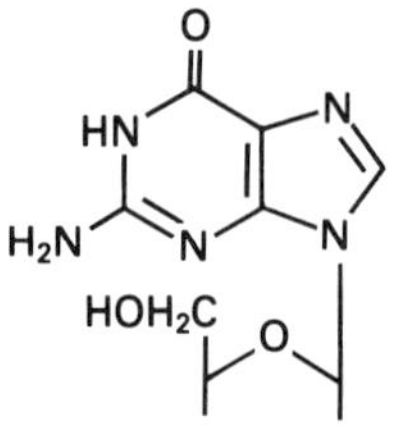

Acyclovir is active to a similar extent against types 1 and 2 virus with 50% inhibition of viral plaque formation at concentrations of 0.06 to 1.8×10^{-6} M for type 1 and 0.65 to 1.8×10^{-6} M for type 2 (Parris and Harrington, 1982). Since there are two enzymes involved in the antiviral activity of acyclovir resistance can and has occurred either (a) in thymidine kinase where the enzyme either is missing entirely or has an altered sensitivity to substrates and to the drug (Larder and Darby, 1982) or (b) in the ultimate target, DNA polymerase (Parris and Harrington, 1983). Mutants of herpesvirus 1 without thymidine kinase are not apparently as virulent as the wild-type, although those that still have the enzyme, albeit altered, are just as virulent (Veerisetty and Gentry, 1983).

Vidarabine (arabinosyladenine or Ara-A) has both anti-tumour and anti-viral activity. The drug is active against keratinization of the cornea and encephalitis caused by herpesvirus 1 in adults and neonates, and against herpes zoster infections in immunocompromised patients (reviewed in Whitley *et al.*, 1980). The nucleoside is phosphorylated to the monophosphate by both host and virus-induced thymidine kinase; further enzymes convert it to the triphosphate. Ara-ATP is more inhibitory to herpesvirus than to host DNA polymerase; a K_i of 1.4×10^{-7} M has been reported for herpes hominis DNA polymerase from infected cells (K_m for dATP was 1.37×10^{-5} M) while DNA polymerase α from rabbit kidney cells was inhibited with a K_i of 7.4×10^{-6} M (K_m for dATP was 6.4×10^{-6} M) (Muller *et al.*, 1977).

All these inhibitions were competitive and were measured with activated DNA as the template. Clearly, low doses of the drug will inhibit viral replication while higher doses will inhibit the mammalian enzyme, and may lead to toxicity as the drug is not absolutely selective for viral replication (Muller *et al.*, 1977). Furthermore, Ara-A is incorporated into viral DNA, and it is possible that this effect contributes to its anti-viral action.

The development of Ara-A as an anti-tumour agent has been greatly limited by its rapid deamination to the inactive metabolite arabinosylhypoxanthine by adenosine deaminase, an enzyme that is widely distributed in body tissues and fluids. The drug is, therefore, insufficiently active in the clinic for practical use. Attempts have been made to circumvent this deamination by the use of 2-deoxycoformycin, a powerful adenosine deaminase inhibitor, which enhances and prolongs the activity of Ara-A in man (Major *et al.*, 1983), but this practice has not yet found general acceptance.

2.3.2 Mammalian DNA polymerase

Arabinosylcytosine (Ara-C) is a drug that is commonly used for the treatment of acute leukaemia, both myelogenous and lymphocytic, and non-Hodgkin's lymphoma (Rubin, 1981). It is specific for the S-phase of the cell cycle. In order to exert its anti-tumour effect, Ara-C must first be converted to the monophosphate. This is performed by deoxycytidine kinase; subsequently deoxycytidylate kinase and nucleoside diphosphate kinase catalyse the conversion of the nucleotide to Ara-CTP. This metabolite interferes with DNA synthesis by inhibiting DNA polymerase competitively with deoxy-CTP (K_i for Ara-CTP is 8.7×10^{-6} M and K_m for dCTP is 9.0×10^{-6} M (Graham and Whitmore, 1970).

Resistance to the drug can occur either by virtue of reduced deoxycytidine kinase levels or by deamination to inactive uracil metabolites by cytidine deaminase. In addition, it has been shown that Ara-C is incorporated into DNA not so much at terminal, but rather at internucleotide linkages. This suggests that chain elongation is not stopped by the presence of the false sugar (Graham and Whitmore, 1970). More recent work with higher concentrations of drug has shown that a greater proportion of Ara-C may be found at the 3'-termini, which implies that chain elongation is greatly slowed by the drug (Major *et al.*, 1982). This is understandable since the 2'-hydroxyl of arabinose is *trans* to the 3'-hydroxyl, in contrast to ribose, and will cause steric hindrance to the rotation of the pyrimidine base about the nucleoside bond. The bases of polyarabinonucleotides cannot, therefore, stack in the same way as do deoxyribonucleotides, and slowing of DNA elongation could well be the consequence.

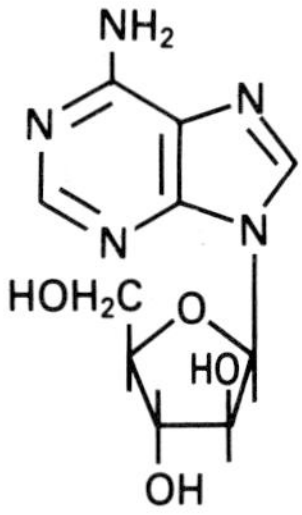

Ara-A

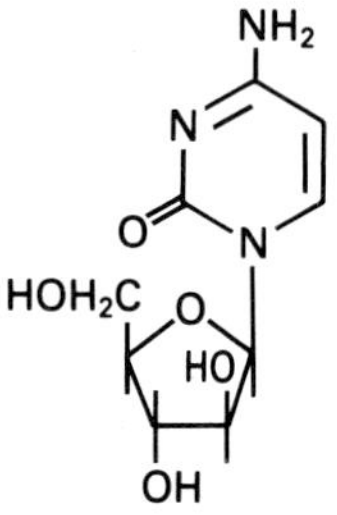

Ara-C

2.3.3 Bacterial topoisomerase II (DNA gyrase)

Often in the history of medicine a drug has been used for many years without any precise knowledge of its mode of action. Nalidixic acid and its analogues form a case in point. They first went on the market in the early 1960s for the treatment of urinary tract infections caused by gram-negative organisms such as *Escherichia coli* and Proteus species. The drug has the drawback that resistance rapidly appears in previously sensitive organisms. Recently, the advent of analogues with a greatly broadened spectrum of action which includes gram-positive organisms has occurred, notably norfloxacin (Neu and Labthavikul, 1982) and ciprofloxacin (Barry *et al.*, 1984); the clinical use of these 'quinolones' has been reviewed by Hooper and Wolfson (1985).

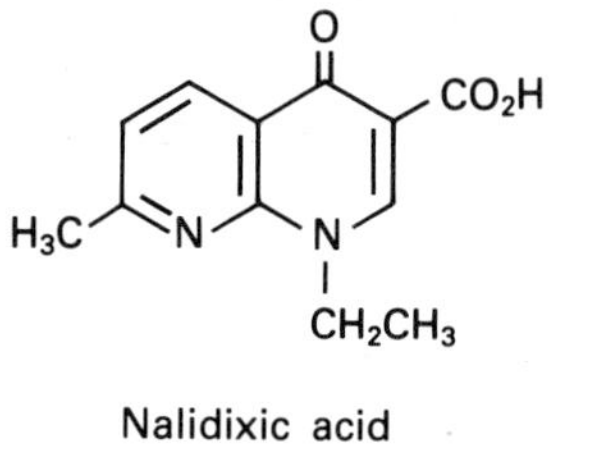

Nalidixic acid

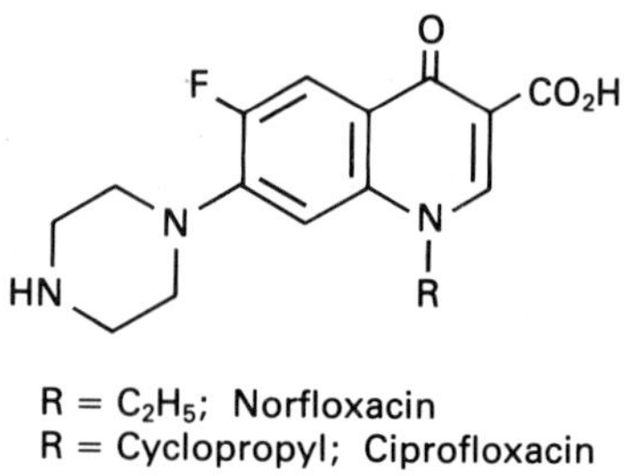

R = C_2H_5; Norfloxacin
R = Cyclopropyl; Ciprofloxacin

It became clear early on that nalidixic acid was interfering with DNA synthesis in some unidentified fashion, but it required the discovery of DNA gyrase before the precise target was identified (Gellert *et al.*, 1976a; reviewed in Gellert, 1980). DNA as normally extracted from the bacterial cell, whether plasmid or chromosomal, is supercoiled, i.e. over and above the double helical structure there are a number of supertwists which work in an opposite sense to the double helix and are thus termed negative. Many of the activities that bacterial DNA undergoes, including recombination, conjugation, replication and repair, require the DNA to be in the negatively supercoiled form. The enzyme that catalyses this supercoiling, known as DNA gyrase, clearly plays a crucial role in the bacterial cell.

The enzyme takes double-stranded closed circular DNA as a substrate and, in the presence of magnesium and ATP, introduces negative supercoils, two at a time, into the DNA. It does this by first introducing a double loop into the DNA; both strands of the DNA are broken, the loop is passed through and the original break is resealed. ATP is hydrolysed to ADP in the process. Thus the enzyme merely changes the topology of the DNA without any net bond breaking (for a detailed description of the process see Cozzarelli, 1980), hence the name topoisomerase II (the II refers to the fact that both strands of the DNA are cleaved initially). The temporary link between enzyme and split ends appears to be a phosphotyrosine bond (Klevan and Tse, 1983). If sodium dodecyl sulphate is added with nalidixic acid, enzyme and DNA, the break can be 'frozen', i.e. the enzyme remains attached to the broken DNA strand.

DNA gyrase is composed of two subunits in the form of a tetramer, A_2B_2. Nalidixic acid binds to the A subunit which also binds the DNA (Sugino *et al.*, 1977). There is still some doubt, however, as to whether this entirely explains the mode of action of the quinolone antibiotics. Normally one would expect a drug to inhibit a target enzyme at a lower concentration than an organism because less of the drug might be expected intracellularly, as the drug has to penetrate the cell wall and membrane to reach the enzyme. In the case of nalidixic acid and its analogues the reverse is the case. The overall level of supercoiling in the bacterial cell is also not affected by these agents which is consistent with moderate enzyme inhibition (Engle *et al.*, 1982).

Suggestions have been made that the DNA gyrase operating near the precise point where the DNA is dividing ready for replication (the replication fork) is much more sensitive to the drug than is the bulk enzyme measured *in vitro*, but there appears at present to be no clearcut solution of this difficulty. In sharp contrast, the genetic data indicate quite clearly that mutations in subunit A confer resistance to nalidixic acid (Sugino *et al.*, 1977). This discrepancy does highlight the potential difficulty in the extrapolation from the 'test-tube' to the whole organism.

Novobiocin

At the same time, the 4-hydroxycoumarin group of antibiotics, of which novobiocin is the only one to have reached the market, were shown to interfere with the B subunit which binds the adenine nucleotides (Gellert *et al.*, 1976b). In this instance, the evidence for the identification of DNA gyrase as the target enzyme is more consistent, as the overall level of supercoiling in the cell is reduced by these drugs (Engle *et al.*, 1982) and the drug inhibits the enzyme at a lower concentration than is required to kill the bacterium (Hooper *et al.*, 1982).

2.3.4 Mammalian DNA topoisomerase II

No enzyme exactly equivalent to DNA gyrase has yet been found unequivocally in mammalian cells, probably because mammalian DNA is negatively supercoiled by being wrapped around histones. There is no need therefore, to 'wind up' the nucleic acid, merely to unwind it with the help of ATP. A well-known class of anti-tumour agents may interfere with mammalian topoisomerases, namely the anthracycline antibiotics, such as doxorubicin (adriamycin), produced by *Streptomyces peucetius var. caesius*. Early studies suggested that the drug exerted its antitumour effect by binding tightly to DNA, possibly by slotting (intercalating) in between the bases like actinomycin, and thus interfering with DNA and RNA biosynthesis (DiMarco, 1975).

More recently, however, it has been found that use of these drugs results in single and double strand cleavage of the DNA. These breaks are associated with a covalently bound protein on the 5'-end of a DNA strand (Tewey *et al.*, 1984) in a similar fashion to DNA gyrase and bacterial DNA. As in the bacterial case, this effect can be 'frozen' by precipitation of the complex with sodium dodecyl sulphate. Some non-intercalative antitumour drugs of the epipodophyllotoxin class have also been shown to act in a similar way (Chen *et al.*, 1984). Studies with closely related groups of structures have indicated a relationship between the ability to cleave DNA and cytotoxicity. The mechanism, however, by which the cleavage results in cell death is not fully clear, although it may have to do with the requirement for intact DNA for mitosis, and the inability of the cell treated with these drugs to repair the damage to DNA. Nevertheless, it is interesting to see how an enzyme can cooperate in the action of a drug in a suicidal fashion (see Ross, 1985, for a review).

The anthracyclines are effective in the treatment of acute leukaemias and malignant lymphomas and also in a variety of solid tumours. Doxorubicin is part of a cocktail for the treatment of carcinoma of the ovary, breast and oatcell carcinoma of the lung, and is also of value against a wide range of sarcomas including osteogenic and soft-tissue sarcomas (Blum and Carter, 1974). Sarcoma is the term given to tumours arising in bone, connective tissue or muscle, while carcinomas arise in covering or lining membranes. The latter are more commonly encountered than sarcomas.

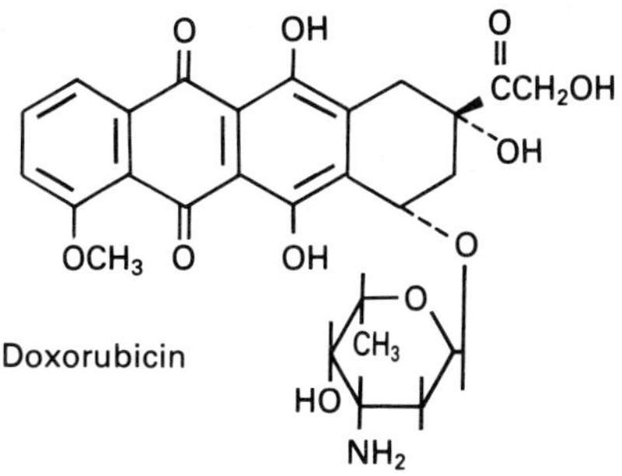

Doxorubicin

2.3.5 Reverse transcriptase (RNA-primed DNA polymerase)

A number of RNA viruses that can infect man and animals are known as retroviruses because they use their genomic RNA to prime an enzyme to synthesize complementary DNA. This enzyme is called reverse transcriptase and it has both RNA and DNA-dependent DNA polymerase activity. The nucleic acid of the viral genome is replicated via a mixed RNA/DNA duplex. DNA synthesized then acts as a template for the synthesis of viral RNA, both messenger and genomic, that subsequently ensures the synthesis of viral proteins in order to form new viral particles.

If ever the medicinal scientist fraternity needed to be reminded that micro-organisms are capable of changing in such a way as to present new threats to the health of man, the development of acquired immune deficiency syndrome (AIDS) as a result of virus infection has provided such a reminder. The disease is caused by the virus known as human immunodeficiency virus (HIV) (earlier names were human T-cell lymphotropic virus III (HTLV-III) and lymphadenopathy-associated virus (LAV)), which appears to be a recent import into humans possibly from the African green monkey. The viral genome may initially integrate into the host DNA as a provirus which is subsequently reactivated to produce virus particles.

The virus infects the T-cells of the immune system and thus the host defences are seriously weakened. Consequently, the AIDS patient frequently dies of other diseases that are normally controlled by the T-cells, for example pneumonia or an otherwise rare form of cancer, Kaposi's sarcoma (Shaw *et al.*, 1985). Furthermore, there may be degenerative changes in nervous tissue as a consequence of infection, but it is probably too early to be sure how serious these changes are.

At present only one drug has so far shown any ability to prolong the life of AIDS sufferers, and that is 3'-azido-3'-deoxythymidine (generic name azidothymidine). The drug is converted to the triphosphate in virus infected cells. The false nucleotide is used by the DNA polymerase coded for by the virus as a false substrate, thereby introducing the azidothymidine into the growing chain of DNA and causing it to terminate prematurely. The triphosphate also inhibits the enzyme directly at concentrations below 10^{-7} M (Mitsuya *et al.*, 1985). Because this is a reverse transcriptase it is different from DNA polymerases present in mammalian cells and as such is a selective target — mammalian enzymes do not readily use azidothymidine triphosphate as a false substrate. Studies in mice have shown that azidothymidine protects against retrovirus infection and clinical trials in man are showing promise (Ruprecht *et al.*, 1986).

Azidothymidine

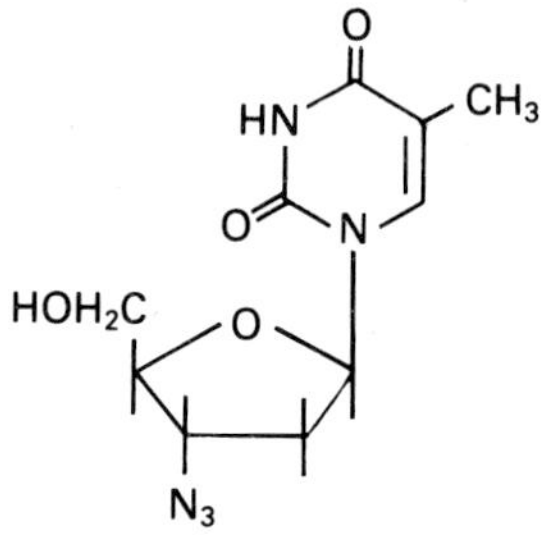

2.4 Nucleotide catabolism

2.4.1 Adenosine deaminase

Adenosine deaminase catalyses the deamination of adenosine and deoxyadenosine to inosine and deoxyinosine respectively. This enzyme is important in lymphocytic function and its absence through genetic defect leads to severe immune deficiency (Thompson and Seegmillar, 1980). 2-Deoxycoformycin is the most potent known inhibitor of the enzyme and acute lymphoblastic leukaemia of T-cell origin responds to such inhibition probably because T-cells are particularly rich in adenosine deaminase, and the enzyme levels are raised even further in various forms of acute lymphocytic leukaemia (Smyth *et al.*, 1980). The enzyme plays an important role in the salvage pathway for purines since inosine may be phosphorylated to IMP which can, as we have seen, be converted into GMP. Inhibition of this salvage pathway in times of high purine requirement is likely to be critical to the cell.

The action of 2-deoxycoformycin as a powerful adenosine deaminase inhibitor may also be indirect. A build-up of dATP, derived by phosphorylation from adenosine and adenylate kinases, can inhibit ribonucleotide reductase which is particularly sensitive to the triphosphate balance. Secondly, accumulated adenosine can drive S-adenosylhomocysteine hydrolase into reverse to form S-adenosylhomocysteine (Hershfield *et al.*, 1983). This agent is a powerful inhibitor of the methylation of RNA and DNA, possibly by antagonising the methyl transfer activity of S-adenosylmethionine.

2-Deoxycoformycin is a fermentation product of *Streptomyces antibioticus*. The mode of action is reviewed in Agarwhal (1982). The drug is slightly more tightly bound than its ribose analogue, coformycin (K_i of 2.5×10^{-12} M compared with 1×10^{-11} M). Since a similar relationship occurs with the substrates with adenosine (K_m 2.5×10^{-5} M) and deoxyadenosine (7×10^{-6} M), it is likely that the enzyme has a greater affinity for deoxyribose than ribose.

The drug is a close structural analogue of the transition state of the enzyme reaction (see Chapter 1 for discussion on transition states). The six-membered ring of the purine nucleus has been expanded to a seven-membered ring by the addition of a carbon atom. There are two double bonds in that ring so the aromaticity is lost and the ring becomes puckered; the hydroxyl group is a secondary alcohol, and so the keto–enol tautomerism is also lost. This leads to a structure similar to the transition state intermediate in which a hydroxyl group has been added at the 6-position to give a tetrahedral carbon atom (Wolfenden *et al.*, 1977). The reaction mechanism is shown in Figure 2.7.

This type of inhibition leads to some rather unusual kinetic results. The binding of the inhibitor is to the form of the enzyme present in the transition not to the ground state, and so the inhibition takes time to develop. The initial velocity plots show only relatively moderate inhibition if the inhibitor is

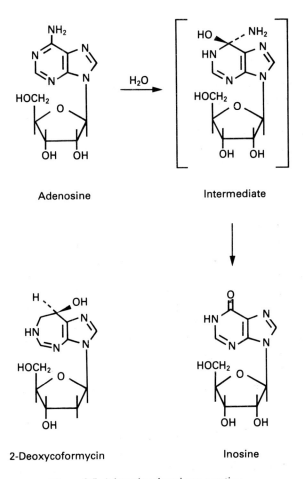

Figure 2.7 Adenosine deaminase reaction.

not pre-incubated with the enzyme. A double reciprocal plot demonstrates characteristics of competitive inhibition with a K_i of 1×10^{-8} M. With pre-incubation, however, the plot obtained indicated a non-competitive inhibitor with a K_i of $2 \cdot 5 \times 10^{-12}$ M. The situation is similar to what is obtained with irreversible time-dependent binding to an enzyme, namely that a proportion of the enzyme is put out of action by the inhibitor because the enzyme–inhibitor complex takes so long to dissociate (the half-life for dissociation is 25 to 30 hours). The activity observed is that of the remaining uncomplexed enzyme, and so the K_m remains the same while V_{max} is reduced — a classical example of non-competitive behaviour. The rates of association and dissociation of this complex are slow, possibly as a result of a slow conformational change in the enzyme as it is transformed from the ground state to the activated transition state (Agarwhal, 1982).

2.4.2 Xanthine oxidase

Xanthine oxidase is the last enzyme on the breakdown pathway of purine bases in primates (Figure 2.4), and it catalyses the conversion of hypoxanthine to xanthine and of xanthine to uric acid. The latter is normally excreted, although quantities of the other purines may also find their way into the urine. In some diseases, notably gout, the production of purines can be increased as a primary cause of the disease (Scott, 1980). Enzyme deficiencies with a genetic origin may play a part. One such case is deficiency of the salvage enzyme hypoxanthine phosphoribosyltransferase (HPRT) which leads to an elevated level of hypoxanthine phosphoribosylpyrophosphate. The latter stimulates *de novo* purine biosynthesis at the initial rate-limiting step of the formation of phosphoribosylamine (see Figure 2.2).

The consequence of increased purine synthesis is an increased throughput down the catabolic pathway to uric acid. When levels of the latter rise above saturation, crystals of monosodium urate form in the synovial fluid. The characteristic symptoms of gout derive from an inflammatory response to these crystals, and thus closely resemble the painful joint swellings in rheumatoid arthritis. This may occur in one joint only or in several (Kelley and Wyngarden, 1974). In advanced gout, deposits (tophi) of sodium urate form on or near joints or tendon sheaths, which are soft initially but eventually harden (Scott, 1980).

For therapy the major need is to lower serum uric acid levels, although anti-inflammatory drugs will relieve the symptoms on a short-term basis. One of the most useful drugs in effecting a long-term cure is allopurinol. Xanthine oxidase is the target of the drug, and so serum and urine hypoxanthine and xanthine levels are raised whilst, more importantly, uric acid levels are lowered. Additionally, the drug is useful when given in combination with anti-leukaemic drugs since serum urate levels can rise sharply as the leukaemia cells die. This is an example of secondary gout, secondary in that uric acid formation is increased as a consequence of other changes. In this case, the danger is not only that acute episodes of gout may occur, but also that sodium urate crystals may form in the distal tubule of the kidney (Scott, 1980).

Clearly, if a drug is metabolized by xanthine oxidase, its action is likely to be potentiated by allopurinol. A case in point is 6-mercaptopurine (a drug used for the treatment of leukaemia) which is metabolized by xanthine oxidase to 6-thiouric acid, an inactive metabolite (Elion, 1967). The dose of mercaptopurine required when given in conjunction with allopurinol must therefore be reduced to avoid widespread toxicity which would otherwise occur if higher mercaptopurine levels were sustained for longer periods of time.

Xanthine oxidase is a complex enzyme containing in effect a transport system involving molybdenum, flavin nucleotide and four iron–sulphur centres which convey electrons to oxygen to yield superoxide anion (O_2^-). Allopurinol inhibits the enzyme in a complex fashion, and may be regarded as one of the earliest examples of a suicide substrate (Spector and Johns, 1970; Massey *et al.*, 1970 — see Chapter 1 for a discussion on suicide substrates). If the inhibition is studied without pre-incubation of enzyme and inhibitor, allopurinol behaves as though it were a competitive inhibitor with a K_i of 7×10^{-7} M. With pre-incubation in the presence of air, the inhibition increases and is no longer competitive with substrate. Allopurinol is also a substrate for xanthine oxidase and the product of the reaction, oxipurinol (alloxanthine), is also an inhibitor. In the presence of xanthine as substrate and oxygen, or anaerobically without substrate, the enzyme is inactivated by oxipurinol. If the oxidation of xanthine, which requires the enzyme to cycle between reduced and oxidized forms, and for the enzyme to be in an anaerobic environment, both result in enzyme inactivation by oxipurinol, it is likely that the reduced form of the enzyme is sensitive to oxipurinol (Massey *et al.*, 1970). The dissociation constant of the oxipurinol–enzyme complex is 5.4×10^{-10} M (Spector and Johns, 1970). Inhibition can be reversed by prolonged dialysis or by allowing the complex to be reoxidised in the presence of air, thus confirming that it is the partly reduced form of the enzyme that is receptive to oxipurinol inhibition.

Allopurinol   Oxipurinol

The inactivation of reduced xanthine oxidase by oxipurinol follows first order kinetics by appearing to be dependent on the concentration of reduced enzyme. This may be the result of an internal rearrangement of the enzyme/

inhibitor complex in a time-dependent fashion (Spector and Johns, 1970). The similarity between the tight or stoichiometric binding of oxipurinol to xanthine oxidase, and coformycin to adenosine deaminase was noted by Cha *et al.*, 1975.

Allopurinol has been found to be effective in the treatment of kala-azar (leishmaniasis) on clinical trial (Kager *et al.*, 1981). In this instance the drug is acting as a false substrate for the parasite's hypoxanthine phosphoribosyl-transferase (see Figure 2.8) — much more efficiently than for the human

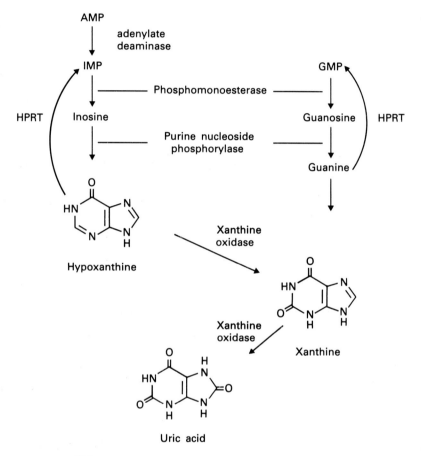

HPRT: Hypoxanthine phosphoribosylamine transferase

Figure 2.8 Purine nucleotide catabolism and salvage.

erythrocyte enzyme (Tuttle and Krenitsky, 1980). Subsequent enzymes convert the ribonucleotide into an analogue of ATP which is then incorporated into a faulty RNA (Nelson *et al.*, 1979).

References

Agarwhal, R.P. (1982). *Pharmacol. Ther.*, **17**, 399–429.

Ariel, I. (1970). *Cancer*, **25**, 705–714.

Ashton, W.T., Karkas, J.D., Field, A.K. and Tolman, R.L. (1982). *Biochim. Biophys. Res. Commun.*, **108**, 1716–1721.

Barry, A.L., Jones, R.N., Thornberry, C., Ayers, L.W., Gerlach, E.H. and Sommers, H.M. (1984). *Antimicrob. Ag. Chemother.*, **25**, 633–637.

Block, E.R., Jennings, A.E. and Bennett, J.E. (1973). *Antimicrob. Ag. Chemother.*, **3**, 649–656.

Blum, R.H. and Carter, S.K. (1974). *Ann. Intern. Med.*, **80**, 249–259.

Bonadonna, G. and Valagussa, P. (1981). *New Engl. J. Med.*, **304**, 10–15.

Cha, S., Agarwhal, R.P. and Parks, R.E. (1975). *Biochem. Pharmacol.*, **24**, 2187–2197.

Chang, T.-W. and Heel, R.C. (1981). *Drugs*, **22**, 111–128.

Chen, G.L., Yang, L., Rowe, T.C., Halligan, B.D. Tewey, K.M. and Liu, L.F. (1984). *J. Biol. Chem.*, **259**, 13560–13566.

Churchill, W.S. (1951). In: The Hinge of Fate, Vol 4 in the Second World War, (London: Cassell), p. 582.

Cohen, J. (1982). *Lancet*, **ii**, 532–537.

Cozzarelli, N.R. (1980). *Science*, **207**, 532–537.

De Clerq, E. and Walker, R.T. (1984). *Pharmacol Ther.*, **26**, 1–44.

Diasio, R.B., Bennett, J.E. and Myers, C.E. (1978). *Biochem. Pharmacol.*, **27**, 703–707.

Di Marco, A. (1975). *Cancer Chemother. Rep.*, **6**, 91–106.

Elford, H.L., Wamplen, G.L. and Van't Riet, B. (1979). *Cancer Res.*, **39**, 844–851.

Elford, H.L., Van't, Riet B., Wamplen, G.L., Lin, A.L. and Elford, R.M. (1981). *Adv. Enzyme Regul.*, **19**, 151–168.

Elion, G.B. (1967). *Fed. Proc.*, **26**, 898–904.

Engle, E.C., Manes, H.S. and Drlica, K. (1982). *J. Bacteriol.*, **149**, 92–99.

Erikkson, S. Thelander, L. and Akerman, M. (1979). *Biochemistry*, **18**, 2948–2972.

Ferone, R., Burchall, J.S. and Hitchings, G.H. (1969). *Molec. Pharmacol.*, **5**, 49–59.

Gellert, M. (1980). *Annu. Rev. Biochem.*, **50**, 879–910.

Gellert, M., Mizuuchi, K., O'Dea, M.H. and Nash, H.A. (1976a). *Proc. Nat. Acad. Sci. (U.S.A.)*, **7**, 3872–3876.

Gellert, M., O'Dea, M.H., Itoh, T. nd Tomizawa, J.I. (1976b). *Proc. Nat. Acad. Sci. (U.S.A.)*, **73**, 4474–4478.

Gilbert, B.E and Knight, V. (1986). *Antimicrob. Ag. Chemother.*, **30**, 201–205.

Graham, F.L. and Whitmore, G.F. (1970). *Cancer Res.*, **30**, 2636–2644.

Hammond, C.B., Weed, J.C., Barnard, D.E. and Tyray, L. (1981). *Cancer*, **31**, 322–332.

Heidelberger, C., Danenberg, P.V. and Moran, R.G. (1983). *Adv. Enzymol.*, **54**, 57–119.

Hershfield, M.S., Kredich, N.M., Koller, C.A., Mitchell, B.S. Kurtzberg, J., Kinney, T.R. and Falletta, J.M. (1983). *Cancer Res.*, **43**, 3451–3456.

Hitchings, G.H. and Smith, S.L. (1980). *Adv. Enzyme Regul.*, **18**, 349–371.

Hooper, D.C. and Wolfson, J.S. (1985). *Antimicrob. Ag. Chemother.*, **28**, 716–721.

Hooper, D.C. Wolfson, J.S. McHugh, G.L., Winters, M.B. and Swartz, M.N. (1982). *Antimicrob. Ag. Chemother.*, **22**, 662–671.

Hurly, M.G.D. (1959). *Trans. R. Soc. Trop. Med. Hyg.*, **53**, 412–413.

Kager, P.A., Rees, P.N., Wellde, B.T., Hockmeyer, W.T. and Lyerley, W.T. (1981). *Trans. R. Soc. Trop. Med. Hyg.*, **75**, 556–559.

Kelley, W.N. and Wyngaarden, J.B. (1974). *Adv. Enzymol.*, **41**, 1–33.

Kennedy, B.J. and Yarbro, J.W. (1966). *J. Am. Med. Assoc.*, **195**, 1038–1043.

Kisner, D.L. Schein, P.S. and Macdonald, J.S. (1981). *Rec. Results Cancer Res.*, **79**, 28–40.

Klevan, L. and Tse, Y.-C. (1983). *Biochim. Biophys. Acta*, **745**, 175–180.

Larder, B.A. and Darby, G. (1982). *J. Virol*, **42**, 649–658.

Lucas, A.O., Hendrickse, R.G., Okubadse, O.A., Richards, W.H.G., Neal, R.A. and Kofie, B.A.K. (1969). *Trans. R. Soc. Trop. Med. Hyg.*, **63**, 216–229.

Major, P.P., Agarwhal, R.P. and Kufe, D.W. (1983). *Cancer Chemother. Pharmacol.*, **10**, 128–138.

Major, P.P., Egan, E.M., Herrick, D.J. and Kufe, D.W. (1982). *Biochem. Pharmacol.*, **31**, 2937–2940.

Mar, E.-C. and Huang, E.-S. (1979). *Intervirol.*, **12**, 73–83.

Massey, V., Komai, H., Palmer, G. and Elion, G.B. (1970). *J. Biol. Chem.*, **245**, 2837–2844.

Matthews, D.A., Alden, R.A., Bolim, J.T., Filman, D.J., Freer, S.T., Hamlin, R., Wim, G.J.H., Kislink, R.L., Pastore, E.J., Plante, L.T., Xuong, N. and Krant, J. (1978). *J. Biol. Chem.*, **253**, 6946–6954.

McClung, H.W., Knight, V., Gilbert, B.E., Wilson, S.Z., Quarles, J.M. and Divine, G.W. (1983). *J. Am. Med. Assoc.*, **249**, 2671–2674.

Miller, W.H. and Miller, R.L. (1982). *Biochem. Pharmacol.*, **31**, 3879–3884.

Mitsuya, H., Weinhold, K.J., Furman, P.J., St. Clair, M.H., Nusinoff-Lehrman, S., Gallo, R.C., Bolognese, D., Barry, D.W. and Broder, S. (1985). *Proc. Nat. Acad. Sci. (U.S.A.)*, **82**, 7096–7100.

Moody, M.R. and Young, V.M. (1975). *Antimicrob. Ag. Chemother.*, **7**, 836–839.

Moore, E.C. (1969). *Cancer Res.*, **29**, 291–295.

Muller, W.E.G., Zahn, R.K., Brittlingmaier, K. and Felke, D. (1977). *Ann. N.Y. Acad. Sci.*, **284**, 34–48.

Nelson, D.J., LaFon, S.W., Tuttle, J.V., Miller, W.H., Miller, R.L., Krenitsky, T.A., Elion, G.B., Berens, R.L. and Marr, J.J. (1979). *J. Biol. Chem.*, **254**, 11544–11549.

Neu, H.C. and Labthavikul, P. (1982). *Antimicrob. Ag. Chemother.*, **22**, 23–37.

Parris, D.S. and Harrington, J.E. (1982). *Antimicrob. Ag. Chemother.*, **22**, 71–77.

Ramachandran, S., Godfrey, J.J. and Lionel, N.D.W. (1978). *J. Trop. Med. Hyg.*, **81**, 36–39.

Ross, W.E. (1985). *Biochem. Pharmacol.*, **34**, 4191–4195.

Rubin, R.N. (1981). *Clin. Ther.*, **4**, 74–94.

Ruprecht, R.M., O'Brien, L.G., Rossoni, L.D. and Nusinoff-Lehrman, S. (1986). *Nature*, **323**, 467–469.

Scott, J.T. (1980). *Brit. Med. J.*, **281**, 1164–1166.

Shaw, G.M., Harper, M.E., Hahn, B.H., Epstein, L.G., Gadjusek, D.C., Price, R.W., Navia, B.A., Petito, C.K., O'Hara, C.J., Groopman, J.E., Cho, E.-S., Oleske, J.M., Wong-staal, F. and Gallo, R.C. (1985). *Science*, **227**, 177–182.

Shepard, C.C., Ellard, G.A., Levy, L., Upromolla, V., Pattyn, S.R., Peters, J.H., Rees, R.J.W. and Waters, M.F.R. (1976). *Bull. WHO*, **53**, 425–433.

Smyth, J.F., Paine, J.M., Jackman, A.L., Harrap, K.R., Chassin, M.M., Adamson, R.H. and Johns, D.G. (1980). *Cancer Chemother. Pharmacol.*, **5**, 93–101.

Spector, T. and Johns, D.G. (1970). *J. Biol. Chem.*, **245**, 5079–5086.

Streeter, D.G., Witkowski, J.T., Khare, G.D., Sidwell, R.W., Bauer, R.J., Robins, R.K. and Simon, L.N. (1973). *Proc. Nat. Acad. Sci. (U.S.A.)*, **70**, 1174–1178.

Sugino, A., Peebles, C.L., Kreuzer, K.N. and Cozzarelli, N.R. (1977). *Proc. Nat. Acad. Sci. (U.S.A.)*, **74**, 4767–4771.

Tewey, K.M., Rowe, T.C., Yang, L., Halligan, B.D. and Liu, L.F. (1984). *Science*, **226**, 466–468.

Thelander, L. and Reichard, P. (1979). *Annu. Rev. Biochem.*, **48**, 133–158.

Thompson, L.F. and Seegmiller, J.E. (1980). *Adv. Enzymol.*, **51**, 167–210.

Tuttle, J.V. and Krenitsky, T.A. (1980). *J. Biol. Chem.*, **255**, 909–916.

Veerisetty, V. and Gentry, G.A. (1983). *J. Virol.*, **46**, 901–908.

Whelan, W.L. and Kerridge, D. (1984). *Antimicrob. Ag. Chemother.*, **26**, 570–574.

Whitley, R.J., Alford, C., Hess, F. and Buchanan, B. (1980). *Drugs*, **20**, 267–282.

Wolfenden, R.L., Wentworth, D.F. and Mitchell, G.N. (1977). *Biochemistry*, **16**, 5071–5077.

Woods, D.D. (1962). *J. Gen. Microbiol.*, **29**, 687–702.

Wray, S.K., Gilbert, B.E., Noall, M.W. and Knight, V. (1985). *Antiviral Res.*, **5**, 29–37.

Chapter 3

Protein biosynthesis

3.1 Introduction

The synthesis of protein from individual amino acids is an essential feature of living organisms. Major differences exist between the process in eukaryotes and prokaryotes, and these have been exploited serendipitously to produce anti-bacterial agents. Subtle differences are found between higher and lower eukaryotes (for example mammalian and yeast cells), but these have not so far indicated a compound that could be used to kill pathogenic fungi. This chapter, is, therefore, concerned solely with antibacterial agents.

Protein synthesis in eukaryotes and prokaryotes takes place both in ribosomes which are located in the cytoplasm and in mitochondria, but as no drugs are believed to owe their mechanism of action to interference with the mitochondrial process (although chloramphenicol does interfere with this process in a few individuals giving rise to side-effects), only the cytoplasmic system will be considered in this chapter. Protein synthesis can be divided into three stages: chain initiation, elongation and termination. Although almost all the medically useful inhibitors of protein synthesis act on the elongation stage, initiation will also be discussed in order that the whole process may be understood (reviewed in Kozak, 1983; Maitra *et al.*, 1982).

An amino acid first has to be activated by combination with a transfer RNA that is specific for that amino acid. This is done in two stages both catalysed by the same specific aminoacyl–tRNA ligase or synthetase: the amino acid first reacts with ATP and Mg^{2+} to form an enzyme-bound amino acid adenylate and then the amino acid is transferred to the appropriate tRNA, with the concomitant release of AMP. The triplet anticodon on the tRNA recognizes

Table 3.1 Antibacterial drugs discussed in Chapter 3

Section	Drug	Target
3.2	Aminoglycosides	30S ribosomal subunit
3.3	Chloramphenicol	50S ribosomal subunit
3.4	Tetracycline	30S ribosomal subunit
3.5	Erythromycin	50S ribosomal subunit
3.6	Clindamycin	50S ribosomal subunit

the codon for that amino acid on the messenger RNA, this complex formation takes place on the ribosome in the presence of GTP (Figure 3.1).

Protein synthesis occurs on the ribosome, a complex particle containing various proteins and RNA. In bacteria, the overall sedimentation coefficient of the ribosome is 70S, and it readily dissociates into unequal subunits of 30S (containing 21 proteins and 16S RNA) and 50S (34 proteins, and both 23S and 5S RNA) respectively. Ribosomes are normally linked together by an mRNA molecule like beads on a wire (polysomes), with the growing peptide chain at increasing lengths according to the ribosomal position on the mRNA. The ribosome directs the addition of amino acyl groups to a growing peptidyl tRNA by a condensation reaction that forms the peptide bond, and then moves with its growing chain along the messenger RNA to the next codon to be read.

In prokaryotes, protein synthesis is always initiated by formylmethionyl–tRNA (fMet–tRNA) while in eukaryotes this role is carried out by unformylated methionine. Formylation of methionine takes place after the methionine is linked to a form of tRNA specific for chain initiation. Also required are three proteins or initiation factors called IF 1 to 3 and also GTP. IF 1 and 2 are needed to position the mRNA and fMet–tRNA on the ribosome, while the role of IF 3 is to recognize the mRNA. GTP is hydrolysed to GDP in the process.

In order to assemble the active complex, the three initiation factors bind to the 30S subunit of the ribosome. FMet–tRNA and mRNA bind to the complex with a specific start codon, AUG, as part of the binding site on the mRNA. GTP also binds at this stage. A 50S subunit joins the complex to form the functional 70S ribosome, and GTP is hydrolysed to GDP and phosphate by ribosome-bound IF 2. IF 1 and 2 are then released from the complex so that fMet–tRNA is unblocked ready for peptide bound formation.

Chain elongation requires two sites on the ribosome (see Figure 3.1): one called the A-site (Acceptor) where the fMet–tRNA is initially located before it transfers to the P- or Peptide site (nearer the 5′ end of the nucleic acid). The A-site is located on the 3′ side of the P-site and is largely on the 30S subunit. The 50S subunit carries the P-site, although there are clearly interactions between the two subunits as the substrates bind. The incoming aminoacyl–tRNA binds to the A-site, and then the fMet is condensed with the second amino-acid residue to make a dipeptidyl-tRNA; a reaction which requires a protein known as elongation factor 1 together with GTP which is hydrolysed in the process. The reaction is catalysed by a specific protein on the ribosome itself, and the spent tRNA is released from the ribosome.

The de-acylated tRNA vacates the P-site and the dipeptidyl-tRNA is transferred back to the P-site still hydrogen-bonded to its mRNA codon — a process which requires the intervention of another molecule of GTP and elongation factor 2. This allows a third aminoacyl–tRNA defined by the mRNA codon to take up position in the A-site and the condensation process is repeated etc. (see Figure 3.1).

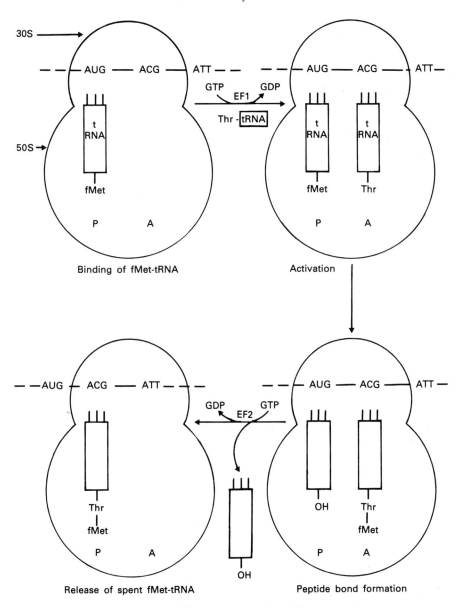

Figure 3.1 Chain initiation and formation of first peptide link.

Protein release factors (RF1 and RF2) are needed in order to terminate the process. They recognize 'stop' codons UAA, UAG and UGA. Termination is stimulated by GTP, and probably requires the hydrolysis of GTP by one of the release factors.

The drugs that are discussed in this chapter, namely the aminoglycosides, tetracyclines, chloramphenicol, erythromycin and clindamycin have their major effect on the chain elongation procedure. They are specific for protein synthesis in bacteria; apart from the tetracycline family which also inhibit eukaryotic protein synthesis. Nevertheless, despite this specificity, without exception they have undesirable side-effects and, in addition, their use over several decades has induced a considerable level of resistance among the bacterial population. For both these reasons, their use has been superseded, except in particular instances, by the much safer β-lactams (penicillins and cephalosporins) discussed in Chapter 5.

Bacteria may be divided into two main classes as to whether they take up the Gram stain; the cells are treated with crystal violet and then with iodine to fix the stain, followed by decolourization with ethanol or acetone. Finally the cells are exposed to safranine, a counterstain. Gram-positive microorganisms stain violet, whilst gram-negative stain red. This difference in response to a histological stain is derived from major differences in the cell wall of the two types of bacteria; gram-positive cells contain teichoic acids, polymers of glycerol or ribitol phosphate, unlike gram-negative cells, which may explain the affinity of the gram-positive cell wall for a basic dye such as crystal violet.

It is, however, not only a matter of chemistry but also of integrity as to whether a cell will take up the Gram stain because old, ruptured or dead gram-positive cells may stain as though they were gram-negative. Furthermore, gram-positive cells contain very little lipid whereas gram-negative are rich in lipid. In sharp contrast, one portion of the wall that both types have in common, but contributes more to the gram-positive cell, is peptidoglycan (also known as murein or glycopeptide — see Chapter 5).

3.2 Aminoglycosides

Aminoglycosides are a family of antibiotics produced mainly by streptomycetes (names ending in -mycin) or micromonospora (ending in -micin). Structurally, they consist of aminosugars connected to a central hexose/aminocyclitol — streptidine in streptomycin or L-deoxystreptamine in the other drugs mentioned in this chapter. Streptomycin was the first to be discovered but now is much less widely used than previously, while tobramycin is frequently regarded as the drug of choice when an aminoglycoside is indicated. Amikacin, a semisynthetic agent, is useful for the treatment of infections caused by aminoglycoside-resistant bacteria (Bailey, 1981).

Aminoglycosides are most useful therapeutically for infections caused principally by gram-negative organisms such as the enterobacteriaceae and pseudomonads, and are largely ineffective against gram-positive agents such as streptococci (DeTorres, 1981). In aerobic environments these drugs are bactericidal, i.e. they kill the bacterium (in certain instances a drug might only prevent the bacterium from growing and multiplying and would be termed bacteriostatic). They require the energy of oxidative phosphorylation mediated by respiratory quinones in the cell-wall to facilitate active transport into the cell because they are highly polar and do not diffuse to any extent across the cell-wall.

Aminoglycosides

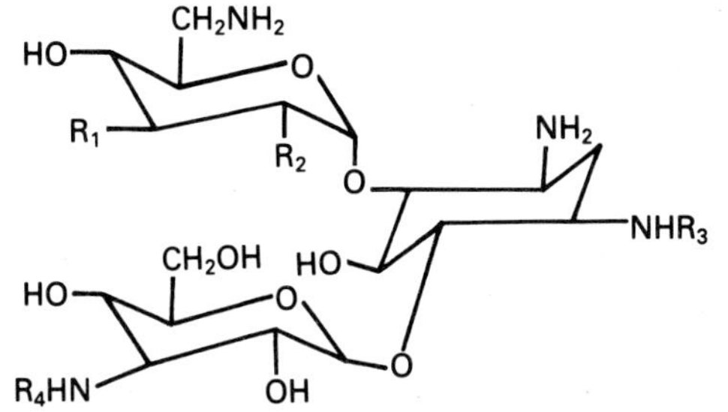

	R_1	R_2	R_3	R_4
Amikacin	OH	OH	COCH(OH)(CH$_2$)$_2$NH$_2$	H
Tobramycin	H	NH$_2$	H	NH$_2$

The primary intracellular site of action of the aminoglycosides is the 30S ribosomal subunit which consists of 21 proteins and a single 16S molecule of RNA. A single substitution of asparagine for lysine in protein S12 (i.e. protein 12 of the smaller ribosomal subunit as numbered by position of the protein on two dimensional gel electrophoresis (Kaltschmidt and Wittman, 1970) of *Escherichia coli* prevents streptomycin binding to the ribosome (Birge and Kurland, 1969) but it is doubtful whether this is the only protein involved (Lando *et al.*, 1976; Davies and Courvalin, 1977).

Dihydrostreptomycin binds to the 30S subunit with a binding constant of $2 \cdot 7 \times 10^{-7}$ M (Lando *et al.*, 1976). Competition is observed between streptomycin and dihydrostreptomycin, not surprisingly, but not with gentamicin; indeed gentamicin appears to bind to the 50S subunit since protein L16 (protein 16 of the large ribosomal subunit) is altered in a gentamicin-resistant mutant (Buckel *et al.*, 1977). Tobramycin appears to bind to both subunits of the ribosome and at several different sites since no saturation curve kinetics can be obtained when increasing concentrations of [^{3}H]tobramycin are used. Nevertheless tobramycin antagonism of acetyltobramycin binding suggests a binding constant of $1 \cdot 42 \times 10^{-6}$ M for the most tightly bound site (LeGoffic *et al.*, 1979).

One major effect of streptomycin binding is to cause misreading of the message so that the wrong amino acids are inserted into the growing polypeptide chain (Tai *et al.*, 1978). The aminoglycosides vary in their ability to cause misreading of the genetic code, presumably as a result of their varying affinity for different ribosomal proteins. This effect is found both in cell-free systems and in whole bacteria. Streptomycin can also interfere with chain

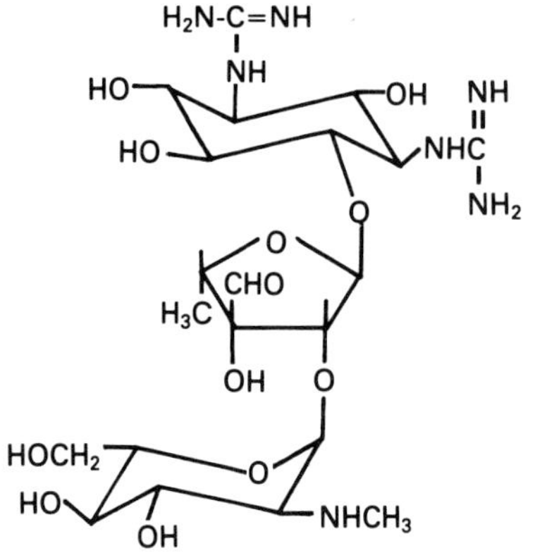

Streptomycin

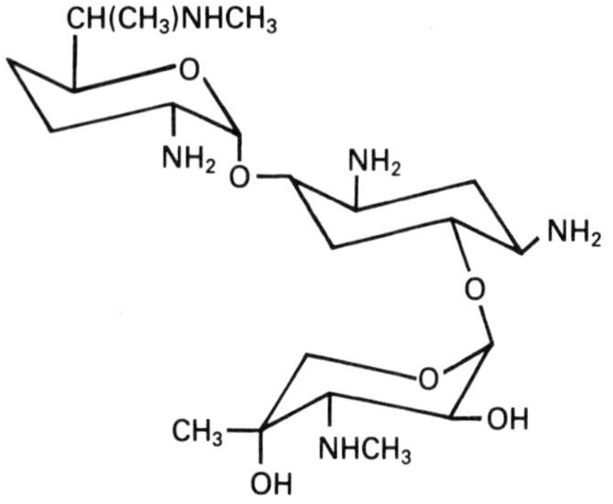

Gentamicin C$_1$

elongation by inhibiting, in a partial and reversible manner, protein synthesis by the full complex of ribosomes attached to messenger RNA (polysomes).

The drug does not, however, affect the process of initiation. It merely ensures that the initiation complex formed in its presence is less stable and less likely to engage in appreciable protein synthesis. The presence of the full complex would appear to mask the major streptomycin binding site, however, as streptomycin will additionally bind irreversibly to uncomplexed ribosomes, and thus cause the ribosomes to fall off the messenger RNA prematurely. The drug-bound ribosomes are impaired in IF 3-dependent dissociation and become engaged in abortive initiation and elongation (Wallace *et al.*, 1973). These two types of effect may cause the observed phenomena of partial inhibition of synthesis at lower concentration and total inhibition at higher concentration leading to cell death, in that at low concentrations streptomycin will not be able to block all the initiation sites in a cell but merely interfere transiently with polysomal complexes (Wallace *et al.*, 1973).

Drug resistance can occur, as one might expect, by alteration of the ribosome. Furthermore in an anaerobic environment where the energy of oxidative phosphorylation is not available to facilitate the active transport of the drug, bacteria sensitive in an aerobic environment become resistant. Moreover, the development of enzymes that can inactivate the drugs by acetylation of amino-groups and phosphorylation or adenylylation of hydroxyl groups allows another type of resistance to emerge. No fewer than 12 metabolizing enzymes have been identified with the genetic information carried on plasmids that require the presence of transfer resistance factors to express the resistance. Amikacin, however, is sensitive to only two of these enzymes (Bailey, 1981; DeTorres, 1981), so it can be used where other aminoglycosides would fail.

The use of aminoglycosides is restricted by a narrow therapeutic margin since they are in varying degrees toxic to the ear and kidney (Phillips, 1982). Streptomycin, in particular, damages the auditory and vestibular branches of the eighth cranial nerve. This drug is still in use for the treatment of tuberculosis although isoniazid and rifampicin are sometimes preferred.

3.3 Chloramphenicol

Chloramphenicol is a natural product isolated from the culture filtrate of *Streptomyces venezuelae*. It is a molecule of low molecular weight characterized by the presence of a nitrobenzene group, unusual both for a natural product and also for a drug. The drug has a broad spectrum of activity against both gram-positive and -negative organisms, and is able to penetrate freely into host tissues. In general, chloramphenicol is believed to be bacteriostatic, although it can be bactericidal to some species — notably *Haemophilus influenzae* which causes meningitis (Turk, 1977).

The drug binds primarily to the 50S subunit of the bacterial ribosome and, *in vitro*, protein L16 appears to be the target (Nierhaus and Nierhaus, 1973). When monoiodoamphenicol is incubated with whole cells, however, proteins S6 and L24 are labelled in addition to L16, suggesting that all three proteins are involved with the receptor site (Pongs and Messer, 1976). The binding constant for chloramphenicol binding to free ribosomes is 6×10^{-7} M (Fernandez-Munos *et al.*, 1971).

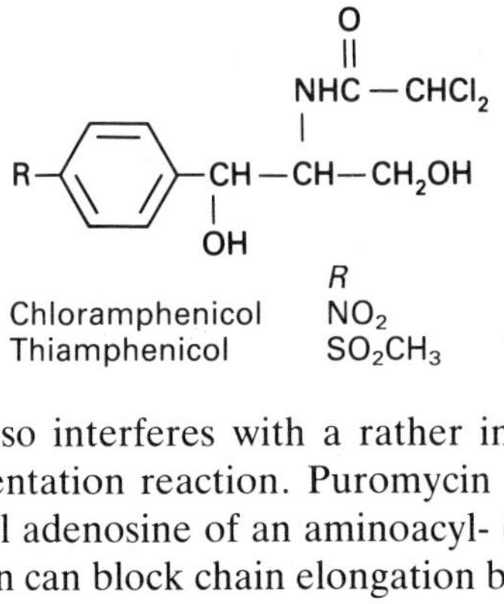

$$R\text{—}\underset{\overset{|}{OH}}{CH}\text{—}\underset{\underset{\overset{|}{NHC}\text{—}CHCl_2}{\overset{O}{\parallel}}}{CH}\text{—}CH_2OH$$

	R
Chloramphenicol	NO_2
Thiamphenicol	SO_2CH_3

Chloramphenicol also interferes with a rather interesting reaction called the puromycin fragmentation reaction. Puromycin is a natural product that resembles the terminal adenosine of an aminoacyl- (or peptidyl-) tRNA (see Figure 3.2). Puromycin can block chain elongation both by substituting for an incoming aminoacyl-tRNA at the A-site, and by accepting the peptidyl-tRNA bound at the P-site. The peptidyl-puromycin thus formed is incapable of further reaction as it contains an amide link attached to the ribose ring rather than the expected ester; the peptide chain consequently falls off the ribosome. All of the peptides so formed have the correct amino acid at the N terminus but puromycin at the C terminus.

The overall process is known as the puromycin fragmentation reaction and requires 50S subunits, a peptidyl- or aminoacyl-tRNA fragment, puromycin and K^+, Mg^{2+} and the presence of 33% ethanol. This reaction is inhibited by other antibiotics that bind at or near the A-site of the ribosome, notably chloramphenicol with a K_i of approximately 10^{-5} M (Pestka, 1970).

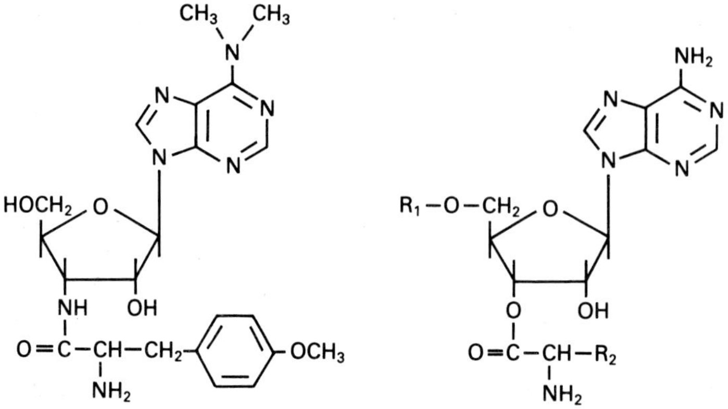

Figure 3.2 The resemblance of puromycin to an aminoacyl adenosine.

Puromycin   Terminal adenosine of an
                             aminoacyl (or peptidyl)-tRNA

R_1 represents transfer RNA
R_2 is the side-chain of an amino-acid or a growing peptide.
Puromycin resembles methyltyrosine sufficiently closely to accept the growing peptide chain. No futher change can take place because of the amide link at the 3' position of the sugar ring in puromycin instead of the expected, and more easily broken, ester linkage.

Chloramphenicol was extremely useful in earlier decades for combatting a wide range of infections, but it soon became clear that the drug was responsible for haematological side-effects, notably bone marrow suppression and, more seriously, aplastic anaemia. One in 25 000 patients who take the drug succumb to the latter condition, which is frequently fatal (Kucers, 1982). Bone marrow suppression is a consequence of mitochondrial injury which appears to result from the inhibition of mitochondrial protein synthesis. Aplastic anaemia, however, may result from reduction of the nitro group of the drug to less chemically stable intermediates since thiamphenicol, which has an SO_2CH_3 group instead of the nitro, does not show the same toxicity (Yunis *et al.*, 1980).

The recognition, however, that various serious infections, notably with anaerobic bacteria such as *Bacteroides fragilis*, are sensitive to chloramphenicol has to some extent brought back the use of the drug — particularly when, as in this case, the safer β-lactams are rendered less effective by the bacterial production of β-lactamases. Nevertheless, chloramphenicol therapy should be limited to those infections for which the benefit of the drug outweighs the risks of the side-effects. If other drugs are available that are equally effective but are less toxic, they should be used instead (see Kucers, 1982).

One condition where chloramphenicol appears to be the antibiotic of choice is for the treatment of meningitis and brain abscess caused by *Haemophilus influenzae* type b (Kucer, 1982) although in America chloramphenicol is recommended only for those strains of organism that are ampicillin resistant (AMA Drug Evaluation, 1983). *B. fragilis* infection of the central nervous system and systemic infections by *Salmonella* species (e.g. typhoid fever) are often treated by chloramphenicol (AMA Drug Evaluation, 1983).

3.4 Tetracyclines

Various streptomycetes elaborate members of the tetracycline family of anti-biotics, notably oxytetracycline from *S. rimosus* and chlortetracycline from *S. aureofaciens*. Later members of the family, such as minocycline and doxy-cycline, have been synthesized chemically and have clinical advantages in being better absorbed by host tissues.

For drugs which have been used for so many years, there is still a surprising amount of disagreement as to which protein is the target of their action. Tetracycline binds to the 30S ribosomal subunit in a 1:1 fashion. Specific targets are primarily proteins S4 and S18 with S7, S13 and S14 as minor targets as shown by photoincorporation studies with 70S ribosomes (Goldman *et al.*, 1980). There is only one strong binding site for tetracyclines on the ribosome and it is likely that the occupancy of this site is sufficient to block protein synthesis. The binding constant is 2×10^{-5} M (Tritton, 1977). Tetra-cycline binding stabilizes the ribosome thermally and raises the melting temperature of the ribosomal RNA. This effect correlates moderately well with the inhibition of polyphenylalanine synthesis for a series of tetracyclines (Tritton, 1977).

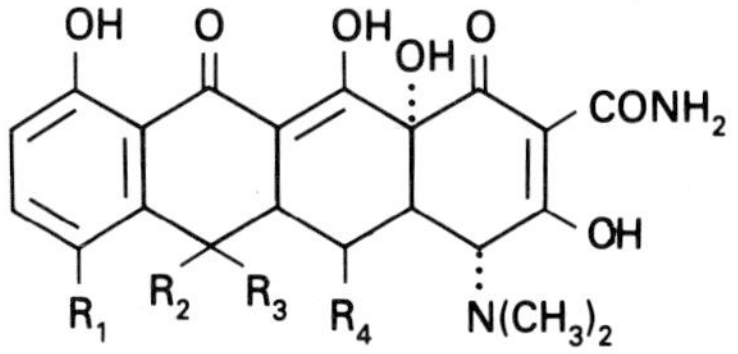

	R_1	R_2	R_3	R_4
Chlortetracycline	Cl	OH	CH_3	H
Oxytetracycline	H	OH	CH_3	OH
Doxycycline	H	H	CH_3	OH
Minocycline	$N(CH_3)_2$	H	H	H

As a consequence of this binding, aminoacyl-tRNA is prevented from binding to the A-site, and so no peptide bond can be formed. Elongation factor, EF_1, dependent hydrolysis of GTP is not affected by tetracycline (Modolell *et al.*, 1971). The interaction between the ribosome and EF_1 may be interrupted directly, or indirectly by interfering with the process whereby EF_1 recognizes the anti-codon on the tRNA (Smythies *et al.*, 1972).

Some interesting studies have been carried out on the process by which tetracyclines gain access to the cell (reviewed in Chopra and Howe, 1978). In gram-negative species, passive diffusion of the non-ionized form of the drug takes place through hydrophilic pores in the outer cell membrane, specifically through protein 1A, one of three proteins situated there. In contrast, mino-cycline and doxycycline, which are more lipophilic than the original natural products, pass more readily through the lipid bilayer rather than the pore. The result is better uptake, as shown by studies on *E. coli* in which the uptake of minocycline was 10 to 20 times more rapid than that of tetracycline (McMurray *et al.*, 1982). The second process involves an energy-dependent active transport system that pumps tetracycline through the cytoplasmic membrane. Less is known about transport into gram-positive bacteria but it also requires an energy dependent process. Plasmids which confer resistance to tetracyclines in *E. coli* and possibly other species code for proteins which are located in the cell envelope and probably interfere with the uptake of the drug (Chopra and Howe, 1978).

The tetracyclines are active against a wide range of gram-positive and -negative bacteria but a number of organisms such as *Pseudomonas aeruginosa*, and staphylococcal and streptococcal species have acquired resistance possibly as a result of indiscriminate use of these drugs. For the treatment of Chlamydial infections notably urethritis, and pneumonia induced by *Mycoplasma pneumoniae* the tetracyclines are still regarded as the drugs of choice. The tetracyclines are primarily bacteriostatic at low concentrations and bactericidal at high (Wilson and Cockerill, 1983).

3.5 Erythromycin

Erythromycin is a larger molecule than the other antibiotics considered in this chapter, with a molecular weight of 734. It consists of a 14-membered lactone ring with two sugar molecules attached, and was first detected as a product of *Streptomyces erythreus*. Erythromycin can be either bacteriostatic or bactericidal, depending on both the organism and the concentration of the drug. In general it is not active against most aerobic gram-negative bacilli. It is active against some gram-positive cocci such as *Streptococcus pneumoniae*, and is particularly useful in patients that show an allergy to penicillins. Erythromycin can also be helpful against staphylococcal strains that are resistant to penicillins. It is still the drug of choice in Legionnaire's disease caused by *Legionella pneumophila* and for the treatment of diphtheria (Kucers, 1982;

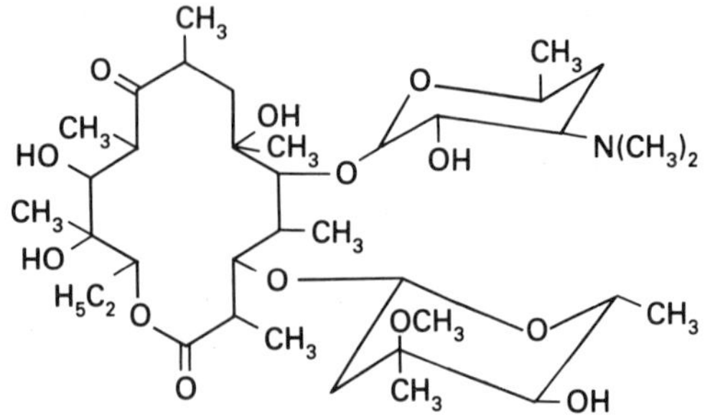

Erythromycin

Blagg and Gleckman, 1981). Unfortunately, erythromycin usage is not free from side-effects, notably liver toxicity, reversible hearing loss and pseudo-membraneous colitis (Blagg and Gleckman, 1981) — see under clindamycin.

Protein L16 is the binding site for erythromycin since the loss of this protein, carried out by washing with ammonium chloride and ethanol, parallels the loss of drug binding to the ribosome (Bernabeu *et al.*, 1977). In *E. coli*, an erythromycin-resistant mutant has been identified in which L22 is altered but binding of the drug to the ribosome still takes place, whereas alteration of L4 causes drug binding to the ribosome to be greatly decreased (Pardo and Rosset, 1977). These authors suggest, in view of other work which indicates that alterations in the 30S subunit can restore binding of a 50S subunit previously made resistant, that the large subunit proteins involved may be close to the small subunit and that both subunits may participate in the mode of action of erythromycin.

The drug blocks protein synthesis soon after initiation at the stage of elongation of the oligopeptide chain. It does this from its probable binding site on the large subunit by sterically hindering the A-site where the sequential addition of the amino acids takes place. The optimum length of nascent polypeptide chain for erythromycin to block appears to be two to five residues (Contreras and Vazquez, 1977). Accordingly, erythromycin blocks peptide bond formation in the puromycin fragment assay when the donor oligopeptidyl-tRNA is a maximum of five residues in length. The binding of the drug is reversible and can take place only if the 50S subunit does not contain tRNA molecules with growing peptide chains. Production of small peptides can continue in the presence of the antibiotic, but the production of larger peptides is blocked. The consequence is premature release of oligopeptidyl-tRNA from polysomes; rather like the effect of puromycin except that the antibiotic does not attach itself to the growing polypeptide chain (Otaka and Kaji, 1982).

3.6 Clindamycin

Clindamycin is the 7-deoxy, 7-chloro analogue of lincomycin, which was initially obtained from *Streptomyces lincolnensis*. The former is less toxic, however, and its use has superseded that of lincomycin. Broadly speaking, the spectrum of activity of clindamycin resembles that of erythromycin, and the former is of considerable use in the treatment of infections caused by anaerobic bacteria, notably *B. fragilis* (Dhawan and Thadepalli, 1982). Central nervous system infections are best managed by chloramphenicol which crosses the blood-brain barrier more effectively (Fass *et al.*, 1973). More recent work suggests that metronidazole and cefoxitin are at least as effective as clindamycin *in vivo* (Dubreil *et al.*, 1984).

Antibiotics of the lincomycin family inhibit protein synthesis in whole bacteria, and in cell-free systems, by interacting with the 50S subunit at a site very close to that occupied by erythromycin and chloramphenicol. This is shown by a parallel loss of binding for all three antibiotics when protein L16 is washed out of the ribosome by ammonium chloride and ethanol treatment (Bernabeu *et al.*, 1977). Furthermore, lincomycin inhibits the binding of chloramphenicol and erythromycin and *vice versa* (Fernandez-Munos *et al.*, 1971). One molecule of lincomycin binds per ribosome preventing peptide bond formation by blocking binding of the 3′ terminal end of the substrate to the A-site on the ribosome (Pestka, 1970). The binding constant for linco-mycin binding to *E. coli* ribosomes was $3 \cdot 4 \times 10^{-5}$ M as measured by direct binding of radiolabelled drug by equilibrium dialysis. Under conditions of the puromycin fragment assay with 33% ethanol the binding was increased to give a figure of $1 \cdot 7 \times 10^{-6}$ M but this is clearly not a physiological situation.

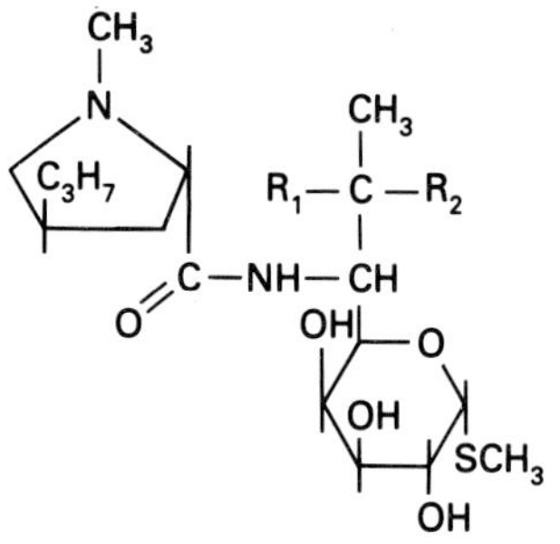

	R_1	R_2
Lincomycin	OH	H
Clindamycin	H	Cl

Clindamycin suffers from the drawback that diarrhoea frequently accompa-nies treatment, which can occasionally degenerate into pseudomembraneous colitis, sometimes a lethal disease which is caused by *Clostridium difficile* and characterized by diarrhoea, fever and abdominal pain (Dhawan and Thadepalli, 1982).

References

AMA Drug Evaluation (1983). Prepared by A.M.A Drug Division 5th edition (Philadelphia, U.S.A.: W.B. Saunders Co) p. 1255.

Bailey, R.R. (1981). *Drugs*, **22**, 321–327.

Bernabeu, C., Vazquez, D. and Ballesta, J.P.G. (1977). *Eur. J. Biochem.*, **79**, 469–477.

Birge, E.A. and Kurland, C.G. (1969). *Science*, **166**, 1282–1284.

Blagg, N.A. and Gleckman, R.A. (1981). *Postgrad. Med.*, **69**, 59–67.

Buckel, P., Buchberger, A., Bock, A. and Wittman, H.G. (1977). *Molec. Gen. Genetics*, **158**, 47–54.

Chopra, I. and Howe, T.G.B. (1978). *Microbiol. Rev.*, **42**, 707–724.

Contreras, A. and Vazquez, D. (1977). *Eur. J. Biochem.*, **74**, 539–547.

Davies, J. and Courvalin, P. (1977). *Am. J. Med.*, **62**, 868–872.

DeTorres, O.H. (1981). *Clin. Ther.*, **3**, 399–412.

Dhawan, V.K. and Thadepalli, H. (1982). *Rev. Infect. Dis.*, **4**, 1133–1153.

Dubreil, L., Devos, J., Neut, C. and Romond, C. (1984). *Antimicrob. Ag. Chemother.*, **25**, 764–766.

Fass, R.J., Scholand, J.F., Hodges, G.R. and Saslaw, S. (1973). *Ann. Intern. Med.*, **78**, 853–859.

Fernandez-Munos, R., Monro, R.E., Torres-Pinedo, R. and Vazquez, D. (1971). *Eur. J. Biochem.*, **23**, 185–193.

Goldman, R.A., Cooperman, B.S., Strycharz, W.A., Williams, B.A. and Tritton, T.R. (1980). *FEBS Letters*, **118**, 113–118.

Kaltschmidt, E. and Wittman, H.G. (1970). *Proc. Nat. Acad. Sci. (U.S.A.)*, **67**, 1276–1282.

Kozak, M. (1983). *Microbiol. Rev.*, **47**, 1–45.

Kucers, A. (1982) *Lancet*, **ii**, 425–428.

Lando, D., Cousin, M.A., Ojasoo, T. and Raynaud, J.P. (1976). *Eur. J. Biochem.*, **66**, 597–606.

LeGoffic, F., Capmau, M.L. Tangy, F. and Baillarge, M. (1979). *Eur. J. Biochem.*, **102**, 73–81.

McMurray, L.M., Cullinane, J.C. and Levy, S.B. (1982). *Antimicrob. Ag. Chemother.*, **22**, 791–799.

Maitra, U., Stringer, E.A. and Chaudhuri, A. (1982). *Annu. Rev. Biochem.*, **51**, 869–900.

Modolell, J., Cabrer, B., Parmeggiani, A. and Vazquez, D. (1971). *Proc. Natl. Acad. Sci (U.S.A.)*, **68**, 1796–1800.

Nierhaus, D. and Nierhaus, K.H. (1973). *Proc. Natl. Acad. Sci. (U.S.A.)*, **70**, 2224–2228.

Otaka, T. and Kaji, A. (1982). *Arch. Biochem. Biophys.*, **214**, 846–849.

Pardo, D. and Rosset, R. (1977). *Molec. Gen. Genetics*, **156**, 267–271.

Pestka, S. (1970). *Arch. Biochem. Biophys.*, **136**, 80–88.

Phillips, I. (1982). *Lancet*, **ii**, 311–314.

Pongs, O. and Messer, W. (1976). *J. Mol. Biol.*, **101**, 171–184.

Smythies, J.R., Benington, F. and Morin, R.D. (1972). *Experientia*, **28**, 1253–1254.

Tai, P.-C., Wallace, B.J. and Davis, B.D. (1978). *Proc. Natl. Acad. Sci. (U.S.A.)*, **75**, 275–279.

Tritton, T.R. (1977). *Biochemistry*, **16**, 4133–4138.

Turk, D.C. (1977). *J. Med. Microbiol.*, **10**, 127–131.

Wallace, B.J., Tai, P.-C., Herzog, E.L. and Davis, B.D. (1973). *Proc. Natl. Acad. Sci. (U.S.A.)*, **70**, 1234–1237.

Wilson, W.R. and Cockerill, F.R. (1983). *Mayo Clin. Proc.*, **58**, 92–98.

Yunis, A.A., Miller, A.M., Salem, Z. and Arimura, G.K. (1980). *Clin. Toxicol.*, **17**, 359–373.

Chapter 4

Carbohydrate metabolism in anaerobic microorganisms

4.1 Introduction

Metronidazole is a drug which is used widely for the treatment of infections caused by anaerobic organisms, either (a) protozoa such as *Trichomonas vaginalis* which provokes vaginal discharges; and *Entamoeba histolytica* causing amoebic dysentery or (b) bacteria as in, for example, abscesses or tissue necroses caused by *Bacteroides fragilis* and antibiotic-induced pseudomembraneous colitis (*Clostridium difficile*). For a general review on metronidazole, see Muller (1981). Although there is agreement that the drug has to be reduced to show activity (metronidazole is in fact a pro-drug) the ultimate target may be DNA in bacteria and, possibly, carbohydrate metabolism in protozoa. In view of this uncertainty, therefore, it has been placed in a chapter on its own.

Carbohydrate metabolism in anaerobes appears to be similar to that in aerobes — at least as far as the conversion of glucose to pyruvate is concerned. Although anaerobes subsequently transform pyruvate to acetyl coenzyme A, they use a very different type of enzyme from the mammalian or bacterial pyruvate dehydrogenase. Pyruvate decarboxylation in anaerobes is linked to the reduction of hydrogen ion by ferredoxin (Fd), a protein which contains four iron/sulphur centres and is a very powerful reducing agent. This reaction is catalysed by hydrogenase, to give hydrogen gas, and the reduction of molecular oxygen to give hydrogen peroxide. Pyruvate dehydrogenase (pyruvate: ferredoxin oxidoreductase) appears to have been a very early product of evolution as it is also present in archaebacteria which are thought to have existed initially in a highly reduced environment (Kerscher and Oesterhelt, 1982). The overall reaction is shown below:

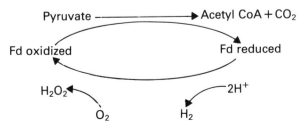

In trichomonads, acetyl coenzyme A is converted to acetate directly by an enzyme which simultaneously transfers the coenzyme A moiety to succinate. Succinyl coenzyme A is then hydrolysed to succinate and free coenzyme A coupled with phosphorylation of ADP to ATP (this is known as substrate level phosphorylation in contrast to oxidative phosphorylation in which an electron transport chain is required). In prokaryotes, acetate is still the end-product but an intervening step of acetylphosphate formation is included (sometimes referred to as the phosphoroclastic reaction). In trichomonads, but not in bacteria, pyruvate decarboxylase and the other enzymes involved in the later stages of carbohydrate metabolism are enclosed in a cytoplasmic organelle called the hydrogenosome (anaerobic protozoa do not have mito-chondria and peroxisomes) (Lindmark *et al.*, 1975).

4.2 *Metronidazole action*

Metronidazole is the original and most widely used of a family of alkyl-substituted 5-nitroimidazoles. It is inactive against organisms that grow aero-bically and have reducing activity equivalent to a redox potential of -350 mV whereas anaerobic organisms possess a reducing system sufficiently powerful to reduce the nitro group ($E_o = -0.47$ mV; O'Brien and Morris, 1972). These products are likely to be the nitro group radical-ion for one-election reduction, nitroso for two-electron reduction, reduction to a hydroxylamino group for four while a 6-electron reduction would give rise to products at the oxidation level of the amino group. Whether all or only some of these products occur in the intact cell is not established but the instability of some of them makes their detection extremely difficult. The chemical instabil-ity of one product at least ensures that reaction takes place with a sensitive constituent of the cell leading to cell death (Muller and Gorrell, 1983). Intracellular reduction of the drug causes more to be transported into the cell to maintain equilibrium. The net effect is that accumulation of the drug occurs in sensitive cells (Muller, 1981).

Metronidazole

$$O_2N \diagdown \quad \diagup CH_2CH_2OH$$
$$N$$
$$N \diagdown CH_3$$

The powerful reducing agent required to carry out this reduction is prob-ably the iron–sulphur protein ferredoxin ($E_o = -0.64$ mV; O'Brien and Mor-ris, 1972). Ferredoxins contain equal numbers of iron and sulphur atoms at the active site (i.e. 4 Fe: 4 S in mammalian systems), capable of acting as redox agents in a respiratory chain because of the ability of the iron to change valency. They are characterized by the presence of acid-labile sulphur atoms — so-called because treatment with acid liberates H_2S. Ferredoxin-linked reduction of metronidazole has been demonstrated in cell-free extracts of *Clostridium pasteurianum*, and indeed it appears that the drug acts as an electron sink, siphoning reducing equivalents away from other cellular reductive processes. Methylviologen is less effective, while the pyridine nucleotide coenzymes are totally ineffective. The data suggest that a 4-electron reduction takes place, consistent with earlier studies (Lockerby *et al.*, 1984).

Although the model is perhaps rather too simplistic to be regarded as complete, rapid reduction of metronidazole occurs in the presence of cysteine and ferrous iron (Willson and Searle, 1975). The radical-ion is formed first by one-electron reduction followed by further reduction to a nitroso moiety, as shown by electron spin resonance (e.s.r.) measurements (Mason and Josephy, 1985).

Experimentally, almost the earliest effects detectable after drug treatment are the cessation of carbon dioxide and hydrogen evolution, as seen with *C. welchii* for example (Edwards *et al.*, 1973). The radical-ion produced by metronidazole acting as an electron acceptor in place of hydrogen ion has been detected by e.s.r. measurements both in hydrogenosomes from *T. vaginalis* (Lloyd and Kristensen, 1985) and in the intact organism (Chapman *et al.*, 1985). The radical-ion is then reduced further to the active species that attacks crucial cell constituents. If the reduced product(s) were chemically stable it is likely that the effect of the drug would be reversible, in the sense that when all the drug had been reduced hydrogen ion would continue as the electron acceptor, and hydrogen evolution would re-start.

In aerobic situations oxygen can accept the electron from the radical-ion because the redox potential of the O_2/O_2^- is greater than that of RNO_2/RNO_2^-. Superoxide ion and the uncharged drug molecule are produced — a futile metabolic cycle. Such a cycle has been observed in rat liver microsomes in the presence of NADPH (Perez-Reyes *et al.*, 1979), and in hydrogenosomes from *Tritrichomonas foetus*, the cattle pathogen, in the presence of oxygen (Moreno *et al.*, 1984). In the latter case, pyruvate and CoA addition stimulated the production of the radical-ion, but NADPH was totally ineffective, as one would expect.

The precise nature of the protozoal target for the reduced active species is not known. Chapman *et al.* (1985) showed that progressive damage occurs to the radical generating system as a consequence of metronidazole reduction and considered that the radical-ion is likely to be reactive enough not to travel very far in the whole cell. Furthermore, Lloyd and Kristensen (1985) proposed that a metabolite of the drug may irreversibly interfere with part of the system for hydrogen production, since the level of gas release falls progressively with time in the presence of metronidazole. The likelihood, therefore, of a reative metabolite crossing both the hydrogenosomal and nuclear membranes to attack the DNA is not very high.

In bacteria, where there are neither nuclear nor hydrogenosomal membranes, DNA synthesis has been proposed as the target for the active species. Although protein and RNA synthesis in *B. fragilis* are unaffected by treatment with 10 μg/ml metronidazole, DNA synthesis, as measured by [³H]thymidine incorporation, halted 20 minutes after treatment. The DNA remained structurally intact, as shown by subsequent extraction and gel electrophoresis, and it is unlikely that the level of supercoiling was affected, thus ruling out an effect on DNA gyrase (see Chapter 2). The precise target may be in a multi-enzyme replication step since DNA polymerase also was not inhibited by drug in sonicated cells, and DNA was clearly capable of serving as a template for RNA synthesis (Sigeti *et al.*, 1983). Earlier studies had shown that if metronidazole is reduced in the presence of DNA, the reduced products will bind covalently to the nucleic acid but without interfering with the sedimentation velocity or melting temperature (LaRusso *et al.*, 1978).

The development of resistance to 5-nitroimidazoles does not occur at all frequently. Two possibilities that in theory can occur with an anti-infective agent are, (a) reduced transport into the cell, which would be very unlikely with such a small molecule as metronidazole or (b) metabolism by enzymes, which have not been discovered. Resistance modes normally give a clue as to the mode of action of a given drug, but in this case the evidence is very thin. Oxygen detoxification of metronidazole has been suggested as a possible method of inducing resistance (Morent *et al.*, 1984), since oxygen can react with the radical-ion to reform uncharged drug. Indeed, if sensitivity tests to 5-nitroimidazoles are carried out under aerobic conditions the microorganisms often appear more or less resistant. In studies with *T. vaginalis* isolates that markedly differed in their susceptibility to the drug *in vivo*, the availability of oxygen was shown to be an important factor in that NADH oxidase, which operates to keep oxygen tensions low, was reduced in one strain (Clackson and Coombs, 1982).

Mutations in pyruvate: ferredoxin oxidoreductase could be a possible mode of resistance, as shown for *Clostridium perfringens* by mutagenesis with *N*-methyl-*N'*-nitro-*N*-nitrosoguanidine (Sindar *et al.*, 1982). In these mutants pyruvate and lactate levels rose whereas acetate and carbon dioxide levels fall during growth, and so the enzyme activity was not totally abolished. Muller and Gorrell (1983), however, discounted the importance of this enzyme since activity can be quite low even in susceptible strains. Furthermore, a number of clinically resistant isolates defied identification of their locus of resistance (Muller and Gorrell, 1983). The overall difficulty in assigning the resistance to a given site is that the organism, by virtue of its ferredoxin or other powerful reducing agent, is almost certain to be able to reduce metronidazole to chemically unstable intermediate(s). The nature of the ultimate target could vary from species to species depending on the nature of the nearest sensitive molecule.

4.3 Medical use of metronidazole

Metronidazole was originally introduced on to the market in 1959 for the treatment of protozoal infections. The development of the drug for the treatment of infections caused by anaerobic bacteria coincided with, and played a major part in, the recognition of the importance of these organisms in infectious disease (Finegold, 1981; Bartlett, 1982). Metronidazole is regarded as the drug of choice in the treatment of these conditions, particularly those likely to be caused by *B. fragilis*. In one study on the susceptibility of anaerobic bacteria isolated from patients in several French hospitals, metronidazole and cefoxitin were found to be equally effective while clindamycin was less so. Metronidazole was considered to be the more suitable drug partly because fewer resistant strains have been recorded and, in addition, a smaller dose is normally required than for cefoxitin (Dubreil *et al.*, 1984).

Metronidazole is particularly effective *in vivo*, being readily absorbed from the gastrointestinal tract and widely disseminated in body fluids and tissues. Furthermore, in the considerable period of the use of the drug, very few genuinely resistant strains have been reported. Another potential problem with the use of the drug has been the possibility of mutagenicity as indicated by the Ames test (Ames *et al.*, 1975). This test measures the ability of a compound to reverse a mutation in *Salmonella typhimuria* that lacks an enzyme in the histidine biosynthetic pathway (phosphoribosyl adenosine triphosphate synthetase). The bacterium is unable to grow in histidine-deficient medium unless a reverse mutation occurs. Since the microsomal mixed function oxidase system which involves cytochrome P–450, is often responsible for the metabolism of inactive precursors to frank carcinogens, liver microsomes and a NADPH generating system are also included. Metronidazole is positive in this test.

It should be noted that the Ames test is not the only test used for potential carcinogens and that some of its results are controversial. In practice over nearly 30 years the use of metronidazole has not given rise to an abnormal incidence of tumours (Beard *et at.*, 1979). The drug is also very cheap, at least in oral tablet form.

References

Ames, B.N., McCann, J. and Yamasaki, E. (1975). *Mutat. Res.*, **31**, 347–364.

Bartlett, J.G. (1982). *Lancet*, **ii**, 478–481.

Beard, C.M., Noller, K.L., O'Fallon, W.M., Kurland, L.T. and Dockerty, M.B. (1979). *New Engl. J. Med.*, **301**, 519–521.

Chapman, A., Cammack, R., Linstead, D. and Lloyd, D. (1985). *J. Gen. Microbiol.*, **131**, 2141–2144.

Clackson, T.E. and Coombs, G.H. (1982). *J. Protozool.*, **29**, 636.

Dubreil, L., Devos, J., Neut, C. and Romond, C. (1984). *Antimicrob. Ag. Chemother.*, **25**, 764–766.

Edwards, D.I., Dye, M. and Carne, H. (1973). *J. Gen. Microbiol.*, **76**, 135–145.

Finegold, S.M. (1981). *Scand. J. Infect. Dis.*, Suppl **26**, 9–13.

Kerscher, L. and Oesterhelt, D. (1982). *Trends in Biochem Sci.*, **3**, 371–374.

LaRusso, N.F., Tomasz, M., Kaplan, D. and Muller, M. (1978). *Antimicrob. Ag. Chemother.*, **13**, 19–24.

Lindmark, D.G., Muller, M. and Shio, H. (1975). *J. Parasitol*, **61**, 552–554.

Lindmark, D.G., and Muller, M. (1976). *Antimicrob. Ag. Chemother.*, **10**, 476–482.

Lloyd, D. and Kristensen, B. (1985). *J. Gen. Microbiol.*, **131**, 849–853.

Lockerby, D.L., Rabin, H.R., Bryan, L.E. and Laishley, F.J. (1984). *Antimicrob. Ag. Chemother.*, **26**, 665–669.

Mason, R.P. and Josephy, P.J. (1985). *J. Inorg. Biochem.*, **24**, 161–165.

Moreno, S.N.J., Mason, R.P. and Decampo, R. (1984). *J. Biol. Chem.*, **259**, 8252–8259.

Muller, M. (1981). *Scand. J. Infect. Dis.*, Suppl. **26**, 31–41.

Muller, M. and Gorrell, T.E. (1983). *Antimicrob. Ag. Chemother.*, **24**, 667–673.

O'Brien, R.W. and Morris, J.G. (1972). *Arch. Mikrobiol.*, **84**, 225–233.

Perez-Reyes, E., Kalyanaraman, B. and Mason R.P. (1979). *Molec. Pharmacol.*, **17**, 239–244.

Sigeti, J.S., Guiney, D.G. and Davis, C.E. (1983). *J. Infect. Dis.*, **148**, 1083–1089.

Sindar, P., Britz, M.L. and Wilkinson, R.G. (1982). *J. Med. Microbiol.*, **15**, 503–509.

Willson, R.L. and Searle, A.J.F. (1975). *Nature*, **255**, 498–500.

Chapter 5

Cell wall biosynthesis

5.1 Introduction

Fungi and bacteria both require a structure external to their cell membrane to maintain the integrity of the organism — preventing a high internal osmotic pressure from rupturing the cell in a hypotonic external environment. Drugs that interfere with the biosynthesis of the cell wall might therefore be expected to be of value in combating infection by these organisms, particularly since mammalian cells do not require such a structure. The β-lactams (penicillins and cephalosporins) and vancomycin fall into this category.

β-Lactams, indeed, have been on the market for many decades; the first observation that led to the identification of the precursor penicillin was made in 1929, although it took nearly 20 years for the drug to reach the market. In addition, a number of agents are known to interfere with the biosynthesis of the fungal cell wall, such as nikkomycins and polyoxins that block the biosynthesis of chitin; the latter have been used for the control of fungal diseases in plants, such as rice blast in Japan (Gooday, 1977). No compound with this mode of action, however, has yet been found effective in the treatment of human mycoses, and so it is only antibacterial agents with which we are concerned in this chapter.

Peptidoglycan is a polymeric molecule which consists of parallel polysaccharide chains covalently joined by peptide cross-links. The whole structure can be regarded as a single molecule encompassing the entire cell in a sort of rigid bag. The glycan chain is a polymer consisting of alternating pyranose residues of N-acetylglucosamine and N-acetylmuramic acid connected in a β-(1–4) linkage (similar to the fungal cell wall polysaccharide chitin) with D-lactyl groups attached to alternate sugar residues. Attached via amide links through the carboxyl groups of the D-lactate are tetrapeptide chains (Figure 5.1). The sequence of this peptide is L-ala-γ-D-glu-L-lys (or *meso*-diaminopimelic acid)-D-ala (*meso*-diaminopimelic acid is found in gram-negative organisms, lysine in the gram-positive *Straphylococcus aureus*).

The bifunctional amino acid in the chain (lysine for example) acts as a cross-link to accept the terminal carboxyl group of a D-ala residue from an adjacent chain. This linkage may be either direct as in *Escherichia coli*, or

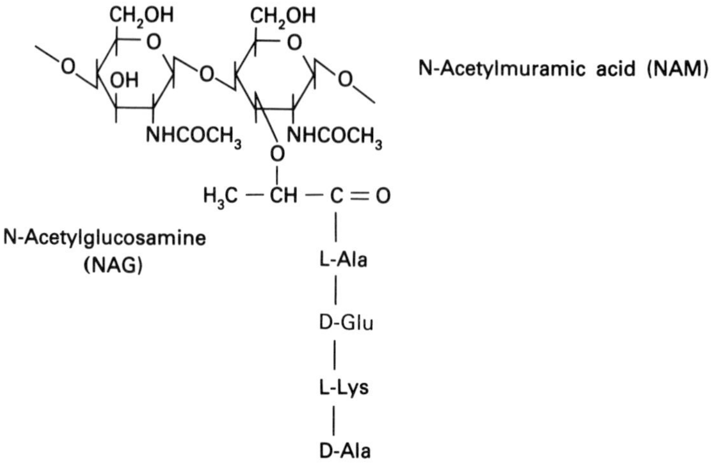

The γ-carboxyl group of D-Glu is used for peptide bond formation.

Figure 5.1 Peptidoglycan repeating unit (*S. aureus*).

through a short connecting peptide such as the pentaglycine found in *Staphylococcus aureus* (Figure 2.5). This structure is resistant to peptidases which cannot attack peptides containing D-amino acids. The enzyme lysozyme, however (found in tears and egg-white) is able to split the polysaccharide back-bone at the β-(1–4) link, to yield disaccharides of the basic repeating unit to which peptide chains are still attached. This breaks the strength of the peptidoglycan so that it can no longer contain the osmotic pressure inside the cell, and so the membrane ruptures with loss of cell contents.

The biosynthesis of the peptidoglycan involves the action of about 30 enzymes, starting with the formation of active precursors by soluble enzymes in the cytoplasm. The first major precursor or monomer is the pentapeptide attached to uridine diphosphate: UDP-MurNac-L-ala-γ-D-glu-L-lys-D-ala-D-ala (MurNac is short hand for *N*-acetylmuramic acid). Synthesis of the D-ala-D-ala fragment, which is added last, occurs by conversion of L-alanine to D-alanine (epimerization) followed by condensation of two D-alanine molecules. The monomer is then transferred to a C_{55}-terpene alcohol with a terminal phosphate group, known as bactoprenol, which acts as a membrane-bound carrier instead of UDP, and is linked to a phospholipid in the cell membrane via a pyrophosphate bridge. Condensation of UDP-*N*-acetylmuramyl-pentapeptide with UDP-*N*-acetylglucosamine is effected. In the case of the gram-positive organism *Staphylococcus aureus*, five glycine residues are transferred from glycinyl-tRNA to form a chain attached at one end to the ε-amino group of the L-lysine residue in the side-chain. The unit is then inserted into the cell wall and simultaneously the bactoprenol is released as a pyrophosphate. This is the reaction inhibited by vancomycin. The pyrophosphate then has to be converted to a monophosphate before being used again.

The third and final stage involves the formation of the cross-link. The

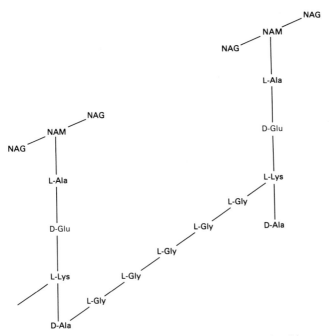

NAG is *N*-acetylglucosamine; NAM is *N*-acetylmuramic acid.
The γ-carboxyl group of D-Glu is used for peptide bond formation.

Figure 5.2 Mode of peptidoglycan cross-linking (*S. aureus*).

terminal glycine is linked to the penultimate D-alanine residue of a neigh-bouring chain (a reaction catalysed by a transpeptidase) and at the same time releasing the terminal D-alanine. It is this last step in peptidoglycan biosyn-thesis that was originally thought to be the only target for the β-lactam antibiotics, probably because of their structural resemblance to the D-ala-D-ala dipeptide (Tipper and Strominger, 1965). The overall process is reviewed in more detail in Gale *et al.* (1981).

5.2 β-lactams

The significant feature of the penicillins and cephalosporins is a 4-membered cyclic amide, known as a β-lactam ring. This is fused to a six-membered thiazine ring in the cephalosporins, a five membered ring in the penicillins or, in the case of the more recently discovered monobactams, no ring at all (Sykes *et al.*, 1981). The activity of penicillin was first observed in 1929 by Alexander Fleming. The antibiotic is elaborated by a fungus, *Penicillium notatum*, and the extraction and identification of the antibiotic from broth cultures was carried out by Chain, Florey and Abraham starting in 1939. The drug became available for human use during the Second World War. A multitude of semi-synthetic penicillins have been produced with a variety of substituents on the 6-amino group.

Penicillins

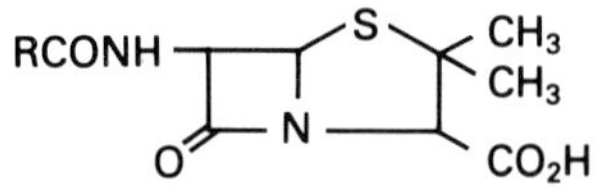

Benzylpenicillin : R = (phenyl)$-CH_2-$

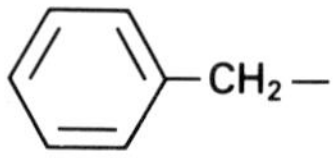

Amoxycillin : R = HO$-$(phenyl)$-CH(NH_2)-$

Methicillin : R = (phenyl with OCH_3 and OCH_3)$-$

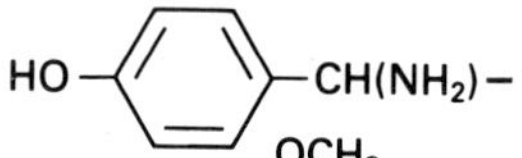

Ampicillin : R = (phenyl)$-CH-$ with NH_2

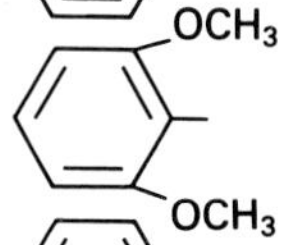

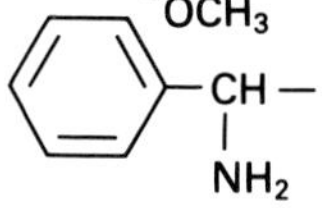

Cephalosporins

	R_1	R_2	R_3
Cefoxitin	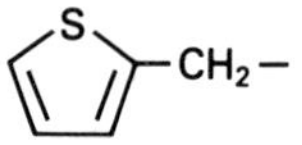 (thiophene)$-CH_2-$	$-CH_2OCONH_2$	$-OCH_3$
Cephalosporin C	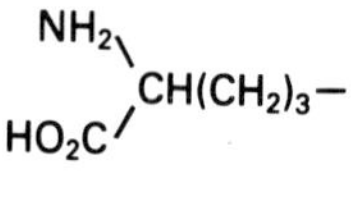 NH_2, HO_2C $CH(CH_2)_3-$	$-CH_2OCOCH_3$	$-H$
Cefuroxime	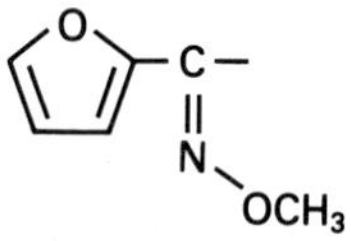 (furan)$-C-$ with $\parallel$ $N-OCH_3$	$-CH_2OCONH_2$	$-H$
Ceftazidime	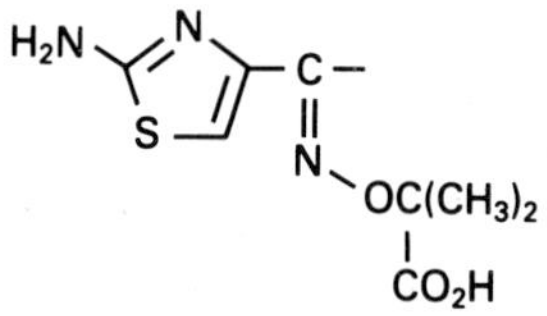 H_2N-(thiazole)$-C-$ with $\parallel$ $N-OC(CH_3)_2-CO_2H$	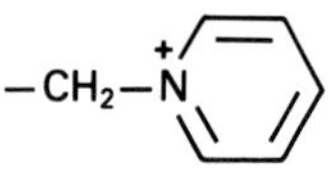 $-CH_2-N^+$(pyridine)	$-H$

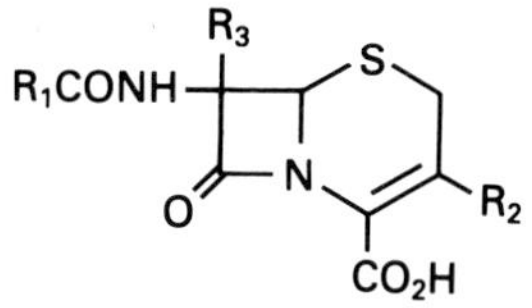

Cephalosporium acremonium was the first source of the cephalosporin family of antibiotics, and was isolated in 1948 from a sewer outlet off the Sardinian coast. One of the earliest cephalosporins identified was cephalosporin C which can be hydrolysed to yield 7-aminocephalosporanic acid; this has allowed the preparation of a vast range of semi-synthetic agents. The cephamycins are a closely related family of antibacterials which are derived from Streptomyces species, in particular *Streptomyces lactamdurans*. The latter produces cephamycin C, which has a methoxyl group at position 7 of the β-lactam ring of the 7-aminocephalosporanic acid nucleus. Recently, the monobactam series of structures has been developed of which one of the main compounds of interest is aztreonam (see Neu, 1983).

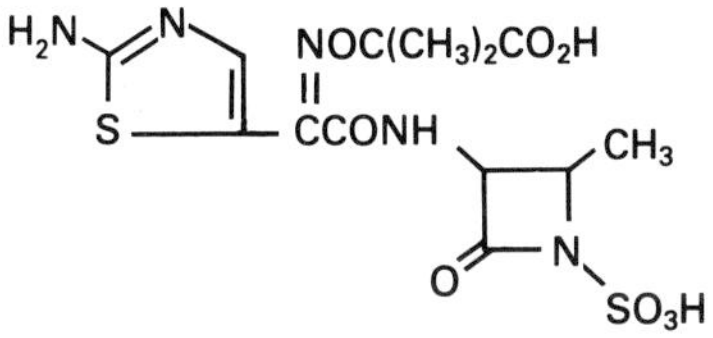

Aztreonam

Originally the mode of action of the penicillins was attributed to inhibition of the transpeptidase that catalyses the final cross-linking of the peptide side-chain of the nascent peptidoglycan (Tipper and Strominger, 1965). In *Staphylococcus aureus* treated with penicillin these authors detected large amounts of non-crosslinked peptidoglycan units, confirmed in later studies by the detection of soluble linear peptidoglycan in penicillin-treated cells (Waxman *et al.*, 1980).

More recent studies have shown, however, that the situation is not so clear-cut (reviewed in Spratt 1980, 1983). In addition to the transpeptidase, both a D-alanine carboxypeptidase and a peptidoglycan endopeptidase have been found that are sensitive to penicillin, and it has become clear that bacteria possess a number of enzymes that are β-lactam sensitive. It appears, however, that the carboxypeptidase reaction at least is not crucial for the organism as 6-aminopenicillanic acid inactivates at least 95% of the enzyme at non-lethal concentrations in *Bacillus subtilis*, whereas cephalothin kills the bacterium at concentrations that do not affect the enzyme (Blumberg and Strominger, 1971).

5.2.1 Penicillin-binding proteins

In contrast, more recent approaches have been to study the pattern of binding when radiolabelled benzylpenicillin is incubated with bacterial cells, the membrane portion is isolated and solubilized, and the proteins are separated by gel electrophoresis (Curtis *et al.*, 1979a; Spratt and Pardee, 1975). A number of proteins to which the β-lactams will bind have been identified by this process and are known as penicillin-binding proteins (PBP) — numbered 1–6 in decreasing order of apparent molecular weight.

Antibiotics that bind to these proteins produce a variety of effects: in *E. coli*, blockade of PBP1a and 1b leads to cell lysis although it appears that the loss of either one alone is not lethal, presumably because the other protein can compensate; PBP2 inhibition leads to cessation of growth and change of shape from rod to sphere and eventual lysis; PBP3 appears to be responsible for the formation of the cross-wall to divide the two daughter cells at cell division (septation), and blockade of this protein results in the formation of long filaments and eventual cell death. Binding to the three other penicillin binding proteins does not lead to any obvious growth defects (reviewed in Spratt, 1983).

Other distinctions that correlate with the separation of the penicillin binding proteins into two groups depending on their importance to the bacterium are that PBPs 1–3 have molecular weights between 60 and 92 kDa in *E. coli* for example, and exhibit transglycosylase and transpeptidase activity while PBPs 4–6 have molecular weights 40–49 kDa and show D-alanine carboxypeptidase activity. This work has been carried out both with agents that bind specifically to one or other penicillin-binding protein, and also by the use of mutants that lack, or produce thermolabile varieties of, these proteins (Spratt 1980, 1983).

It is clearly of importance to correlate the binding of an antibiotic to a given penicillin binding protein with the effect on the whole cell. Cefoxitin, for example, exhibits an I_{50} of 0·1 µg/ml for PBP1a and 3·9 µg/ml for PBP1b from *E. coli* K12, values which are close to the minimum inhibitory concentration (M.I.C., the minimum concentration that totally inhibits growth as noted in Chapter 1) (Curtis *et al.*, 1979b), and so it is likely that the penicillin-binding proteins are the targets particularly as the drug causes the bacteria to lyse. Mecillinam is an example of β-lactam that binds particularly effectively to PBP2 (I_{50} <0·25 µg/ml), kills *E. coli* with an M.I.C. of 0·05 µg/ml, producing in the process spherical forms of the organism as we would expect from a binding to PBP2 (see above). In sharp contrast, ceftazidime has a potent affinity to PBP3 in both *E. coli* and *Pseudomonas aeruginosa*, with an I_{50} of 0·06 µg/ml for PBP3 of the former bacterium and M.I.C. of 0·2 µg/ml (Hayes and Orr, 1983). As a consequence filamentation of the organism results.

Mecillinam

Resistance to β-lactams among gram-positive bacteria may often be closely linked with changes in the penicillin-binding proteins. This can arise by a reduction in the quantity or binding affinity of a given protein, or apparently, the addition of an extra low-affinity protein. In gram-negative organisms, likewise, a change in affinity of the penicillin-binding protein may be responsible for the induction of β-lactam resistance (Malouin and Bryan, 1986).

The detailed interaction of the penicillins with transpeptidases has been studied with much interest, albeit using soluble enzymes from streptomyces species that are not of medical importance (reviewed in Frere and Joris, 1985). Initially the antibiotic forms a reversible complex with the enzyme, followed by the irreversible conversion into a complex with a covalent link between enzyme and drug. This complex then dissociates into enzyme and penicilloic acid:

$$E + I \overset{K}{\rightleftharpoons} EI \overset{k_2}{\to} EI^* \overset{k_3}{\to} E + P$$

Provided that k_3 is small the enzyme remains in an inactivated state, but if k_3 is appreciable, we have a recipe for drug detoxification, as could apply to the β-lactamases (see below).

Detailed studies on the reaction mechanism have been carried out by incubation of the protein with radiolabelled β-lactam, followed by digestion with a proteolytic enzyme called pronase and identification of the residue to which the label was attached. It is clear that a serine residue makes a nucleophilic attack on the β-lactam ring of benzylpenicillin. It may be that this is the same serine to which the substrate binds during the normal reaction but this is not yet certain. Figure 5.3 shows the two major pathways of β-lactam metabolism.

Studies on the penicillin binding proteins are at an early stage but the active site region of PBP1 from *E. coli* shows a considerable degree of homology with the transpeptidases discussed above. In particular, serine is the active

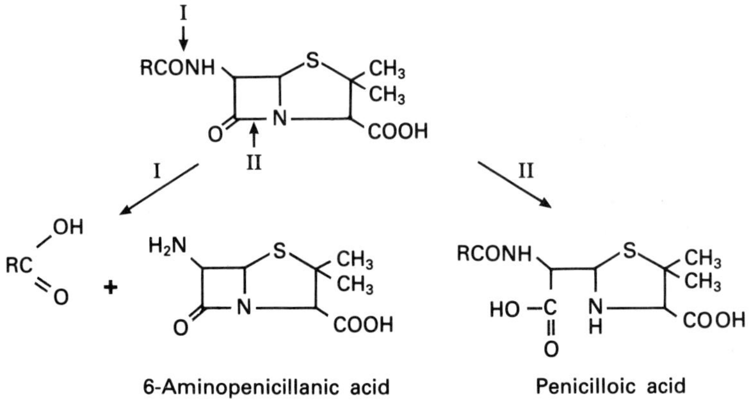

6-Aminopenicillanic acid   Penicilloic acid

I. Penicillin amidase breaks the amide link in the side-chain of the drug. This enzyme is not used by the microorganism for protection against the drug. Penicillin amidase has been used to prepare 6-aminopenicillanic acid from which new semi-synthetic deriva-tives can be made by linking the amino group to new acyl side-chains.

II. β-Lactam hydrolysis of the β-lactam ring proceeds through attack by a nucleophilic serine hydroxyl on the β-lactam carbonyl, yielding a transient acyl-enzyme inter-mediate. It is this enzyme that confers drug resistance upon the bacterium, and organic chemists have been at pains to synthesize less susceptible new compounds.

Figure 5.3 Pathways of β-lactam metabolism

site nucleophile. It is interesting that this sequence is related to that occurring around the active site of the β-lactamases from *E. coli, B. cereus* etc. (Nicholas *et al.*, 1985). It is too early to relate the differences between the two classes of enzyme to detailed changes in amino acid residues. Nevertheless, recent X-ray data suggest that the β-lactamase 1 from *B. cereus* bears a considerable similarity in tertiary structure, particularly around the active site, to the D-alanine-D-alanine carboxypeptidase elaborated by *Streptomyces* R61 that is penicillin-sensitive (Samraoui *et al.*, 1986).

Both β-lactamases and penicillin-binding proteins pass through a reversible enzyme inhibitor complex formation followed by transformation into an acyl-enzyme intermediate with the active site serine. The distinction between β-lactamases and penicillin-binding proteins tends to become blurred to some extent when k_3 becomes appreciable for the breakdown of the penicilloyl-PBP complex into penicilloic acid. Indeed, some penicillin-binding proteins could almost be regarded as β-lactamases (see Spratt, 1983).

5.2.2 β-lactamases

Bacterial resistance to the β-lactams may be derived, as with other antibiotics, by preventing the ingress of the antibiotic into the cell or by altering the enzyme target. Although the former possibility does not appear to play a major part, it is clear that the latter does (Malouin and Bryan, 1986). In addition, another major method for the development of bacterial resistance to the β-lactams is detoxification by enzymes which destroy the β-lactam ring, namely β-lactamases. The genetic coding for the enzyme may be found both on a chromosome or a plasmid. In gram-positive bacteria, β-lactamase production is the major method of resistance in that the bacteria produce a large quantity of the enzyme and secrete it extracellularly.

In gram-negative bacteria the situation is more complex. The organism produces less β-lactamase and secretes it into the periplasmic space, i.e. the space between the peptidoglycan band around the cytoplasmic membrane and the bilayer outer membrane, which is very hydrophobic (construction of the gram-negative cell envelope is reviewed in Costerton and Cheng, 1975). The enzyme is thereby positioned at the site where the peptidoglycan is formed on the outer side of the cytoplasmic membrane, and it can therefore protect the organism to the greatest extent. Furthermore, the outer membrane presents a much more serious obstacle to the transport of the antibiotic in gram-negative bacteria than it does in gram-positive ones. As a consequence, the possession of a β-lactamase correlates better with bacterial resistance for a gram-positive than for a gram-negative bacterium (Sykes and Matthew, 1976).

β-Lactamases have been classified in various ways but probably the most widely accepted classification depends on their amino acid sequence and catalytic properties. Two groups of enzymes with serine at the active site have a preference for penicillins on the one hand (class A), and cephalosporins on the other (class C). The latter are usually chromosomally encoded enzymes derived from gram-negative enterobacteria (Jaurin and Grundstroem, 1981). Class B includes the Zn^{2+} containing enzymes (see Frere and Joris, 1985). The enzymes frequently are extremely active, although the most useful way to describe the activity is by the ratio of catalytic rate constant to Michaelis constant (K_{cat} to K_m). For benzylpenicillin this figure is 9.8×10^5 M/s, which is close to the diffusion limit, i.e. the substrate is hydrolysed almost as fast as it can diffuse to the active site of the enzyme.

The enzyme splits most penicillins by breaking open the β-lactam ring to yield penicilloic acids (see Figure 5.3). Cephalosporins also are hydrolysed by β-lactamases to yield inactive metabolites, in which the β-lactam ring is split open while the six-membered thiazine ring remains intact. The final products of the reaction depend to some extent on the relative stability, or otherwise, of the cephalosporoic acid formed. Studies with β-lactamase from *Staphylococcus aureus* have shown that a good leaving group at the 3' position of the ring is of great advantage in forming an acyl-enzyme intermediate and thereby a more effective antibiotic (Boyd, 1984).

A number of suicide substrates have been described, i.e. drugs which the enzyme converts into products that destroy the enzyme's activity (see Chapter 1). These are frequently derived from fermentation cultures, as in the case of clavulanic acid from *Streptomyces clavuligerus*. Clavulanic acid interacts with β-lactamases in a complex fashion but, in general, the first step involves the formation of a complex that is non-covalent and readily dissociable. Secondly, the ligand acts as a substrate and the β-lactam ring is opened with the active site serine forming an ester link. A true substrate would now dissociate from the enzyme, but the clavulanic acid moiety rearranges to produce a conjugated series of double bonds giving rise to an ultraviolet absorbance spectrum. Subsequently, in some cases, the enzyme/inhibitor complex is reactivated and the enzyme recovers its full activity. In others the complex is stable and the enzyme remains irreversibly inhibited (Labia *et al.*, 1985).

5.2.3 β-lactam therapy

It has been of great importance to synthesize drugs which are resistant to β-lactamase detoxification. The introduction of a 2,6-dimethoxyphenacyl group to form an amide with the amino group at the 6-position of the β-lactam ring sterically hindered β-lactam hydrolysis as in methicillin. Replacing the CH_2 between the ring and the carbonyl group with a methoxyimine side-chain, as in cefuroxime, also conferred stability to β-lactamases. Another approach is to insert a 7α-methoxy group in the β-lactam ring as in the family of cephamycins of which cefoxitin is an example, although larger substituents in this position produced a reduction in activity. Further substitution of the imino grouping, as in ceftazidime, maintains resistance to β-lactamases while increasing intrinsic activity against gram-negative organisms of clinical importance (such as *E. coli* and *Klebsiella pneumoniae*) apparently by virtue of the aminothiazolyl side-chain (see Neu, 1986).

An alternative approach is to use β-lactams that are potent β-lactamase inhibitors, but are not necessarily antibacterial, to act in synergy with β-lactam antibiotics. Various compounds with these properties have been discovered in culture filtrates of streptomycetes. In particular, clavulanic acid has been developed for this purpose. A combination of clavulanic acid with amoxicillin (tradename Augmentin) is effective in treating infections produced by organisms that produce β-lactamase in large quantities, notably *Bacteroides fragilis*. It can be shown *in vitro* that clavulanic acid will enhance the activity of β-lactams against *B. fragilis* by lowering the M.I.C. by two orders of magnitude (Bansal *et al.*, 1985).

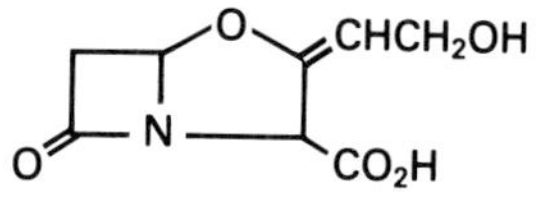

Clavulanic acid

The β-lactams are probably the most widely used antibacterials at the present time, partly because of their broad spectrum of action, and partly because of their very low toxicity. They must, indeed, be about the least toxic drugs available on the market. The third generation of the cephalosporin family of antibiotics have improved the modest activity of earlier members against gram-negative organisms, although Gould and Wise (1985) feel that they are still not sufficiently active to be used against *Pseudomonas* infections that may be picked up in hospital and are frequently difficult to treat (Garzone *et al.*, 1983).

One situation where β-lactams cannot be used with confidence is in the treatment of so-called methicillin-resistant staphylococcal species. This term is used to indicate resistance to all β-lactamase-resistant penicillins and cephalosporins. This resistance arises, therefore, not as a consequence of β-lactamase production, but probably because of altered penicillin-binding proteins (Malouin and Bryan, 1986). Vancomycin is often used as the drug of choice in these situations (Kucers, 1984).

5.3 Vancomycin

Vancomycin is a large glycopeptide with a molecular weight in the region of 1500, obtained from the culture filtrates of *Streptomyces orientalis*. It has a complex molecular structure (see Sheldrick *et al.*, 1978) containing an amino-sugar called vancosamine linked to three aromatic rings; several amino acid residues and a bis-resorcinol system. Vancomycin is active primarily against gram-positive bacteria. It is not a new antibiotic, having been discovered in the middle of the 1950s, but while it demonstrated undesirable side-effects it had no advantage over the β-lactams, especially with the advent of the β-lactamase-stable β-lactams.

The structure of vancomycin

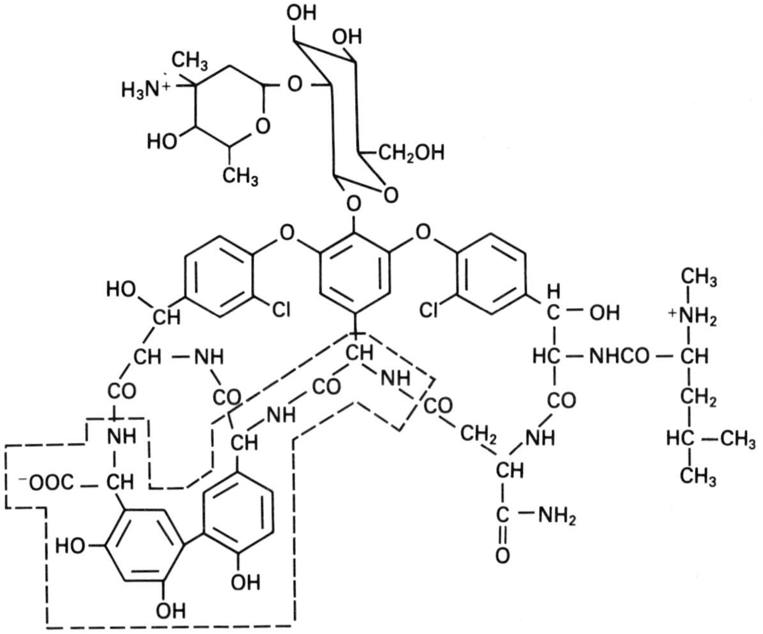

The three-dimensional molecular structure is very compact and excludes water molecules. A hydrophobic cleft is present on the side of the molecule on which the chlorine atoms reside. The D-ala-D-ala portion of the muramyl pentapeptide binds in this cleft. The groups postulated to take part in the interaction are surrounded by a dotted line. See Figure 5.4 for the detailed interaction proposed by Sheldrick *et al.*, 1978.

Latterly, however, the importance of methicillin-resistant *Staphylococcus aureus* infections has indicated the need for a tried and tested antibiotic such as vancomycin. In addition, pseudo-membraneous colitis responds well to vancomycin. This condition is induced by the action of other antibiotics which 'scour' the gut leaving it almost empty of bacteria thus allowing *Clostridium difficile* to colonize. Furthermore, more modern preparations appear to be 'cleaner' and, therefore, the occurrence of side-effects is markedly reduced compared with impure preparations of earlier years (Kucers, 1984; Watanakunakorn, 1984).

The mode of action of vancomycin involves interference with the biosynthesis of cell wall peptidoglycan, but in a different way to the β-lactams. Vancomycin binds tightly to UDP-pentapeptides containing D-alanine-D-alanine at the free carboxyl end, thus causing UDP-*N*-acetylmuramyl-peptide precursors to accumulate. The tightest binding was observed with UDP-*N*-acetylmuramylglycyl-D-glutamylhomoseryl-D-alanyl-D-alanine, with a binding constant of 6×10^{-7} M, as measured by ultraviolet difference spectroscopy. The smallest fragment to which binding was detected was acetyl-D-ala-D-alanine (Perkins, 1969). The molecule carries a marked hydrophobic cleft on one side to which the peptide is bound as shown by nuclear magnetic resonance studies. The binding is initiated by the interaction between the peptide carboxylate group and a protonated amine on the antibiotic (see Williamson and Williams, 1984, and Figure 5.4).

The primary mode of action of vancomycin is inhibition of cell wall biosynthesis, and consequently results in cell lysis. The drug may also alter the permeability of the cell membrane and, in addition, interfere with ribonucleic acid synthesis (see Watanakunakorn, 1984).

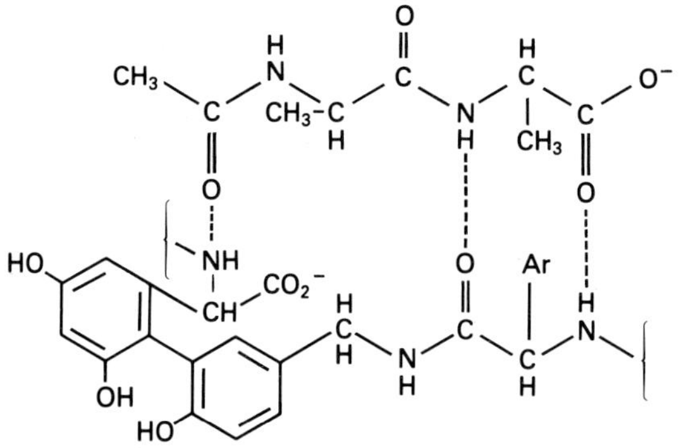

Interaction with acetyl-D-Ala-D-Ala is shown above. Ar represents the trihydroxylated benzene ring in the middle of the structure. Dotted lines indicate hydrogen bonds which are postulated to be the major force in holding the complex together. The portion of the molecule that interacts with the dipeptide moiety is outlined in the structure on previous page.

Figure 5.4 Vancomycin interaction with D-ala-D-ala moiety of pentapeptide

References

Bansal, M.B., Church, S.K., Onjema-Lalobo, M. and Thadepalli, H. (1985). *Chemotherapy* (*Basel*), **31**, 173–177.

Blumberg, P.M. and Strominger, J.L. (1971). *Proc. Natl. Acad. Sci.* (*U.S.A.*), **68**, 2814–2817.

Boyd, D.B. (1984). *J. Med. Chem.*, **27**, 63–66.

Costerton, J.W. and Cheng, K.J. (1975). *Bact. Rev.*, **38**, 87–110.

Curtis, N.A.C., Brown, C., Boxall, M. and Boulton, M.G. (1979a). *Antimicrob. Ag. Chemother.*, **15**, 332–336.

Curtis, N.A.C, Orr, D., Ross, G.W. and Boulton, M.G. (1979b). *Antimicrob. Ag. Chemother.*, **16**, 533–539.

Frere, J.M. and Joris, B. (1985). *CRC Crit. Rev. Microbiol.*, **11**, 299–396.

Gale, E.F., Cundliffe, E., Reynolds, P.E., Richmond, M.H. and Waring, M.J. (1981) Eds. *The Molecular Basis of Antibiotic Action*, 2nd Ed. (London: John Wiley) Ch. 3.

Garzone, P., Lyon, J. and Yu, V.L. (1983). *Drug. Intell. Clin. Pharm.*, **17**, 615–622.
Gooday, G.W. (1979). *J. Gen. Microbiol.*, **99**, 1–11.
Gould, I.M. and Wise, R. (1985). *Brit. Med. J.*, **290**, 878–879.
Hayes, M.V. and Orr, D.C. (1983). *J. Antimicrob. Chemother.*, **12**, 119–126.
Jaurin, B. and Grundstroem, T. (1981). *Revs. Infect. Dis.*, **8**, Suppl. 3, S237–S259.
Kucers, A. (1984). *J. Antimicrob. Chemother.*, **14**, 564–567.
Labia, H., Barthelemy, M. and Peduzzi, J. (1985). *Drugs Exptl. Clin. Res.*, **11**, 765–770.
Malouin, F. and Bryan, L.E. (1986). *Antimicrob. Ag. Chemother.*, **30**, 1–6.
Neu, H.C. (1986). *Revs. Infect. Dis.*, **8**, Suppl. 3, S237–S259.
Nicholas, R.A., Suzuki, H., Hirota, Y. and Strominger, J.L. (1985). *Biochemistry*, **24**, 3448–3553.
Perkins, H.R. (1969). *Biochem. J.*, **111**, 195–205.
Samraoui, B., Sutton, B.J., Todd, R.J., Artymiuk, P.J., Waley, S.G. and Phillips, D.C. (1986). *Nature*, **320**, 378–380.
Sheldrick, G.M., Jones, P.G., Kennard, O., Williams, D.H. and Smith, G.A. (1978). *Nature*, **271**, 223–225.
Spratt, B.G. and Pardee, A.B. (1975). *Nature*, **254**, 516–517.
Spratt, B.G. (1980). *Phil. Trans. R. Soc. Lond. B.*, **289**, 273–283.
Spratt, B.G. (1983). *J. Gen. Microbiol.*, **129**, 1247–1260.
Sykes, R.B. and Matthew, M. (1976). *J. Antimicrob. Chemother.*, **2**, 115–157.
Sykes, R.B., Cimarusti, C.M., Bonner, D.P., Bush, K., Floyd, D.M., Georgopapadakou, N.H., Koster, W.H., Liu, W.C., Parker, W.L., Principe, P.A., Rathnum, M.L., Slusarchyk, W.A., Trejo, W.H. and Wells, J.S. (1981). *Nature*, **291**, 489–491.
Tipper, D.J. and Strominger, J.L. (1965). *Proc. Natl. Acad. Sci. (U.S.A.)*, **54**, 1133–1141.
Watanakunakorn, C. (1984). *J. Antimicrob. Chemother.*, **14**, Suppl. D, 7–18.
Waxman, D.J., Yu, W. and Strominger, J.L. (1980). *J. Biol. Chem.*, **255**, 11577–11587.
Williamson, M.P. and Williams, D.H. (1984). *Eur. J. Biochem.*, **138**, 345–348.

Chapter 6

Steroid biosynthesis and action

Table 6.1 Drugs discussed in Chapter 6

Section	Target	Drug	Use
6.2 Sterol biosynthesis			
6.2.1	β-Hydroxy-β-methyl glutaryl CoA reductase	Compactin Mevinolin	Hypercholesterolaemia Hypercholesterolaemia
6.2.2	Squalene epoxidase	Naftifine Terbinafine Tolnaftate	Dermatophyte infections Dermatophyte infections Dermatophyte infections
6.2.3	Lanosterol 14α-demethylase	Ketoconazole	Fungal infection
6.3 Steroid biosynthesis			
6.3.1	Steroid 17,20-lyase	Ketoconazole	Hormone dependent prostate tumours
6.3.2	3β-Hydroxysteroid dehydrogenase	Epostane	Pregnancy terminator
6.3.3	Steroid 11β-hydroxylase	Metyrapone	Hypercortisolaemia
6.3.4	Aromatase	Aminoglutethimide	Hormone-dependent breast cancer
6.4 Steroid receptor ligands			
6.4.1	Oestrogen receptor agonists	Ethinyloestradiol Mestranol	Oral contraceptive Oral contraceptive
	Progesterone receptor agonists	Norethisterone Norgestrel	Oral contraceptive Oral contraceptive
6.4.2	Oestrogen receptor antagonist	Tamoxifen	Hormone-dependent breast cancer
	Oestrogen receptor antagonist	Clomiphene	Fertility stimulant
6.4.3	Progesterone receptor antagonist	RU-486	Pregnancy terminator
6.4.4	Aldosterone receptor antagonist	Spironolactone	Diuretic

6.1 Introduction

The major sterols to be discussed in this chapter are ergosterol and cholesterol, which are essential constituents of cell membranes of yeasts and fungi, and of mammals respectively. Both sterols are synthesized in a similar fashion starting from acetyl coenzyme A (see Figure 6.1). The synthesis takes

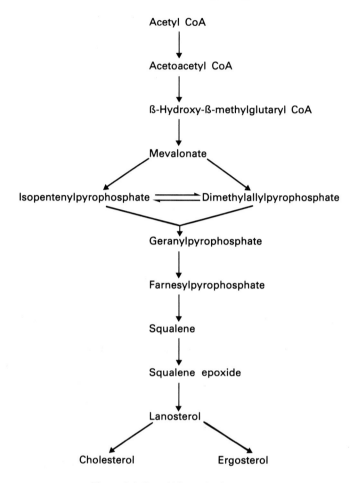

Figure 6.1 Sterol biosynthetic pathway.

place in different parts of the cell: acetate to mevalonate is carried out in the microsomal fraction while mevalonate to squalene is a cytoplasmic function; squalene conversion to ergosterol and cholesterol is performed back in the microsomes and this is the first part of the pathway to require oxygen. The pathways to ergosterol and cholesterol are common up to lanosterol. Furthermore, the removal of the methyl group attached to the carbon atom at position 14 of lanosterol is also common, although in yeasts and fungi it does not necessarily occur next in sequence (Figure 6.3). Not all the enzymes have been purified, but it appears that, although yeast and mammalian enzymes catalyse the same interconversions, they are not necessarily identical, and thus lend themselves to selective inhibition.

Cholesterol can undergo a variety of reactions. Part of the side-chain is removed as the first step in the formation of the steroid hormones (see Figure 6.2). This leads to (a) *glucocorticoids* such as cortisol which, amongst other

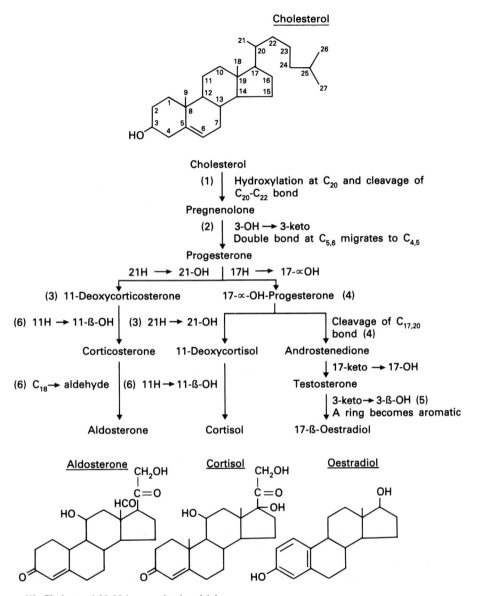

(1) Cholesterol 20,22-lyase, mitochondrial.
(2) 3-β-Hydroxysteroid dehydrogenase, mitochondrial.
(3) 21-Steroid hydroxylase, microsomal.
(4) 17-α-Steroid hydroxylase and 17,20-lyase, microsomal.
(5) Aromatase, microsomal.
(6) 11β-steroid hydroxylase, mitochondrial.
All the enzymes except 3 are cytochrome P–450s, similar to the liver mixed function oxidases, and thus require NADPH and molecular oxygen to function. Unlike the latter, however, they are selective for their particular substrates.

Figure 6.2 Main pathways of steroid hormone biosynthesis.

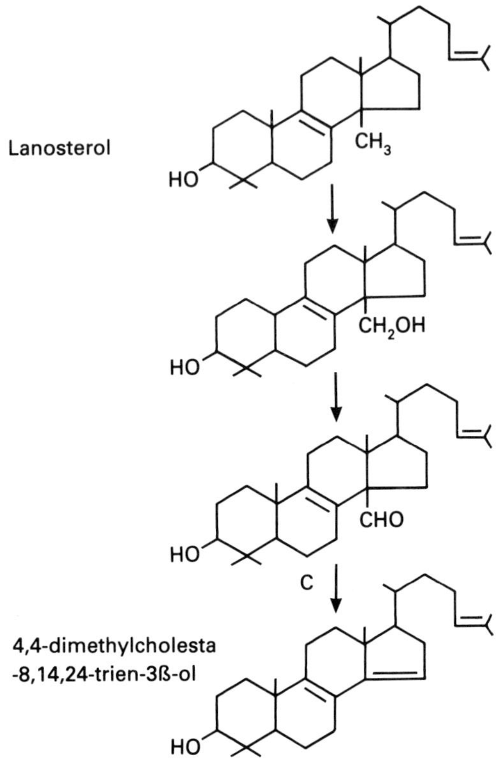

Lanosterol

CH₃

HO

CH₂OH

HO

CHO

HO

C

4,4-dimethylcholesta
-8,14,24-trien-3ß-ol

HO

The 14α-methyl group of lanosterol is sequentially hydroxylated in the presence of a cytochrome P–450 linked enzyme, the flavoprotein cytochrome P–450 reductase, NADPH and molecular oxygen. The final oxidation yields formic acid, together with a sterol double-bonded at the 14–15 position.

Figure 6.3 Lanosterol 14α-demethylation.

actions, increase gluconeogenesis and stimulate lipolysis, (b) *mineralocorticoids* such as aldosterone that cause sodium to be reabsorbed in the kidneys thus ensuring that sufficient water is retained to maintain the body's osmotic balance and (c) the final removal of the side-chain yields the *sex steroids* such as testosterone in the male and oestradiol in the female. Although the throughput down these pathways is fairly small, the effects of the hormones produced is very marked.

The greatest proportion of cholesterol is converted into bile acids which are released into the gut to emulsify triglycerides so that they can be transported into the circulation. Cholesterol can also be esterified at the 3-hydroxy position with long-chain fatty acids donated from either triacyl glycerols such as lecithin (catalysed by the plasma enzyme lecithin cholesterol acyl-transferase: LCAT) or their respective coenzyme A esters (tissue-bound acyl CoA acyltransferase: ACAT). The esters so formed act as a storage depot for cholesterol, and also in the former case to keep cholesterol within the circulating lipoprotein and prevent it from transferring into the tissues.

The cholesterol nucleus cannot be broken down directly to carbon dioxide and water, and so if cholesterol levels rise there is no simple metabolic pathway that can remove it. Elevated cholesterol, when attached to very low density lipoproteins, gives rise to one of the major diseases facing the Western world today (see Section 6.2.1).

Ergosterol, strangely enough, is only known to undergo (a) esterification and (b) conversion to ergocalciferol (Vitamin D_2) under the influence of ultraviolet light. At present there are no other metabolic transformations described, although by analogy with cholesterol we might expect some conversions, perhaps to a yeast hormone. On the other hand, like cholesterol, the ergosterol nucleus cannot be broken down directly to carbon dioxide and water.

The sterol biosynthetic pathway has not, until recently, attracted much attention from the pharmaceutical point of view, expect perhaps from those seeking to lower cholesterol levels. Much effort went into this direction in the 1960s after the important connection between cholesterol and coronary heart disease was found. There was considerable success in reducing the flow of metabolites through this pathway, but the agents led instead to the formation of toxic intermediates and were of no value as drugs. Recently, however, the discovery of non-toxic agents that interfere with the rate-limiting step in the early part of the pathway, β-hydroxy-β-methylglutaryl CoA reduction to mevalonate, has greatly improved the prospects for the treatment of hypercholesterolaemia and, hopefully, coronary heart disease.

More recently, the part of the pathway leading from lanosterol to ergosterol has been spotlighted as the target of various anti-fungal agents, some useful as agricultural fungicides and others in human medicine. This interest is likely to turn what was previously rather a backwater, inhabited only by organic chemists interested in the mechanism of squalene epoxide cyclization, for example, into one of the most fruitful areas in biochemistry.

The conversion of cholesterol into various steroid hormones has been of major pharmaceutical importance for many years because of the development of oral contraceptives. Although the use of oestrogens and progestins to prevent conception is not to combat disease, it is nonetheless one of the major uses of drugs. Possibly because the use of drugs has greatly increased the life expectancy of the population of a large part of the world, the use of drugs to restrain population growth may be regarded as a moral imperative for the industry. Alternatively, it may be argued that the industry is merely responding to a need in that women increasingly desire to regulate their reproductive function. Whichever view is taken, there is still considerable effort being expended to find other means of achieving the same end, particularly in view of the uncertainty as to whether the pill can cause breast cancer (Anon, 1986).

Another pharmaceutical outlet for inhibitors of steroid biosynthesis, which are often hydroxylations carried out by cytochrome P–450 dedicated to a particular reaction, has been to lower the levels of sex hormones, glucocorticoids etc., in conditions where these pathways are hyperactive, and in situations where tumours are particularly dependent on sex steroids. Antagonists of mineralocorticoid action are also useful as diuretics.

6.2 Sterol biosynthesis

6.2.1 β-Hydroxy-β-methylglutarylCoA (HMG CoA) reductase

The lowering of plasma cholesterol levels, whether by inhibition of biosynthesis or reduction in gastrointestinal absorption, has been a major pharmaceutical target for many years, since the establishment of a strong causative connection between plasma lipid, the deposition of fat and other material in the arteries or atherosclerosis and coronary heart disease. In outline, atherosclerosis results in the narrowing of the arteries in the following way. Cholesterol is transported via the plasma in three types of lipoprotein complexes characterized by their densities: very low (VLDL) which has the highest lipid content, low (LDL) which is intermediate and high (HDL) with the least lipid. These complexes are of high molecular weight, with a core of cholesteryl esters, triglycerides and free cholesterol surrounded by phospholipid and lipoprotein and they provide the transport mechanism for hydrophobic lipids in an aqueous environment.

At the surface of the cells in the lining of the arteries (e.g. the aortic intimal cells) are receptors for the protein component of LDL, which is the lipoprotein whose levels correlate most closely with the development of atherosclerosis. LDL binds and the complex is taken up by the cell. There, any remaining free cholesterol is converted to esters by acyl CoA cholesterol acyltransferase (ACAT). These esters can be laid down intracellularly to produce what is known as a 'fatty streak', and at this stage the deposition can be reversed. Subsequently lipids begin to accumulate extracellularly with a contribution in the latter stages from LDL. Ultimately the lipid deposit triggers a formation of fibrin around the fatty streak and collagen is also deposited to form a fibrous plaque so that the artery begins to be blocked. At this point the lesion is no longer reversible. High plasma cholesterol levels, particularly associated with LDL, are correlated with plaque formation, and with the development of coronary heart disease.

In earlier years, many compounds were discovered that were able to block the later stages in cholesterol biosynthesis, notably triparanol, an inhibitor of Δ^{24}-sterol reductase, the final step in the pathway, which leads to elevated levels of desmosterol (24,25-dehydrocholesterol) (Avigan *et al.*, 1960) and dodecylimidazole which causes 2,3-epoxysqualene to accumulate (Atkin *et al.*, 1972). The metabolites so accumulated turned out to be toxic to the organism, and the use of these agents was discontinued. Recently, β-hydroxy-β-methylglutaryl CoA reductase has attracted much more attention, partly because it has been established as the rate-limiting step in the early part of the pathway, and it is subject to feedback inhibition by dietary cholesterol — in other words intake of increased dietary cholesterol leads to reduction of the enzyme level in the liver (see Fears, 1983).

Furthermore, there is a pathway for removal of the HMG CoA that might have accumulated when the enzyme is blocked. Frequently when competitive inhibitors of an enzyme are discovered *in vitro* their effect *in vivo* is greatly reduced or even nullified by a build-up of substrate. In this case, however, substrate may be further metabolized to acetoacetate and β-hydroxybutyrate by HMG CoA lyase and acetoacetate dehydrogenase, although the fact that the breakdown pathway takes place in the mitochondrion and cholesterol synthesis in the cytoplasm requires an exchange between the pools of the HMG CoA.

Two extremely powerful inhibitors of HMG CoA reductase have both been isolated from microbial culture filtrates. Compactin was found in *Penicillium brevicompactum*, while its 6-methyl analogue, mevinolin, was isolated from the broth of *Aspergillus terreus* (Alberts *et al.*, 1980). In each case, the compound is isolated as the cyclic lactone but the open chain hydroxy-acid is the active form. Both compounds are competitive inhibitors of the enzyme which catalyses the conversion of HMG CoA to mevalonate using NADPH as a cofactor. The K_i for compactin is 1.4×10^{-9} M, while the addition of a methyl group increases the binding for mevinolin, which has a K_i of 6×10^{-10} M (Alberts *et al.*, 1980). Rogers and Rudney (1982) have suggested that the inhibition is not simply competitive; the inhibitor may bind to regions of the enzyme other than the catalytic site inducing a change in protein conformation that is essentially irreversible.

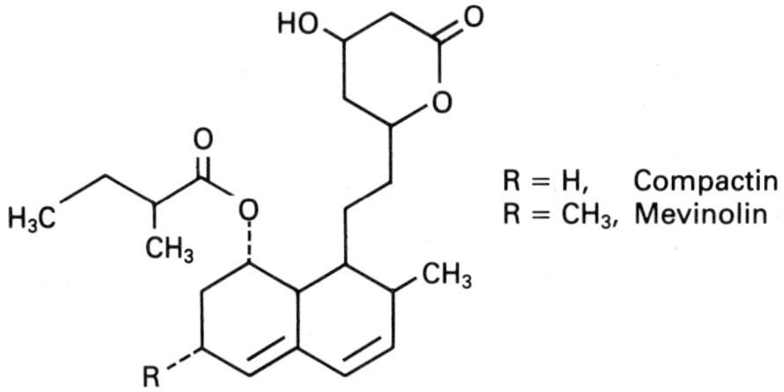

R = H, Compactin
R = CH₃, Mevinolin

Studies in cell culture have shown that inhibition of HMG CoA reductase can lead to increased enzyme synthesis (Luskey *et al.*, 1982) thus nullifying the effect of the inhibitor. Indeed these results are consistent with experiments in rats, where only a transient inhibition of cholesterol biosynthesis is observed. In dogs and man, however, compactin and mevinolin do, in fact, lower serum cholesterol levels by between 20–40% during treatment, both from normal and elevated levels (reviewed in Fears, 1983). This effect is specific for LDL in man. This effect may be supplemented by a reduction in cholesterol synthesis which causes a greater uptake of LDL-bound cholesterol from the plasma. The B receptor, specific for the protein component of LDL is induced in order to facilitate this uptake (Fears, 1983).

Mevinolin is on clinical trial in the United States, and has shown itself to be particularly useful for the long-term treatment of patients with familial high levels of cholesterol (a genetically determined condition) who are particularly at risk for atherosclerosis. They have a complete or partial defect of the LDL receptor on the cell membranes of skin fibroblasts amongst other cells, which is associated with a consequent loss of feedback regulation of HMG CoA reductase. This failure of regulation of cholesterol formation, and a reduction in the breakdown of LDL because of defective receptor uptake, are thought to be the major causes of high levels of cholesterol in the disease (Mabuchi *et al.*, 1981). The drug has been given in conjunction with cholestipol — a resin which binds to the bile acids in the intestine preventing their re-absorption and forcing the body to synthesize more from cholesterol to make up the loss (Illingworth, 1984). The main hope is that coronary heart disease will be reduced in these patients, but it is still too early to be sure.

There was always a concern that ubiquinone levels might suffer as a result of HMG CoA reductase inhibition, since ubiquinone is derived from farnesyl pyrophosphate (Figure 6.1), but this does not apparently occur *in vivo* (Mabuchi *et al.*, 1981). Likewise a reduction in cholesterol levels might lead to a reduction in the synthesis of steroid hormones (Figures 6.2), but fears on this score turned out similarly to be groundless (Tobert *et al.*, 1982).

6.2.2 Squalene epoxidase

The intermediates immediately before squalene in the sterol biosynthetic pathway are rendered soluble by the presence of pyrophosphate groups. Squalene is the first substrate that is insoluble in aqueous media — indeed it must be one of the most, if not the most, insoluble compounds in the cell. It appears that a sterol carrier protein has evolved to handle this situation. In addition, squalene epoxidase is of considerable interest because it is responsible for the first oxidative step in the pathway.

The enzyme is present in the microsomes of eukaryotic cells, and requires the presence of cytoplasmic fraction to show activity, presumably because the latter contains the sterol carrier protein. The cofactors are NADPH or NADH and FAD, but the enzyme is not a cytochrome P–450. Although the enzyme catalyses the placement of an oxygen atom across the 2 and 3 positions of the squalene ring to form an epoxide, it may also be able to put a second one across the 22,23 positions, because conditions that induce an accumulation of the mono-epoxide, such as inhibition of the next step in the pathway (squalene epoxide cyclase) by chloroquine (Chen and Leonhardt, 1984) yield a proportion of the di-epoxide as well. The enzyme has been purified from rat liver (Ono *et al.*, 1982) but not so far from any fungal or yeast source, although the *Candida albicans* enzyme has requirements for activity similar to the rat liver enzyme (Ryder and Dupont, 1984).

Recently a new class of antifungal agents, the allylamines, has been found to exert its antifungal effect by virtue of powerful inhibition at this step (reviewed by Petranyi *et al.*, 1984). Naftifine was the first member of this series to be synthesized but a more recent one, terbinafine, is more potent at inhibiting both enzyme activity and fungal growth. Terbinafine shows non-competitive inhibition of the enzyme from *Candida albicans* with a K_i of 3×10^{-8} M (Petranyi *et al.*, 1984). This was measured without prior incubation of inhibitor and enzyme, however, and it would be of interest to know whether the inhibition is time-dependent and irreversible. The structure of terbinafine, with its conjugated double and triple bonds, suggests a covalent interaction with the enzyme.

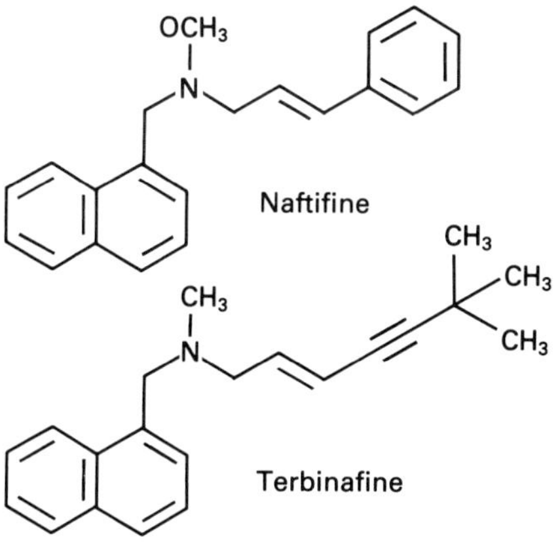

Another curious feature of these compounds is their ability to interfere with the fungal enzyme without always inhibiting fungal growth in the same way. The action of terbinafine on *C. albicans* appears to be fungistatic and much less marked, whereas against the dermatophyte *Trichophyton* species the drug is far more potent and fungicidal. There is no suggestion that the difference lies in the enzyme inhibition, but rather that different fungi respond differently to large increases in squalene and decreases in available ergosterol. The enzyme from rat liver is some 3 to 4 orders of magnitude less sensitive to the drug, and this selectivity appears to be truly reflected *in vivo* (Petranyi *et al.*, 1984). Terbinafine is orally active against dermatophytes — unlike naftifine which only shows promise topically against *Tinea* species (Ganzinger *et al.*, 1982).

Another family of compounds which have been in use for some time against topical infections is exemplified by tolnaftate. Recent studies have shown that it too inhibits squalene epoxidation. Although active against the *Candida* enzyme, the drug apparently cannot reach its site of action in these yeasts (Barrett-Bee *et al.*, 1986).

6.2.3 Lanosterol demethylation

One of the crucial rate-limiting steps in the later part of the pathway of cholesterol biosynthesis is the removal of the C32 methyl group from the 14α position of lanosterol with a concomitant introduction of a double bond at the 14–15 position (Figure 6.3). These transformations appear to be carried out by a single microsomal enzyme that is cytochrome P–450 dependent, and requires NADPH, molecular oxygen and cytochrome P–450 reductase, a flavoprotein, for activity (reviewed by Coulson *et al.*, 1984). The methyl group is sequentially hydroxylated to yield $-CH_2OH$ and $-CHO$, and then removed as formic acid as the double bond is inserted at the 14–15 position. Carbon monoxide which binds to the haem iron of cytochrome P–450 and interferes with a number of P–450 linked reactions, inhibits the first step but not the latter two (Gibbons *et al.*, 1979), but that may merely reflect tighter binding of the oxidized intermediates to the enzyme, i.e. carbon monoxide may be able to displace lanosterol, but not the oxidized intermediates, at the binding site. More recent work has confirmed that the same cytochrome P–450 will catalyse all three steps in the process (Aoyama *et al.*, 1987).

In fungi, the demethylation of lanosterol is not rate-limiting as it is in liver, but it is nevertheless essential for the formation of ergosterol and other sterols that are acceptable materials for the fungal cell membrane. Lanosterol and its analogues which retain the methyl group do not fit properly into the membrane, causing it to become permeable to protons and eventually burst (Nes *et al.*, 1978; Thomas *et al.*, 1983). The 14α-methyl group of lanosterol is axial, protruding from the otherwise planar face. Van der Waals interactions of the sterol underside with the fatty acyl groups in the phospholipid bilayer of the membrane are therefore very much less favourable (Bloch, 1979).

A number of medical and agrochemical antifungal agents in use today owe their activity to potent inhibition at this step. The majority are substituted nitrogen heterocycles. The presence of a lone pair of electrons which can be donated from the nitrogen to the iron of the haem appears to be mandatory. The binding of ligand to enzyme gives rise to a shift in the ultraviolet spectrum whereby the absorbance of the haem is shifted from 410 nm to 420–430 nm (see Figure 6.4). This results in what is known as a Type II difference spectrum characterized by a maximum in the region of 420–430 nm and a minimum around 390–400 nm. Substrates of cytochrome P–450-linked reactions such as lanosterol give a totally reversed type of difference spectrum, called Type 1, where a peak is found at 380–390 nm and a trough at 410–420 nm.

Foremost among inhibitors of lanosterol demethylation in medical use is ketoconazole, a substituted imidazole, which gives a characteristic type II spectrum with yeast microsomes (Van den Bossche *et al.*, 1984). The inhibitory effect of ketoconazole and other antifungal 'azoles' is usually measured by the inhibition of incorporation of ^{14}C-acetate or -mevalonate into ergosterol, which is more convenient to use than a specific assay for lanosterol

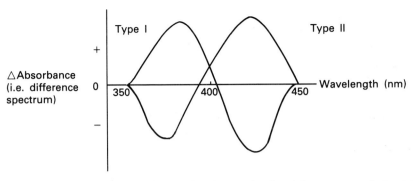

A difference spectrum is used because the changes in ultraviolet spectrum during inter-action of cytochrome P–450 with a ligand are often too small to be detected by reading the direct spectrum. The spectrum of cytochrome and ligand uncomplexed is subtracted from the spectrum of the complex. This can be done electronically by subtracting spectra stored in memory or, experimentally, by the use of split compartment cells.

A Type I difference spectrum is characteristic of a substrate binding to a cytochrome P–450 whereas Type II is given by a poorer substrate or an inhibitor. The absorbance maximum of the cytochrome is shifted to longer wavelength from about 410 nm by such inhibitors to give the Type II spectrum, while a substrate causes the wavelength maximum to move to shorter wavelength. Ketoconazole gives a Type II difference spectra. This subject is also discussed in Chapter 7, section 7.5.

Figure 6.4 Cytochrome P–450 difference spectra (diagrammatic representation).

demethylation and is more flexible in that inhibition of various steps in the pathway may be identified. It is quite clear in this case from the pattern of inhibition obtained, namely a build-up of lanosterol and a decrease in ergo-sterol, where the inhibited step lies. Ketoconazole gives an I_{50} in the region of 10^{-8} to 10^{-9} M in this system (Van den Bossche *et al.*, 1980; Marriott, 1980).

Ketoconazole

The main action of this class of drug is to disrupt the supply of ergosterol for cell membrane synthesis. This effect is likely to be fungistatic not fungi-cidal because the inhibitor is likely to be effective only when the organism is growing and dividing. Such appears to be the case with ketoconazole.

The advent of ketoconazole has greatly improved the treatment of a variety of fungal infections. The fact that it is orally active, and is the first 'azole' to be so, has allowed it to be used both for infections that are carried throughout the body (systemic), and dermatophyte infections that are confined to the surface of the skin (reviewed by Heel *et al.*, 1982).

The drug is relatively specific in inhibiting the fungal lanosterol demethylase as opposed to the same enzyme from rat liver (Van den Bossche *et al.*, 1980, 1984). Nevertheless, a number of other cytochrome P–450 conversions, particularly those in the interconversion of steroid hormones, are sensitive to inhibition by the drug. Two of the most obvious of these are (a) the rate-limiting step in the production of glucocorticoids and sex steroids, i.e. cholesterol 22,24 side-chain cleavage (Loose *et al.*, 1983) and (b) 17,20-lyase (Sikka *et al.*, 1985; reaction 4 in Figure 6.2). These effects are probably responsible for the occurrence of undesirable side-effects such as the growing of breast tissue in men (gynaecomastia), probably through lowered testosterone levels (Pont *et al.*, 1982). Other problems include teratogenicity and some incidence of serious, indeed fatal, liver toxicity (Lewis *et al.*, 1984).

These inhibitions of cytochrome P–450 linked enzymes may be put to good effect, however, in conditions in which there is overproduction of ACTH leading to overstimulation of the adrenals (Cushing's syndrome); ketoconazole has been found effective in this instance (Angeli and Frairia, 1985). Furthermore, some tumours are dependent on sex steroids and respond to the drug (Amery *et al.*, 1986). In this case, the observation of side-effects has led to the development of the drug for another purpose.

6.3 Steroid biosynthesis

6.3.1 Steroid 17,20 lyase

The conversion of progesterone into 17α-hydroxyprogesterone followed by the removal of the acetyl side-chain to yield androstenedione (reaction 4 in Figure 6.2) is believed to be carried out by one enzyme — but there is still a considerable lack of knowledge about some of the finer details of steroid interconversions (see Takemori and Kominami, 1984, for review). The enzyme is sited in the endoplasmic reticulum of adrenals and testes. It is a cytochrome P–450 and, therefore, requires molecular oxygen and NADPH for activity. The molecular weight of the purified enzyme is 52 000.

The inhibition is presumably effected by the donation of the imidazole nitrogen lone pair of electrons to the haem iron (as in lanosterol 14α-demethylase, noted in Section 6.2.3). (Sikka *et al.* (1985) find an I_{50} of 1×10^{-6} M for ketoconazole inhibition of the 17,20-lyase reaction.) Ketoconazole appears to be relatively unselective in its inhibition of dedicated sterol and steroid P–450s. Other drugs are being developed to be selective for this enzyme system.

As mentioned in the previous section, the antifungal drug ketoconazole produces an occasional side-effect in a few men taking the drug in that it causes them to grow breast tissue (gynaecomastia). Testosterone levels are reduced reversibly by the drug in a dose-related fashion, while at the same time progesterone and 17α-hydroxyprogesterone accumulate. Since androstenedione is reduced to form testosterone, these results can be explained by an inhibition of the steroid 17,20-lyase (reviewed in Amery *et al.*, 1986). It appears that the gynaecomastia occurs in men whose 17β-oestradiol levels are high.

This unfortunate side-effect was, however, noted and put to good use in a male condition where the reduction of androgen levels is the goal of therapy, namely prostate cancer. The hormone stimulates the growth of the tumour. Any reduction in androgen level, and possibly antagonism by increasing levels of progesterone, is an advance on one of the previously used treatments — castration! Ketoconazole is at present being used with encouraging results to treat patients who have relapsed on other therapies (Amery *et al.*, 1986).

6.3.2 3-β-Hydroxysteroid dehydrogenase

A long-term study to find means of contraception that do not show the side-effects of the contraceptive pill of the oestrogen/progesterone type has identi-fied the second enzyme in the pathway of steroid hormone biosynthesis as a suitable target (reaction 2 in Figure 6.2). In addition, inhibition of this enzyme shows promise as a potential target for agents that cause abortion (abortifacient). The rationale for this approach lies in the fact that progester-one is essential for the maintenance of pregnancy. The hormone is initially synthesized in the corpus luteum, but 8 to 9 weeks after implantation of the fertilized ovum, the placenta synthesizes enough hormone to take over (reviewed in Gal, 1972; Csapo and Pulkinen, 1978).

Compounds that inhibit this enzyme have been known for some time. The most recent agent, epostane, has been found effective as an abortifacient in rats and monkeys (Creange *et al.*, 1981) and as a contraceptive in rats since it inhibits ovulation (Snyder *et al.*, 1984). Recently a study has shown the efficacy of the drug as an abortifacient in women (Pattisen *et al.*, 1984). Earlier drugs of this series did not distinguish between adrenal and ovary steroid production, but this may not be much of a drawback as the drug does not have to be given over a prolonged period.

3-β-Hydroxysteroid dehydrogenase catalyses the conversion of pregneno-lone to progesterone with NADH as a cofactor. Epostane is a powerful competitive inhibitor of the enzyme from human placental mitochondria, with a K_i of 1×10^{-9} M (Rabe *et al.*, 1982). The inhibition clearly needs to be of this order of magnitude to block an enzyme that is not rate-limiting by a competitive mechanism. The build-up of pregnenolone could otherwise sweep away the inhibition *in vivo*.

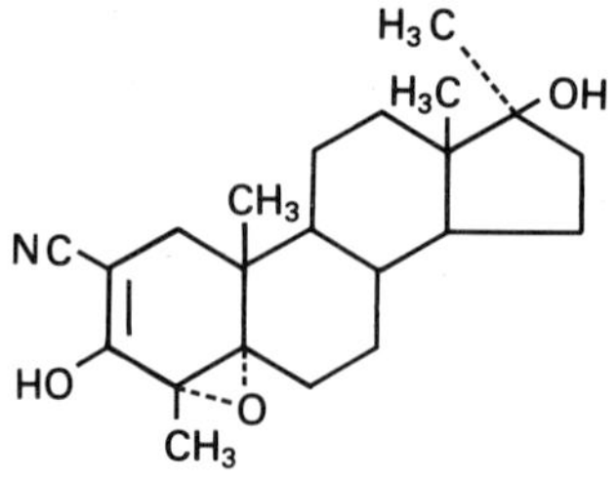

Epostane

6.3.3 11-β-Steroid hydroxylase

This enzyme catalyses the conversion of deoxycorticosterone to corticosterone, and deoxycortisol to cortisol (reaction 6 in Figure 6.2). Like cholesterol 20,22-lyase, the enzyme is found in adrenal mitochondria and requires a powerful reducing agent containing iron–sulphur clusters like ferredoxin (see Chapter 4), known as adrenodoxin, an enzyme to reduce it called adrenodoxin reductase, NADPH and molecular oxygen to form an electron transport chain. Metyrapone inhibits the enzyme competitively with a K_i of 1×10^{-7} M, and shows a type II ultraviolet difference spectrum with adrenal mitochondria (Williams and O'Donnell, 1969). 11-β-Deoxycorticosterone shows a Type I difference spectrum characteristic of a substrate. Both substrate and metyrapone stabilize the enzyme. Furthermore, metyrapone induces a shift in the electron paramagnetic resonance spectrum suggesting that the haem is involved in the interaction (Wilson *et al.*, 1969).

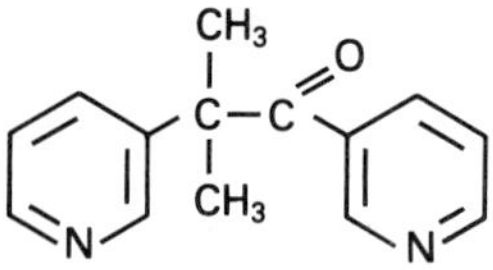

Metyrapone

Metyrapone has been used to treat the hypersecretion of cortisol by adrenal tumours (reviewed by Temple and Liddell, 1970), but on long-term administration the level of ACTH may rise to counteract the effect of the drug. The use of metyrapone alone for this indication is no longer recommended, and aminoglutethimide may be given in conjunction (Gold, 1979).

6.3.4 Aromatase

Another major area of interest concerning the synthesis of steroid hormones is the dependence of certain tumours on steroids for the maintenance of growth. Testosterone appears to be essential for the growth of some tumours of the prostate gland, while certain breast cancers are dependent on oestradiol. At least one third of human breast cancers are hormone-dependent, and regress when the sources of oestrogen are eliminated (Kiang *et al.*, 1978). Inhibition of steroid-producing enzymes is one way of achieving this.

Aminoglutethimide is an inhibitor of both cholesterol 20,22-lyase and aromatase — the latter so-called because it introduces an aromatic ring into the steroid nucleus (ring A, reaction 5 in Figure 6.2) in catalysing the transformation of testosterone and androstenedione to 17-β-oestradiol and oestrone respectively.

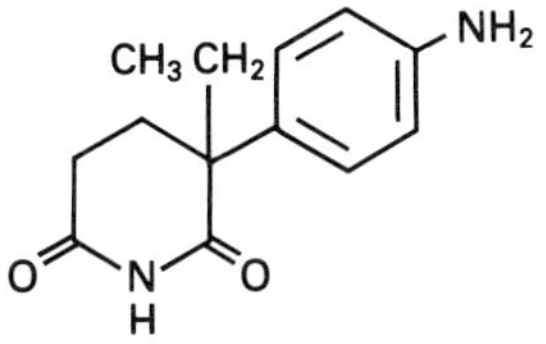

Aminoglutethimide

It is not clear which of these two enzymes is the key target, although the inhibition of aromatase was shown to be at a lower concentration than that of cholesterol 20,22-lyase (K_i for aromatase 6×10^{-7} M — testosterone concentration $1 \cdot 5 \times 10^{-6}$ M; K_i for 20,22-lyase $4 \cdot 5 \times 10^{-6}$ M, cholesterol concentration as normally found in bovine adrenal mitochondria). Aromatase may be the enzyme that is the more clinically relevant since the fall in oestrogens after dosing is immediate, unlike progesterone levels which initially rise (Foster *et al.*, 1983).

These inhibitory effects have promoted the study of aminoglutethimide as a treatment for metastatic breast cancer, against which it is moderately effective (Smith *et al.*, 1978). Attempts to improve on the original structure are under study (Foster *et al.*, 1983). Interestingly, norethisterone, an acetylenic progestin, has been shown to inactivate aromatase in a time-dependent fashion albeit at high concentration (Osawa *et al.*, 1983), but the relevance of this finding to its contraceptive or any other action has not been established. The mechanism of removal of the methyl group bears a formal similarity to the demethylation of lanosterol in that the reaction takes place in three stages: to alcohol, then aldehyde followed by total removal of the fragment as formic acid and the introduction of a double bond into the sterol skeleton. This has prompted the synthesis of some mechanism-based irreversible (suicide) inhibitors (see Chapter 1) (Covey *et al.*, 1981).

6.4 Steroid receptor ligands

6.4.1 Oestrogen/progesterone receptor agonists

One of the major uses of pharmaceutical products is of a prophylactic nature — the avoidance of the condition of pregnancy. Both oestrogen and progesterone receptor agonists are used for this, usually in conjunction, although progestins can be used alone if the patient is believed likely to suffer side-effects on oestrogen treatment. The rationale for the use of such agents is that they inhibit ovulation, with oestrogen mainly responsible for suppressing follicle-stimulating hormone (FSH) secretion, while the progestin suppresses luteinizing hormone (LH) secretion (Briggs *et al.*, 1970) (FSH and LH are known as gonadotrophins). Even if ovulation and subsequent fertilization did occur, it is very unlikely that the fertilized egg could implant into the lining of the womb. Oestrogen produces a cornified or keratinized layer on the surface lining of the womb to promote adhesion of the egg. Progesterone antagonizes this and, in addition, produces a very viscous mucus layer through which the egg would find it difficult to penetrate.

The oral contraceptives inhibit the production of FSH and LH by binding to receptors in the cytosol of hypothalamus and pituitary (see Figure 6.5). The effect on the former site suppresses the release of gonadotrophin releasing hormone (GnRH) whilst the effect on the latter is to reduce gonadotrophin release. This is a negative feedback loop that the endogeneous hormone controls, and is a necessary part of pregnancy (Schally, 1978). It is interesting in this context that the oral contraceptives do not trigger the positive feedback at the level of the pituitary whereby oestrogen together with GnRH stimulates the pre-ovulatory surge of LH (Asch *et al.*, 1983) — see Figure 6.5. The negative feedback closes the cycle since FSH stimulates the granulosa cells of the follicle to produce increasing amounts of oestrogen as the follicle develops. Following ovulation the granulosa cells differentiate into luteal cells

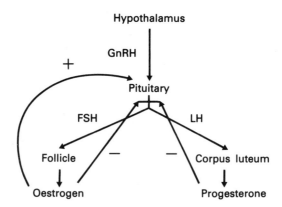

+ indicates a positive feedback
− indicates a negative feedback

The hypothalamus releases GnRH which acts on the pituitary to produce both FSH and LH. FSH stimulates follicles to grow and produce oestrogen which has a negative feedback on LH and FSH release. Eventually, however, an oestrogen level is reached that is sufficient, acting in concert with GnRH, to release LH and FSH in a surge from the pituitary. This converts one follicle to a corpus luteum, synthesizing progesterone that inhibits LH production. Only the negative feedback to the pituitary is shown for the sake of clarity.

Figure 6.5 Hypothalamus – Pituitary – Gonad axis.

which secrete progesterone under the influence of LH in the later part of the cycle (known as the luteal phase) until the corpus luteum begins to degenerate (Murad and Haynes, 1985). Both peptide hormones, LH and FSH, act through adenylate cyclase, after binding to their specific receptors, in a similar fashion to ACTH to stimulate cholesterol 20,22-lyase, cyclic AMP levels and steroidogenesis (see Funkenstein *et al.*, 1983, and references therein).

Two of the most frequently used oestrogens are ethinyloestradiol and its O-methyl analogue, mestranol. The former binds to rat anterior pituitary and hypothalamic cytosol with a potency similar to that of oestradiol, with a K_d from 1 to 1.3×10^{-10} M compared with 2.2 to 3.2×10^{-10} M for oestradiol. Mestranol, on the other hand, is bound much less tightly, with a K_d of 2.9 to 5.5×10^{-8} M, as shown by competitive binding with oestradiol. It has been suggested that mestranol is de-alkylated to the more active ethinyloestradiol by liver mixed-function oxidases — as occurs in the rat (Eisenfeld, 1974).

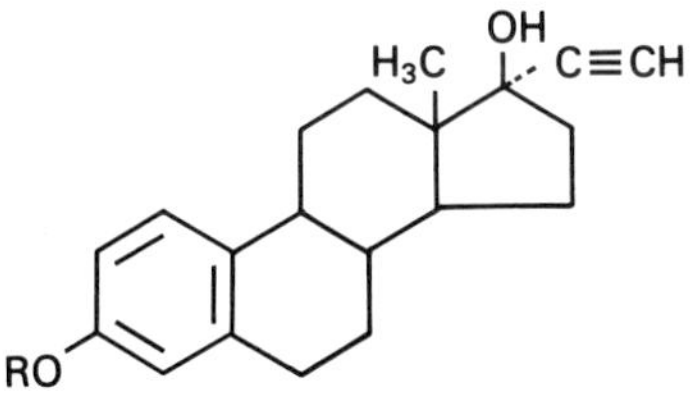

R = CH₃, mestranol
R = H, ethinyloestradiol

After the drug-receptor complex is formed it is converted to a different species by a conformational change and translocated into the nucleus. Specific mRNA and certain proteins are synthesized initially, followed by a general increase in the synthesis of RNA. Specific DNA biosynthesis is a much later event (Murad and Haynes, 1985). How this process actually interferes with the secretion of GnRH is not known at present.

Norethisterone and norgestrel are typical examples of the most widely used progestins. They bind to receptors in the cytosol of human uterus cells with binding constants of 6.8×10^{-9} M and 2.4×10^{-9} M respectively (Kasid *et al.*, 1978; Srivastava, 1978). Both compounds compete with progesterone for these binding sites.

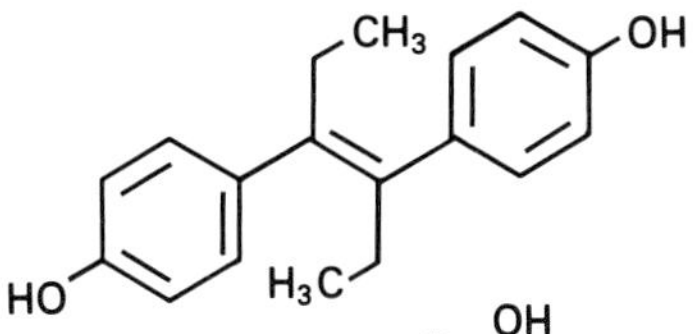

Diethylstilboestrol

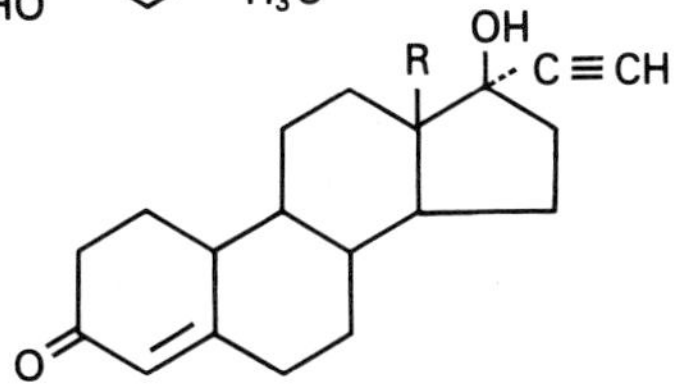

R = CH₃, Norethisterone
R = C₂H₅, Norgestrel

If an oral contraceptive is required after intercourse, diethylstilboestrol, a non-steroid oestrogen is often used. Oestrogens have uses other than oral contraceptives including hormone replacement therapy during the meno-pause (Murad and Haynes, 1985).

6.4.2 Oestrogen antagonists

The treatment of breast cancer has recently undergone a great improvement through the discovery of the existence and biological importance of oestrogen receptors in the tumours. By measuring the tumour oestrogen receptor content, breast cancers can be divided into two distinct groups: receptor-rich and receptor-poor. In the case of receptor-rich tumours, chemotherapy with both cytotoxic and hormonal agents is likely to be more successful than in the receptor-poor group where hormone therapy is likely to be valueless (Kiang *et al.*, 1978). At least one-third of all patients with breast cancer respond to hormone therapy. For the treatment of pre-menopausal women, removal of the ovaries is most frequently used, while tamoxifen, an oestrogen antagonist, is the hormonal agent of choice in post-menopausal women whose ovaries, by definition, have ceased to function (reviewed by Rubin, 1981; Carter and Rubens, 1981). One great advantage of tamoxifen over cytotoxic agents is its relative lack of toxicity.

Tamoxifen is a potent ligand of oestrogen receptors, with a K_d of cytosolic receptors from human mammary tumours of $3 \cdot 7 \times 10^{-9}$ M compared to $1 \cdot 2 \times 10^{-10}$ M for 17-β-oestradiol by direct measurement (Nicholson *et al.*, 1979). A figure of 1–2 orders of magnitude less activity is obtained in competitive binding studies, possibly due to a breakdown of the drug-receptor complex. Using a cell-line of human breast cancer cells (MCF 7), Horwitz *et al.* (1978) found that the action of tamoxifen was biphasic in that at low extracellular levels (10^{-7} M) the drug showed agonist properties in stimulating growth and inducing progesterone receptors. Higher doses (10^{-6} M) were antagonistic. Interference with oestradiol may take the form of blocking the binding of oestradiol to cytosolic receptors, and of reducing the number of receptors available for the natural hormone. Consequently, the natural hormone is unable to maintain tumour growth.

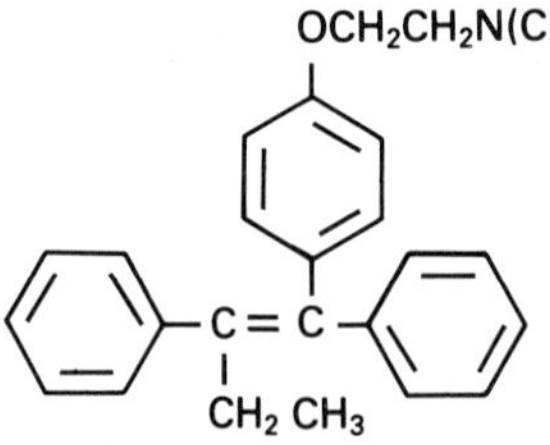

Tamoxifen

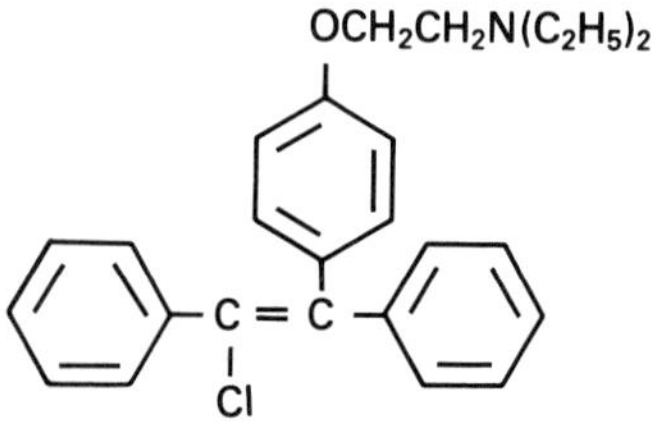

Clomiphene

A close analogue of tamoxifen, clomiphene, has found favour as an agent for the induction of fertility in women who either have no menstrual cycle or cycles in which they do not ovulate (Marshall, 1978). The most obvious physiological effect of this drug is the enlargement of the ovaries, and this hyperstimulation does on occasion give rise to multiple births — approximately 6 to 8% compared with 25% of births when gonadotrophins are given to induce fertility. Clomiphene at high dose is effective for the palliation of breast cancer but tamoxifen is preferred for this indication because of lower toxicity (Marshall, 1978).

Clomiphene acts as an anti-oestrogen showing side-effects consistent with this action, including hot flushes. The major pharmacological effect is to prevent the normal feedback inhibition of oestrogen on the release of GnRH from the hypothalamus (see Figure 6.5). Higher GnRH levels then release more luteinizing hormone and follicle-stimulating hormone from the pituitary and ovulation is stimulated. As with tamoxifen (see above), clomiphene binds to oestrogen receptors and the complex is translocated into the nucleus, thus resulting in a marked diminution of receptors in the cytosol for oestrogen binding.

The binding constant of clomiphene to various rat brain and uterine receptors is in the region of 10^{-8} M compared with 10^{-10} M for 17-β-oestradiol under the same conditions. Unlike the studies with tamoxifen, direct binding and competitive experiments gave similar results (Ginsberg *et al.*, 1977).

6.4.3 Progesterone antagonist

In addition to inhibition of progesterone biosynthesis as an approach to contraception and abortion, receptor antagonism has been found to be effective. Mifepristone (RU-486 or RU-38486) has been developed as a ligand of progesterone receptors, and is apparently considerably more potent than progesterone itself (Schreiber *et al.*, 1983). Mifepristone also binds to glucocorticoid receptors but at markedly higher concentration. The drug also binds to receptors in the hypothalamus which leads to a fall in GnRH secretion and subsequently to luteolysis through a fall in LH levels. A marked fall in progesterone secretion is the result. The drug has been found to be moderately effective at terminating early pregnancy (Kovacs *et al.*, 1984), and its potential use for the induction of menses, abortion and labour has been reviewed by Healy and Fraser (1985).

6.4.4 Aldosterone antagonist

Aldosterone is the steroid (known as a mineralocorticoid because it helps to control the levels of minerals in the blood plasma) which controls reabsorption of sodium ion in the distal portion of the kidney tubule (Figure 6.6). Up to 90% of the sodium is reabsorbed with chloride as the counter-ion, but a portion appears to be linked to the excretion of potassium or hydrogen ion

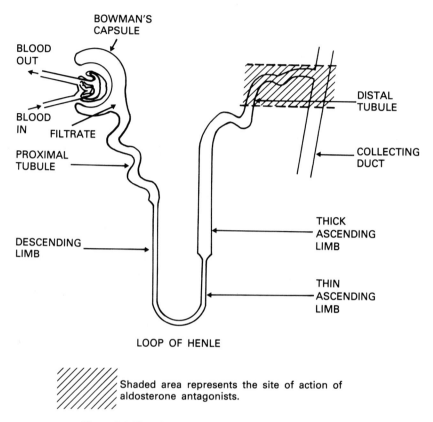

Figure 6.6 Site of aldosterone and spironolactone action.

thereby maintaining electrical neutrality. Water is reabsorbed with the sodium ion; the net effect is an increase in plasma volume and fall in plasma potassium.

Aldosterone is synthesized in the zona glomerulosa of the adrenal cortex. Like the other steroids mentioned in this chapter, the hormone functions by crossing the cell membrane and binding to a cytoplasmic receptor with a K_d of 2×10^{-9} M (see Funder *et al.*, 1974). Two sets of binding sites in the cortex have been proposed by Farman *et al.* (1982), who obtained a biphasic Scatchard plot (see Appendix) for the binding of hormone to preparations from rabbit kidney cortex. Although such a plot does not always imply two sets of sites, and insufficient data points were obtained to reach a definite conclusion about the absolute values of the binding constant, these authors proposed that the higher affffinity binding sites were for mineralocorticoids and the lower for glucocorticoids.

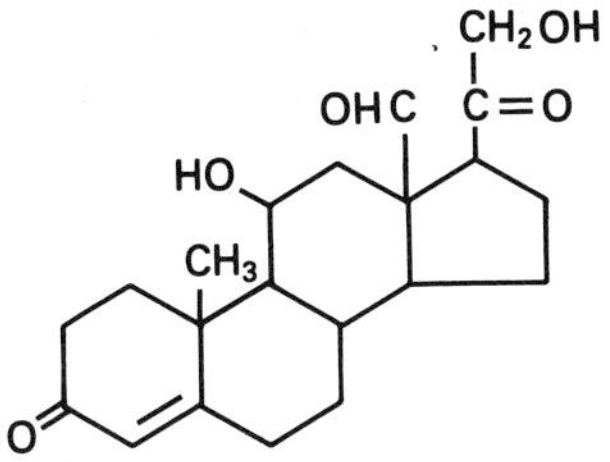

Aldosterone

The steroid/receptor complex undergoes a conformational change before it is translocated into the cell nucleus, where it binds to the DNA thus causing specific mRNAs to be synthesized (see Corvol *et al.*, 1981). Several proteins in the cytosol and plasma membrane of the kidney cell are synthesized in response to aldosterone action — a process which is blocked by cyclohexi-mide — but the precise role each plays in the facilitation of sodium reabsorp-tion is not yet known (Scott *et al.*, 1981).

Steroid antagonists of aldosterone, of which the best known is spironolac-tone, are valuable diuretics in that not only do they antagonize the antidiure-tic action of the steroid by preventing sodium reabsorption, but also they do not cause a loss of potassium (this is referred to as a potassium sparing action) an action not shared by the other types of diuretic. This is because aldoster-one induces a low but measurable exchange of potassium for sodium noted above. Spironolactone and its analogues bind to the aldosterone receptor; the complex so formed does not rearrange into the active conformation and therefore cannot translocate into the nucleus (Funder *et al.*, 1974; Sakauye and Feldman, 1976). These drugs thus act in a fashion different from tamoxifen and its complex with the oestrogen receptor (see section 6.4.3 this chapter).

The binding constant for spironolactone is approximately 4×10^{-8} M when measured competitively against ^{3}H-aldosterone binding to rat kidney (Farman *et al.*, 1974). The correlation of binding to the receptor and physiolo-gical effect has been studied by Rossier *et al.* (1983), who showed that receptor binding in toad bladder correlated with inhibition of sodium trans-port across the membrane induced by aldosterone. These workers also showed that the action of the drugs could be reversed by aldosterone and the drugs were not partial agonists.

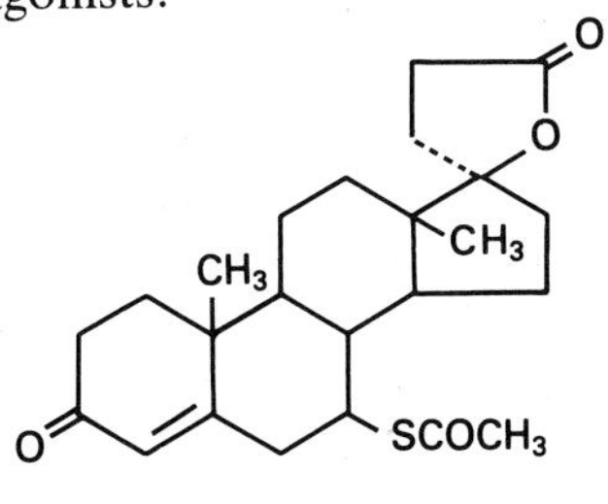

Spironolactone

Spirolactones, as a group, are also able to inhibit the biosynthesis of aldosterone but at rather higher concentrations than are needed to interfere with aldosterone receptor binding. The conclusion is, therefore, that the inhibition of biosynthesis plays little or no part in the antidiuretic action of the drugs (Carvol *et al.*, 1981).

Spironolactone is not, however, selective for aldosterone receptors, and is well able to block androgen receptors, albeit at higher doses. This is clearly a disadvantage when a diuretic effect only is required, because it can lead to the growth of breasts in men (gynaecomastia) and to menstrual disturbances in women. Nevertheless, in some conditions, notably hirsutism in women caused by high androgen levels, an anti-androgen effect is a positive advantage. The drug has been of clinical value in this condition (Shapiro and Evron, 1980).

References

Alberts, A.W., Chen, J. Kuron, G., Hunt, V., Huff, V., Hoffman, C., Rothbrook, J., Lopez, M., Joshua, H., Harris, E., Pitchett, A., Monaghan, R., Currie, S., Stapley, E., Albers-Schonberg, G., Hersens, O., Hirshfield, J., Hoogsteen, K., Liesch, J. and Springer, J. (1980). *Proc. Nat. Acad. Sci. (U.S.A.)*, **77**, 3957–3961.

Amery, W.K., De Coster, R. and Caers, I. (1986). *Drug. Dev. Res.*, **8**, 299–307.

Angeli, A. and Frairia, R. (1985). *Lancet*, **i**, 821.

Anon (1986). *Lancet*, **ii**, 665–666.

Aoyama, Y., Yoshida, Y., Sonoda, Y. and Sato, R. (1987). *J. Biol. Chem.*, **262**, 1239–1243.

Asch, R.H., Balmaceda, J.P., Borghi, M.R., Niesvisky, R., Coy, D.H. and Schally, A.V. (1983). *J. Clin. Endocrinol. Metab.*, **57**, 367–372.

Atkin, S.D., Morgan, B., Baggaley, K.H. and Green, J. (1972). *Biochem. J.*, **130**, 153–157.

Avigan, J., Steinberg, D., Vroman, H.E., Thompson, M.J. and Mosettig, E. (1960). *J. Biol. Chem.*, **235**, 3123–3126.

Barrett-Bee, K.J., Lane, A.C. and Turner, R.W. (1986). *J. Med. Vet. Mycol.*, **24**, 155–160.

Bloch, K.E. (1979). *C.R.C. Crit. Rev. Biochem.*, **7**, 1–5.

Briggs, M.H., Pitchford, A.G., Staniford, M., Barker, H.M. and Taylor, D. (1970). *Adv. Steroid. Biochem. Pharmacol.*, **2**, 111–222.

Carter, S.K. and Rubens, R.D. (1981). *Lancet*, **ii**, 795–797.

Chen, H.W. and Leonhardt, D.A. (1984). *J. Biol. Chem.*, **259**, 8156–8162

Corvol, P., Claire, M., Oblin, M.E., Geering, K. and Rossier, B. (1981). *Kidney Int.*, **20**, 1–6.

Coulson, C.J., King, D.J. and Wiseman, A. (1984). *Trends in Biochem. Sci.*, **9**, 446–449.

Covey, D.F., Hood, W.F. and Parikh, V.D. (1981). *J. Biol. Chem.*, **256**, 1076–1079.

Creange, J.E., Anzalone, A.J., Petts, G.O. and Schane, H.P. (1981). *Contraception*, **24**, 289–299.

Csapo, A.I. and Pulkinen, M. (1978). *Obstet. Gynaecol. Surv.*, **33**, 69–81.

Eisenfeld, A. (1974). *Endocrinology*, **94**, 803–807.

Farman, N., Vandewalle, A. and Bonvallet, J.P. (1982). *Am. J. Physiol.*, **242**, F69–77.

Fears, R. (1983). *Biochem. Soc. Trans.*, **11**, 642–644.

Foster, A.B., Jarman, M., Leung, C.-S., Rowlands, M.G. and Taylor, G.N. (1983). *J. Med. Chem.*, **26**, 50–54.

Funder, J.W., Feldman, D., Highland, E. and Edelman, T.S. (1974). *Biochem. Pharmacol.*, **23**, 1493–1501.

Funkenstein, B., Waterman, M.R., Masters, B.S.S. and Simpson, E.R. (1983). *J. Biol. Chem.*, **243**, 10187–10189.

Gal, I. (1972). *Adv. Fertil.*, **5**, 161–173.

Ganzinger, U., Stephen, A. and Gumhold, G. (1982). *Clin. Trials J.*, **19**, 342–348.

Gibbons, G.F., Pullinger, C.R. and Mitropoulos, K.A. (1979). *Biochem. J.*, **183**, 309–315.

Ginsburg, M., Maclusky, N.J., Morris, I.D. and Thomas, P.J. (1977). *Brit. J. Pharmacol.*, **59**, 397–402.

Gold, E.M. (1979). *Ann. Intern. Med.*, **90**, 829–844.

Healy, D.L. and Fraser, H.M. (1985). *Brit. Med. J.*, **290**, 580–581.

Heel, R.C., Brogden, R.N., Carmine, A., Morley, P.A., Speight, T.M. and Avery, G.S. (1982). *Drugs*, **23**, 1–36.

Horwitz, K.B., Koseki, Y. and McGuire, W.L. (1978). *Endocrinology*, **103**, 1742–1751.

Illingworth, D.R. (1984). *Ann. Intern. Med.*, **101**, 598–604.

Kasid, A., Buckshee, K., Hingorami, V. and Laumas, K.R. (1978). *Biochem. J.*, **176**, 531–539.

Kiang, D.T., Frenning, D.H., Goldman, A.I., Ascensao, V.F. and Kennedy, B.J. (1978). *New Engl. J. Med.*, **299**, 1330–1334.

Kovacs, L., Sas, M., Resch, B.A., Ugocsai, G., Swahu, M.L., Bygdeman, M.M. and Rowe, P.J. (1984). *Contraception*, **29**, 399–410.

Lewis, J.H., Zimmerman, H.J., Benson, G.D. and Ishak, K.G. (1984). *Gastroenterology*, **86**, 503–513.

Loose, D., Kan, P.B., Hirst, A., Marcus, R.A. and Feldman, D. (1983). *J. Clin. Invest.*, **71**, 1495–1499.

Luskey, K.L., Chin, D.J., Macdonald, R.J., Liscum, L., Goldstein, J.L. and Brown, J.S. (1982). *Proc. Nat. Acad. Sci. (U.S.A.)*, **79**, 6210–6214.

Mabuchi, H., Haba, T., Tatami, R., Mujamoto, S., Sakai, Y., Wakasugi, T., Watenabe, A., Keizumi, J. and Takeda, R. (1981). *New Engl. J. Med.*, **305**, 478–482.

Marriott, M.S. (1980). *J. Gen. Microbiol.*, **117**, 253–255.

Marshall, J.R. (1978). *Clin. Obstet. Gynaecol.*, **21**, 147–162.

Murad, F. and Haynes, H.C. Jr, (1985). In: *The Pharmacological Basis of Therapeutics*, 7th Ed., Eds. Goodman, A.G., Gilman, L.S., Rall, T.W. and Murad, F. (New York: Macmillan) Ch. 61.

Nes, W.R., Sekula, B.C., Nes, W.D. and Adler, J.H. (1978). *J. Biol. Chem.*, **253**, 6218–6225.

Nicholson, R.I., Syne, J.S., Daniel, C.P. and Griffiths, K. (1979). *Eur. J. Cancer*, **15**, 317–329.

Ono, T., Nakazone, K. and Kosaka, H. (1982). *Biochim. Biophys. Acta.*, **709**, 84–90.

Osawa, Y., Osawa, Y., Yarborough, C. and Berzynski, L. (1983). *Biochem. Soc. Trans.*, **11**, 656–659.

Pattison, N.S., Webster, M.A., Phipps, S.L., Anderson, A.B. and Gillmer, M.D. (1984). *Fertil. Steril.*, **42**, 875–881.

Petranyi, G., Ryder, N.S. and Stutz, A. (1984). *Science*, **224**, 1239–1241.

Pont, A., Williams, P.L., Azher, S., Reitz, R.E., Rochra, C., Smith, E.R. and Stevens, D.A. (1982). *Arch. Intern. Med.*, **142**, 2137.

Rabe, T., Kiesel, L., Kellerman, J., Weidenhammer, K., Maletz-Kehry, W., Runnebaum, B. (1982). *Acta Endocrinol.*, **99**, Suppl. 246, 24.

Rogers, D.H. and Rudney, H. (1982). *J. Biol. Chem.*, **259**, 10650–10658.

Rossier, B.C., Claire, M., Rafestin-Oblin, M.E., Geering, K., Gaggeler, H.P. and Corvol, P. (1983). *Am. J. Physiol.*, **244**, C24–31.

Rubin, R.N. (1981). *Clin. Ther.*, **4**, 74–94.

Ryder, N.S. and Dupont, M.-C. (1984). *Biochim. Biophys. Acta*, **794**, 466–471.

Sakauye, C. and Feldman, D. (1976). *Am. J. Physiol.*, **231**, 93–97.

Schally, A.V. (1978). *Science*, **202**, 18–28.

Schreiber, J.R., Hsueh, A.J.W. and Barlieu, E.B. (1983). *Contraception*, **28**, 77–85.

Scott, W.N., Yang, C.P., Skipski, I.A., Cobb, M.H., Reich, I.M. and Terry, P.M. (1981). *Ann. N.Y. Acad. Sci*, **372**, 15–28.

Shapiro, G. and Evron, S. (1980). *J. Clin. Endocrinol. Metab.*, **51**, 429–432.

Sikka, S.C., Swerdloff, R.S. and Rajfer, J. (1985). *Endocrinology*, **116**, 1920–1925.

Smith, I.E., Fitzharris, B.M., McKinna, J.A., Fahmy, D.R., Nash, A.G., Neville, A.M., Gazet, J.-C., Ford, H.T. and Powles, T.J. (1978). *Lancet*, **ii**, 646–648.

Snyder, B.W., Beecham, G.D. and Schane, H.P. (1984). *Proc. Soc. Exp. Biol. Med.*, **176**, 238–242.

Srivastava, A.K. (1978). *J. Steroid Biochem.*, **9**, 1241–1244.

Takemori, S. and Kominami, S. (1984). *Trends in Biochem. Sci.*, **9**, 393–396.

Temple, T.E. and Liddle, G.W. (1970). *Annu. Rev. Pharmacol. Toxicol.*, **10**, 199–218.

Thomas, P.G., Haslam, J.M. and Baldwin, B.C. (1983). *Biochem. Soc. Trans.*, **11**, 713.

Tobert, J.A., Bell, G.D., Birtwell, J., James, I., Kukovetz, W.R., Pryor, J.S., Buntinx, A., Holmes, I.B., Chao, Y.-S. and Bolognese, J.A. (1982). *J. Clin. Invest.*, **69**, 913–919.

Van den Bossche, H., Willemsens, G., Cools, W., Cornelissen, F., Lauwers, W.F. and Van Cutsem, J.M. (1980). *Antimicrob. Ag. Chemother.*, **17**, 922–928.

Van den Bossche, H., Lauwers, W., Willemsens, G., Marischal, P., Cornelissen, P. and Cools, W. (1984). *Pesticide Sci*, **15**, 188–198.

Williamson, D.G. and O'Donnell, V.J. (1969). *Biochemistry*, **8**, 1311–1316.

Wilson, L.D., Oldham, S.B. and Harding, B.W. (1969). *Biochemistry*, **8**, 2975–2980.

Chapter 7

Prostaglandin and leukotriene biosynthesis and action

7.1 Introduction

The transformation of arachidonic acid into a multiplicity of lipid intermediates, many of which are extremely active biologically, has been an area of great interest in recent years. The prostaglandins themselves were first isolated and identified by Samuelson and co-workers in the 1960s, but the other agents with quaint, albeit accurate, names like 'rabbit aorta contracting substance' (thromboxane A_2) and 'slow reacting substance of anaphylaxis' (leukotrienes C_4 and D_4) defied identification until the early 1980s, largely because of their instability and low concentrations in biological fluids. The development of sensitive mass spectroscopic and gas chromatographic methods for their isolation was crucial in their identification.

Table 7.1 Drugs discussed in Chapter 7

Target	Drug	Major uses
Phospholipase A_2	Corticosteroids	Asthma, arthritis
Prostaglandin synthetase	Aspirin, Indomethacin	Pain, inflammation in arthritis
Thromboxane synthetase	Dazoxiben, Furegrelate	Anti-thrombosis (under trial) Anti-thrombosis (under trial)
Thromboxane A_2 receptors	AH23.848 BM13.177	Anti-thrombosis (under trial) Anti-thrombosis (under trial)

The details of arachidonic acid metabolism have been elucidated at a remarkable rate over recent years. The present outline of the main pathways is shown in Figure 7.1. The metabolism of arachidonic acid can give rise to four distinct pathways: the cyclooxygenase reaction leads to the prostaglandins D_2, E_2, F_2, G_2 and H_2, the thromboxanes and prostacyclin; 5-lipoxygenase leads to the leukotrienes (see Figure 7.2); a cytochrome P–450 linked reaction gives rise to the eicosatrienoic acids, and 12-lipoxygenase yields another family of hydroxyeicosatrienoic acids (see Taylor and Ritter, 1986). It is the first two pathways with which this chapter is primarily concerned.

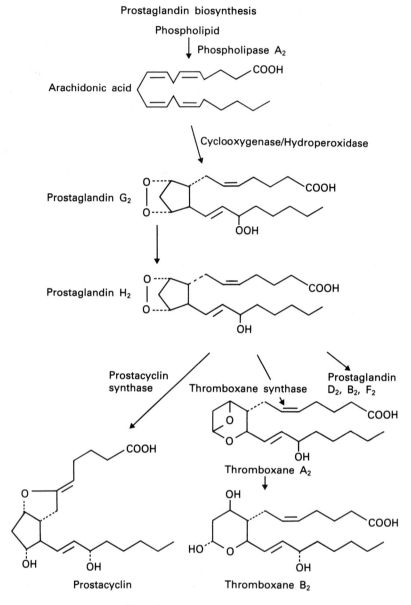

Figure 7.1 Prostaglandin biosynthesis.

7.2 Phospholipase inhibition

The initial step in prostaglandin and leukotriene biosynthesis is the release of arachidonic acid from phospholipids. This may either be direct by phospholipase A_2 hydrolysing the fatty acyl group from position 2 of the phospholipid, or indirect, via the action of phospholipase C that releases the phosphoryl group to form a 1,2-diacylglycerol which is then hydrolysed by another lipase to yield arachidonic acid. The former enzyme, whose action appears to predominate over the latter, is located on the inside of the plasma membrane of the cell and requires calcium ions for activity. Originally it was thought that calcium bound to its intracellular receptor, calmodulin, activated phospholipase A_2 but more recent work has shown that calcium acts directly on the enzyme (Withnall *et al.*, 1984).

Corticosteroids, which have been in use for many years for the treatment of inflammation and asthma, inhibit this reaction in rather a novel way, in that they activate the synthesis of a protein called macrocortin or lipocortin, of molecular weight 15 to 40 kDa depending on the cell type, which inhibits the enzyme completely. For example when cortisol (also known as hydrocortisone) is incubated with peritoneal leukocytes from the rat, macrocortin is released from the cells to the extent of 50% of total release within 30 minutes and completely within 3 hours. It appears that the steroid both causes the release of pre-formed inhibitor and induces fresh synthesis, the latter process being sensitive to cycloheximide and actinomycin D (Blackwell *et al.*, 1980; Hirata *et al.*, 1980). A similar protein inhibitor, known as lipomodulin, is invoked in response to glucocorticoids in the polymorphonuclear leukocyte.

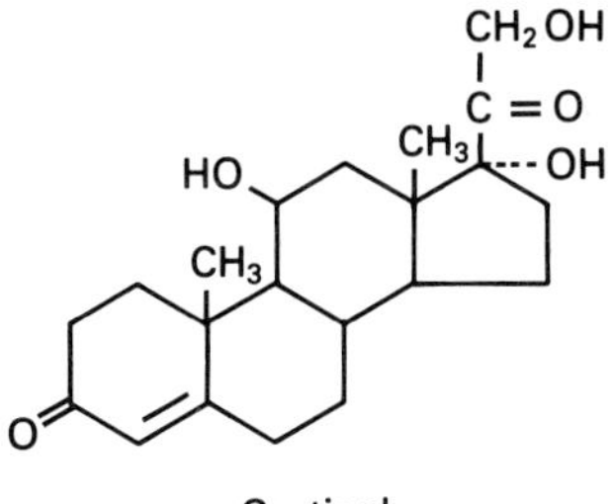

Cortisol

The initial step in the action of the glucocorticoids is the binding to a cytosolic receptor, as with the other steroids mentioned in Chapter 6. The steroid receptor complex is translocated into the nucleus in order to switch on the relevant gene to alter RNA and protein synthesis. Binding of glucocorticoids, not to a purified receptor, but to a homogenate of rat kidney cells, has been studied by the use of ^{3}H-dexamethasone which has a binding constant of $1 \cdot 25 \times 10^{-8}$ M at 37°C (Russo-Marie *et al.*, 1979). Competition of four steroids for the receptor in neutrophils can be correlated with their anti-inflammatory effect (Hirata *et al.*, 1980).

Corticosteroids are still the main therapy for sufferers from severe asthma and other allergic conditions. Although single doses do not normally give rise to side-effects, prolonged use can lead to immune suppression, among other problems.

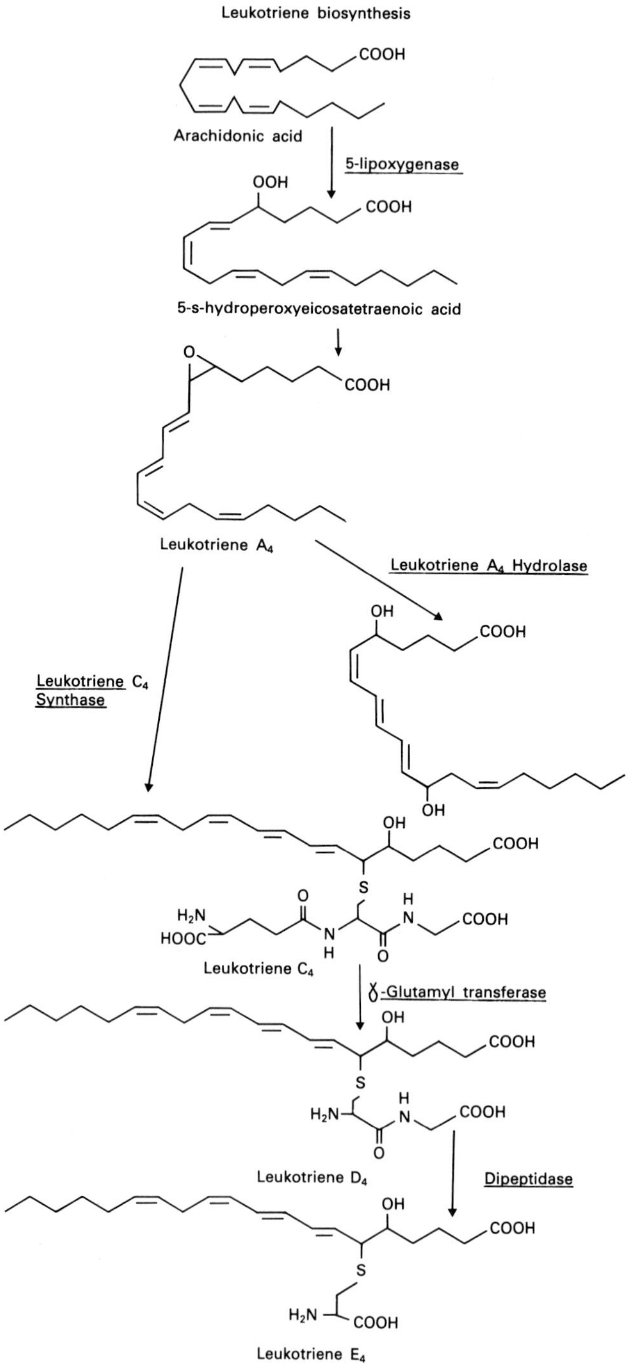

Figure 7.2 Leukotriene biosynthesis.

7.3 Prostaglandin synthetase inhibitors

One of the oldest medications still in use is aspirin or acetylsalicylic acid. The stimulus for the derivation of aspirin was the finding that the bark of the willow tree had a soothing effect on headaches and fevers. The active ingredient turned out to be a glycoside of salicyl alcohol, and subsequently salicylic acid was also found to be effective in reducing temperature (antipyretic action). Acetylation of the phenolic hydroxyl was the next step to yield aspirin (Flower *et al.*, 1985).

A very interesting advance in our understanding of pharmacology arose from the realization that aspirin and other similar drugs, classified as the non-steroidal anti-inflammatory agents, inhibited the synthesis of prostaglandins at the step catalysed by prostaglandin cyclooxygenase (Vane, 1971). There are a considerable number of these drugs which can be broadly grouped into structural families: salicylates such as aspirin, substituted propionic acids like naproxen, pyrazolones such as phenylbutazone, fenamates like mefenamic acid, together with indomethacin and piroxicam which do not readily fall into any of the above structural groupings. Nevertheless, they all show similar pharmacological activity: antipyretic, analgesic and anti-inflammatory properties.

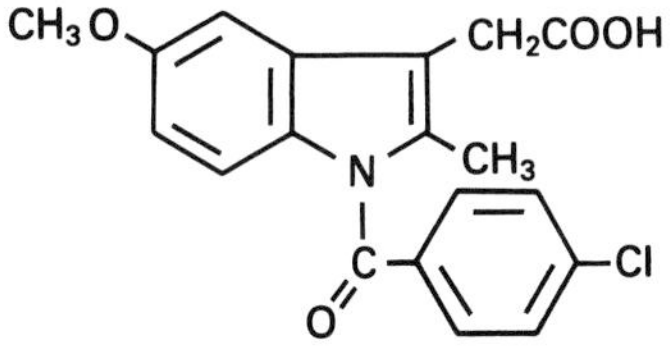

Indomethacin

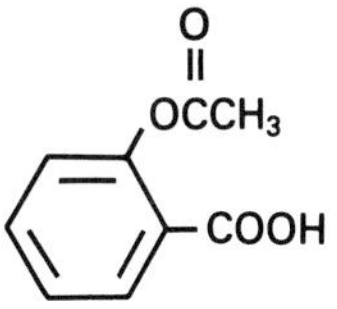

Acetylsalicylic acid

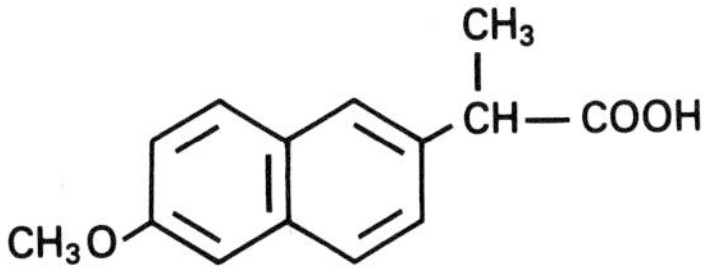

Naproxen

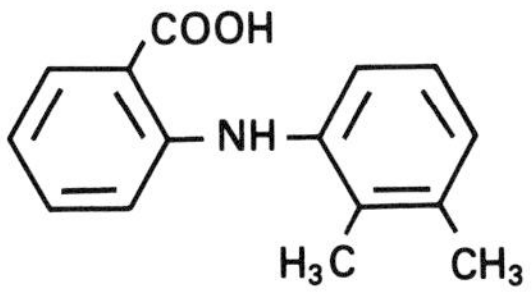

Mefenamic acid

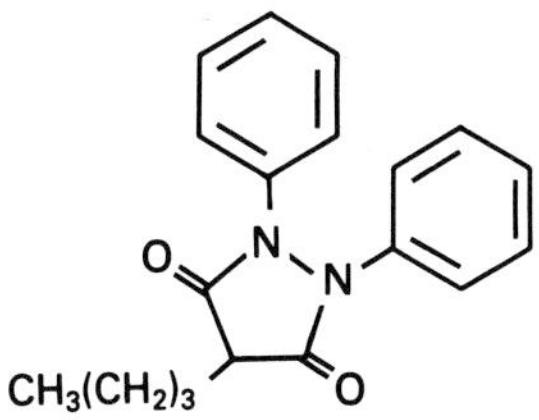

Phenylbutazone

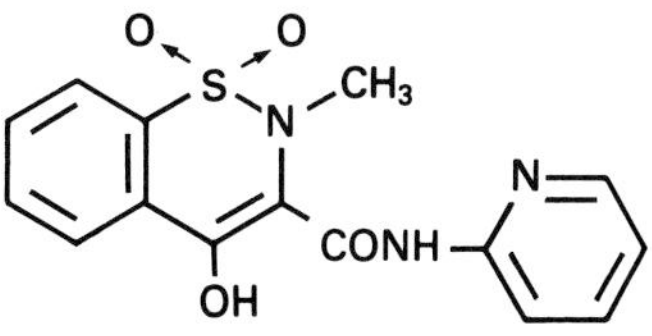

Piroxicam

Prostaglandin cyclooxygenase is a homodimer with a molecular weight of 126 kDa; it is a dioxygenase containing haem as an essential cofactor, membrane-bound in the microsomes and particularly rich in the vesicular gland. It exhibits two activities: adding both atoms of a molecule of oxygen to the carbon atoms at positions 11 and 15 of the arachidonic acid molecule to form a 15-hydroperoxy-9,11-endoperoxide (prostaglandin G_2) followed by reduction of the hydroperoxy group to hydroxy by a hydroperoxidase activity yielding prostaglandin H_2 (see Figure 7.1).

The specific mechanism by which aspirin inhibits prostaglandin cyclooxygenase is the transfer of an acetyl moiety to the amino group of the seryl residue at the amino terminus of the protein. This results in an irreversible inhibition like the inhibition by other agents such as indomethacin, although in a different fashion. It is possible that oxygenase activity resides on one subunit and the subsequent conversion to prostaglandin H2 on the other, since only one acetyl residue is added and the hydroperoxidase activity is left untouched (Roth and Siok, 1978).

Inhibition constants expressed in I_{50} terms for the various non-steroidal anti-inflammatory agents are not readily compared when irreversible inhibition is occurring. Nevertheless, Flower (1974) shows that the rank order of potency taken overall for some of the agents is as follows; meclofenamic acid > indomethacin > mefenamic acid > naproxen > phenylbutazone > aspirin. Many of these drugs have asymmetric structures and are thus optically active; an interesting light is thrown on the inhibition by the finding that the D-isomers of these acids are much more active than the L-form. D-Naproxen, for example, is 70 times more active against the enzyme from bovine seminal vesicles than the L-isomer.

The detailed mechanism, however, of this effect is not entirely clear. The wide variations in structure between the compounds, such that the only common denominator appears to be a hydrophobic moiety and an ionizable proton (except for phenylbutazone) does not make it any easier to understand. Nevertheless, it seems that the drugs bind slowly to a site that is not the active site but is sufficiently close to destroy the catalytic activity. Inhibition by indomethacin does not require the presence of oxygen, unlike the irreversible inhibition by the acetylenic analogue of the arachidonic acid substrate, 5,8,11,14-tetraynoic acid (reviewed in Flower, 1974), i.e. the latter is a suicide substrate (see Chapter 1) while the former is not.

In a more recent study, another four non-steroidal anti-inflammatory drugs besides aspirin, indomethacin and mefenamic acid were confirmed as selective inhibitors of the cyclooxygenase reaction. On the other hand, these agents protected the second activity of the enzyme complex, namely hydroperoxidase, against inactivation by heat treatment or alkali, suggesting that their binding conferred chemical stability on the enzyme (Mizuno *et al.*, 1982). Indeed, another unusual feature of this enzyme is that it is inactivated by the process of catalysis, in that oxygen consumption falls to zero before arachidonic acid is used up (see the comprehensive review by Needleman *et al.*, 1986). The mechanism of this is not understood at present.

Not only do the drugs block the formation of prostaglandins from arachidonic acid, but also it has become clear that prostaglandins themselves are partially responsible for various of the symptoms of rheumatoid arthritis, such as joint inflammation and pain, because the drugs can alleviate these symptoms to some extent. It is now believed that the prostaglandins play a part in the pathogenesis of rheumatoid arthritis. Other important mediators include leukotriene B_4, platelet derived growth factor and interleukin 1 (Harris, 1986).

It should be noted, however, that there are important differences between the drugs; for example, acetaminophen is antipyretic and analgesic but only weakly anti-inflammatory. Salicylate is effective as an anti-inflammatory but is only a weak inhibitor of cyclooxygenase *in vitro*.

7.4 Rheumatoid arthritis

Rheumatoid arthritis is characterized by inflammation of the synovium of particular joints associated with destructive effects, leading to the dissolution of cartilage and bone. It appears to occur in families. Rheumatoid arthritis is not to be confused with osteoarthritis which contains less of an inflammatory component in addition to cartilage breakdown. Both types of arthritis include an element of proliferative growth in different tissues. In osteoarthritis, this can reach the point of excruciating pain as the nodules grind in the joint and require removal, as in the operation for hip replacement for example. As well as the hip, the knee is commonly affected in osteoarthritis. It is a condition more common in later life, although not necessarily a part of the ageing process. Both types of arthritis respond to non-steroidal anti-inflammatory agents.

Rheumatoid arthritis is regarded as an autoimmune disease, in that there is a circulating antibody known as rheumatoid factor to another immunoglobulin molecule (this may also be activated by prostaglandins — see Baker and Rabinowitz, 1986). The nature of the original antigen is unclear — although later an anti-collagen antibody is found, resulting from tissue destruction after the disease has developed. When antigen and antibody combine, complement joins the complex to produce a precipitate, which if deposited in the synovial space can cause the release of factors that attract leukocytes. These cells phagocytose the immune precipitate, releasing a variety of lytic enzymes in the process that gradually destroy the cartilage and bone (Harris, 1986). Early inflammation in the synovium is paralleled by a proliferation of blood vessels in response to factors released by these leukocytes. Activation of lymphocytes in the synovium causes antibody production, including rheumatoid factor.

There remains some doubt about the importance of rheumatoid factor in the progression of the disease, however, because its level does not parallel the severity and indeed is sometimes absent (Harris, 1986). The persistent observation of an immune response in rheumatoid arthritis has led to frequent searches for an infective agent. Mycoplasma were suspected to be the culprits in earlier years, indeed one species of mycoplasma, *M. arthritidis*, can cause a transient condition which resembles arthritis as its name suggests. More recently, however, attention has switched to viruses for the aetiology of the disease, but it is not clear whether the virus is the causative agent or merely a passenger in a damaged synovium.

The problem remains that although the inflammation of the joints can be reduced or eliminated with anti-inflammatory agents in the short term, the progressive deterioration of the synovium, resorption of bone and abnormal growth in response to the degradation still occurs, resulting, in the case of osteoarthritis, in the formation of painful and irregular nodules (Baker and Rabinowitz (1986)). Harris (1986) on the other hand, considers that the drugs interrupt the generation of inflammatory mediators which otherwise assist in the attack on bone and cartilage by attracting hydrolytic enzymes into the joint. The situation is so complex that a subtle regulatory shift may suffice to close down the process and cause remission.

It should be emphasized, however, that the prostaglandins can only contribute to part of the overall picture of inflammation as they are not especially active in the generation of oedema, although they may potentiate the oedema induced by other agents. Furthermore, they are not thought to be directly responsible for the chemotaxis of polymorphonuclear leukocytes. On the other hand, some of the products of the lipoxygenase pathway do have chemotactic activity; 5-hydroperoxyeicosatetraenoic acid (HPETE) (see Figure 7.2) is chemotactic for leukocytes, while leukotriene B_4 is a very potent chemotactic agent and may play an important role in inflammation (see Dahl, 1984).

Many workers over the years have attempted to inhibit putative hydrolytic enzymes that might be responsible for the continuing damage to the joint, such as: lysosomal enzymes with acid pH optima; polysaccharidases (hyaluronidase) and proteases such as cathepsin D; or neutral proteases like collagenase — but so far there is very little to show for this effort.

One recent and novel approach, however, is to look for naturally occurring inhibitors of the hydrolytic enzymes. Collagenase, and other metalloproteinases, are apparently held in check *in vivo* by an excess of a large molecular weight glycoprotein inhibitor. If the control of enzyme activity by this inhibitor were upset in any way, for example by reduced production, then tissue degradation could occur. The gene for the inhibitor has been cloned in *E. coli*, and it is hoped that the primary structure will provide the impetus for the design of new structures to prevent collagen breakdown (Docherty *et al.*, 1985). Nevertheless, at present we have drugs for the treatment of rheumatoid arthritis which are hardly any more effective than aspirin, and are certainly much more expensive.

7.5 Thromboxane synthetase

The next step in the prostaglandin metabolic pathway is conversion of prostaglandin H_2 to thromboxane A_2 (see Figure 7.1). This material is extremely unstable with a half-life of approximately 30 seconds in tissue fluids, being converted by non-enzymic hydrolysis of the epoxide bridge to thomboxane B_2, which is almost biologically inactive. Much interest has centred on thromboxane A_2 in view of its vasoconstrictor activity and its irreversible induction of platelet aggregation which can lead to blockage of blood vessels particularly in the heart. Thromboxane A_2 is generated in vascular smooth muscle and cardiac cells as well as platelets, among other tissues, thus being implicated in angina and acute myocardial infarction.

Although thromboxane A_2 exists only briefly, there is a cascade effect since it induces other platelets to generate more of itself. Much effort, therefore, has gone into inhibition of its generation — partly as it was hoped that prostaglandin equivalents would be diverted into the formation of prostaglandin I_2 or prostacyclin, a vasodilator and extremely powerful inhibitor of platelet aggregation at between 1 and 10×10^{-9} M, which is found in the microsomes of the blood vessel wall or aortic intima. The evidence for this diversion of equivalents, however, is sparse. The problem is that prostacyclin synthetase is not found in the platelet and it is not clear whether prostaglandin H_2 could be transferred from one site to the other. The latter has, in any case, considerable vasoconstricting powers of its own and a build-up may have serious consequences although it too has a short half-life.

Both of these conversions are mediated by cytochrome P–450. Neither, however, is characteristic of the type of mixed function oxidase action of the cytochrome that occurs in the endoplasmic reticulum of, for example, liver and inserts hydroxyl groups into, or removes N-methyl groups from, organic compounds in order to render them more soluble and thus more easily excreted. This reaction requires the presence of the cofactor, NADPH, and a flavoprotein, cytochrome P–450 reductase, to transfer reducing equivalents to molecular oxygen from the cofactor. In contrast, the apoproteins of both prostacyclin and thromboxane synthetases appear to act as a scaffold on which prostaglandin H_2 can be rearranged in two different ways to form either product. In these two reactions, neither cytochrome P–450 reductase, NADPH nor molecular oxygen is required, as no oxidation takes place.

Additional evidence for the involvement of cytochrome P–450 in these reactions lies partly in the nature of the inhibitors such as metyrapone that block the reaction (metyrapone is a characteristic inhibitor of P–450 reactions) but perhaps more importantly in the behaviour and properties of the purified protein itself. It can be purified on an affinity column using a substituted imidazole and N-benzylimidazole acts as the eluant. Imidazoles, as an example of a nitrogen heterocycle with a lone pair of electrons that can be donated to the iron of the haem, are typical inhibitors of cytochrome P–450.

The ultraviolet spectrum of the protein shows a peak at 450 nm when the reduced protein is allowed to react with carbon monoxide, characteristic of cytochromes P–450 (Ullrich and Hauraud, 1983). Furthermore, a number of inhibitors show a classical type of binding spectrum (Type II) when complexed with the protein in that there is a decrease in absorption around 390–400 nm and an increase at 425 nm (see Figure 6.5, Chapter 6). Analysis of the saturation of binding yields a binding constant which correlates well with inhibition of thromboxane synthetase (Ullrich and Hauraud, 1983).

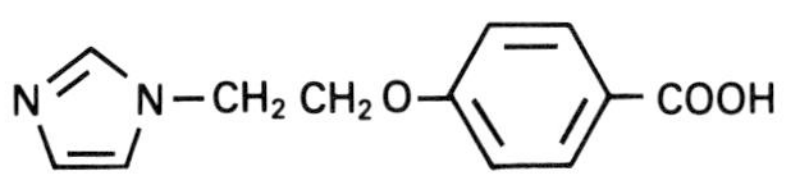

Dazoxiben

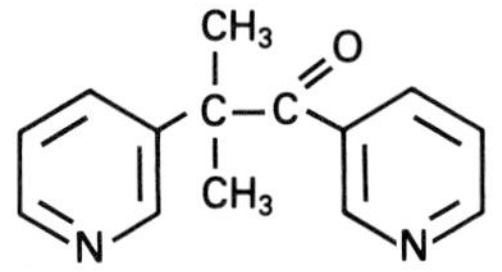

Metyrapone

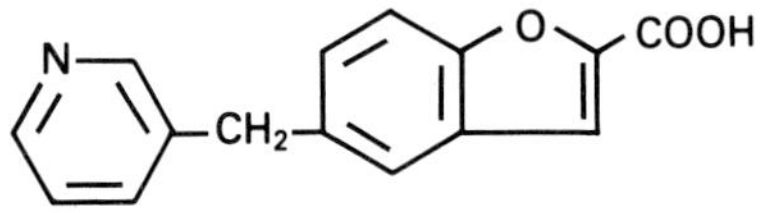

Furegrelate

Starting with the structure of metyrapone and aiming for structures with increased inhibition for thromboxane synthetase, but reduced inhibition for prostacyclin synthetase and the more usual (i.e. oxidative) type of cytochrome P–450 reaction (for example steroid 11-β-hydroxylation) has led to various related structures such as dazoxiben and furegrelate (Lefer, 1986). The latter inhibits the enzyme with I_{50} of $1{\cdot}5 \times 10^{-8}$ M and dazoxiben 5×10^{-8} M. These drugs are now under clinical evaluation for prevention of thombosis and vasoconstriction in coronary heart disease (Lefer, 1986).

7.6 Thromboxane antagonists

Another approach to the antagonism of thromboxane vasoconstriction is by the synthesis of antagonists at the thromboxane receptor. Two such compounds are BM 13.177 and AH 23848. Both compounds have been shown to antagonize thromboxane but in different experimental models: AH 23848 reduces fibrillation in the heart that is induced when the heart is re-perfused after ischaemia. This model is known to cause the release of thromboxane, and the drug did not affect its release, merely its action. It appears that the drug is likely to block the actions of the endoperoxides, such as platelet aggregation and vasoconstriction, which is an additional advantage (Coker and Parratt, 1985).

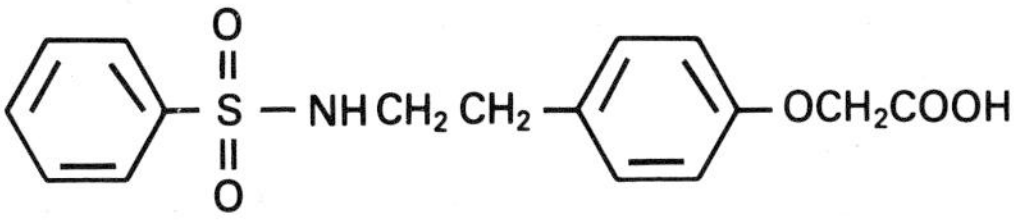

BM 13.177

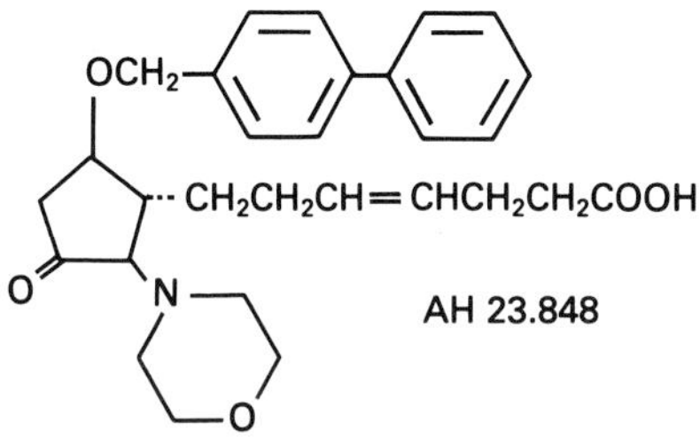

AH 23.848

BM 13.177 was shown to protect against death caused by arachidonic acid administration in rabbits, and also to antagonize the action of a thromboxane mimetic, U 46619, on the contraction of isolated rabbit arterial strips. The drug inhibited platelet aggregation and bronchoconstriction induced by collagen (I_{50} was $3 \cdot 6 \times 10^{-6}$ M) which is believed to be effected through thromboxane A_2. No effect was shown on either thromboxane B_2 levels, unlike aspirin and indomethacin. A short clinical trial against platelet clotting in atherosclerotic patients was shown to be successful and further trials are in progress (Stegmeier *et al.*, 1984). At present it is not clear whether either thromboxane synthetase inhibitors or antagonists, if either, are going to be the most effective drugs.

7.7 Leukotriene biosynthesis

Although no drug has yet reached the market that owes its activity to inhibition of leukotriene biosynthesis, a great deal of effort is going into the search for such a compound. This is largely because of the expectation that such a compound would be useful in the treatment of asthma, since the 'slow reacting substance of anaphylaxis' that is released in asthma and is responsible for much of the bronchoconstriction of the condition, is now recognized to be a mixture of leukotrienes.

The first committed step is peroxidation of arachidonic acid catalyzed by 5-lipoxygenase (see Figure 7.2) which contains a non-haem iron at the active site, producing 5-hydroperoxy-7,9,11,14-eicosatetraenoic acid which can then be converted into leukotriene A_4. This is the branch-point from where leukotriene B_4 may be obtained by hydrolase action or C_4 by the action of a glutathione S-transferase. Leukotrienes D_4 and E_4 result from the sequential action of γ-glutamyl peptidase and a dipeptidase.

It appears that the leukotriene C_4 synthesis, albeit not rate limiting, could be a useful target since leukotrienes C_4 and D_4 are the most potent agents. Alternatively, 5-lipoxygenase as the rate limiting step could be addressed. Glutathione S-transferase is clearly not the ideal target, since it catalyzes the detoxification of a large number of drugs. Whichever is found to be useful, we may soon hope to see a treatment for asthma (Taylor and Clarke, 1986).

References

Baker, D.G. and Rabinowitz, J.L. (1986). *J. Clin. Pharmacol.*, **26**, 2–21.

Blackwell, G.J., Carnuccio, R., Di Rosa, M., Flower, R.J., Parente, L. and Persico, P. (1980). *Nature*, **287**, 147–149.

Coker, S.J and Parratt, J.R. (1985). *Brit. J. Pharmacol.*, **86**, 259–264.

Dahl, S. (1984). *Drug Intell. Clin. Pharmac.*, **18**, 56–58.

Docherty, A.J.P., Lyons, A., Smith, B.J., Wright, E.M., Stephens, P.E. and Harris, T.J.R. (1985). *Nature*, **318**, 66–69.

Flower, R.J. (1974). *Pharmacol. Rev.*, **26**, 33–67.

Flower, R.J., Moncada, S. and Vane, J.R. (1985). In: *The Pharmacological Basis of Therapeutics*, 7th Ed., Eds. Gilman, A.G., Goodman, L.S., Rall, T.W. and Murad, F. (New York: Macmillan) Ch. 29.

Harris, E.D. (1986). *Amer. J. Med.*, **80**, Suppl. 4B, 4–10.

Hirata, F., Schiffman, E., Venkatasubramanian, K., Salomon, D. and Axelrod, J. (1980). *Proc. Nat. Acad. Sci. (U.S.A.)*, **77**, 2533–2536.

Lefer, A.M. (1986). *Drugs of the Future*, **11**, 196–202.

Mizuno, K., Yamamoto, S. and Lands, W.E.M. (1982). *Prostaglandins*, **23**, 743–757.

Needleman, P., Turk, J., Jakschik, B.A., Morrison, A.R. and Lefkowith, J.B. (1986). *Annu. Rev. Biochem.*, **55**, 69–102.

Roth, G.J. and Siok, C.J. (1978). *J. Biol. Chem.*, **253**, 3782–3784.

Russo-Marie, F., Paing, M. and Duval, D. (1979). *J. Biol. Chem.*, **254**, 8498–8504.

Stegmeier, K., Pill, J., Muller-Beckmann, B., Schmidt, F.H., Witte, E.C., Wolff, H.-P. and Patscheke, H. (1984). *Thombosis Res.*, **35**, 379–395.

Taylor, G.W. and Clarke, S.R. (1986). *Trends in Pharmacol. Sci.*, **7**, 100–103.

Taylor, G.W. and Ritter, J. (1986), *Trends in Pharmacol. Sci.*, **7**, March centrepage.

Ullrich, V. and Hauraud, M. (1983). *Adv. Prostaglandin, Thomboxane Leukotriene Res.*, **11**, 105–110.

Vane, J.R. (1971). *Nature New Biology*, **231**, 232–236.

Withnall, M.T., Brown, T.J. and Diocee, B.K. (1984). *Biochem. Biophys. Res. Commun.*, **121**, 507–513.

Chapter 8

Zinc metalloenzymes

8.1 Introduction

Approximately 200 enzymes have so far been discovered that contain at least one zinc atom as part of the integral functioning protein. Vallee and Galdes (1984) have classified the roles that zinc can play in metalloproteins:

(a) Catalytic, in which the metal plays an active part in the catalysis. Notable examples of this are carbonic anhydrase, angiotensin converting enzyme and alcohol dehydrogenase from liver.

(b) Structural, where the zinc is required to maintain the structural integrity of the protein as in β-amylase.

(c) Regulatory, either inhibitory or activatory, as in leucine aminopeptidase.

(d) Other unspecified role(s) which may be regarded as non-catalytic for want of better knowledge.

The complete complement of 10 electrons in the 3d shell of the zinc atom has a screening effect on the positive charge of the nucleus, as compared with calcium which lacks this screen. Consequently, on the one hand, zinc is able to form covalent coordinate complexes in addition to the ionic complexes characteristic of calcium. Ligands can, therefore, easily enter its coordination sphere and donate a pair of electrons to form a coordinate or dative bond. Zinc may have either 4 or 6 such ligands in simple inorganic compounds.

Furthermore, the external shell of zinc's electrons is polarizable and therefore distorts in an asymmetric environment. Characteristic features of the catalytic type of zinc protein are (a) a highly asymmetric environment of the zinc such that the atom can be regarded as either 4- or 5-coordinated and (b) three ligands to the protein amino acid side-chains, while water occupies the fourth position. Non-catalytic and structural zinc is normally bound to the protein by four ligands and is not as distorted as catalytic zinc. The polarizability plays a practical part in the catalysis as it has a higher energy than a non-distorted state and thus the activation energy for the reaction is correspondingly lowered.

Covalent coordinate interactions require a closer inter-atomic approach than is seen in ionic compounds. Furthermore, zinc will accept a variety of

ligands based on oxygen, nitrogen, or sulphur, as exemplified by the formation of sodium zincate (Na_2ZnO_2) and zinc diamine chloride [$Zn(NH_3)_2$]Cl_2.
In this way a variety of reaction types can be catalyzed, with cysteine,
histidine or glutamic acid side-chains acting as anchors for the binding of zinc
to the protein.

Zinc, as an electrophilic cation, can activate the substrate chemically by
behaving as a Lewis acid (meaning that it can readily accept a share in a pair
of electrons from a ligand, with the concomitant formation of a coordinate
covalent bond). This has the effect of enhancing ligand nucleophilicity, as for
example by ionizing a water molecule and binding the hydroxide ion formed,
which can then participate in hydrolysis reactions that the initial water
molecule itself could not achieve. In addition, zinc can facilitate catalysis by
orienting reactants through coordination on the metal as a template (Vallee
and Galdes, 1984).

Another feature of catalytic zinc proteins is the occurrence of acyl-enzyme
intermediates in the catalytic process, as exemplified by acyl-phosphates in
alkaline phosphatase. The nature of the nucleophile which breaks the acyl-
enzyme intermediate is not certain, but in a number of hydrolytic metalloenzymes the proposition of a zinc attached hydroxide ion has been made
(Coleman, 1984). If the variation of enzyme activity with pH is measured (pH
profile) there is a requirement for catalysis of a group with a pK_a of 7 to 7.5
which nuclear magnetic resonance (n.m.r.) studies have shown is not an
imidazole. In effect the zinc is lowering the pK_a of the hydroxide moiety of
the water molecule so that it can act as a nucleophile (see Vallee and Galdes,
1984).

It is the catalytic type of zinc atom with which we are concerned in this
chapter, particularly as the inhibitors of the enzyme actually form complexes
which involve direct attachment of a ligand to the zinc. Inhibitors of two
enzymes containing catalytic zinc have found pharmaceutical use. Carbonic
anhydrase inhibitors are useful both in situations where there is an excess of
fluid (oedema) in various tissues and, therefore, in the treatment of glaucoma, and also when urine needs to be 'alkalinized' because acids like uric
acid might otherwise precipitate. Secondly, angiotensin-converting enzyme
(ACE) inhibitors are now playing a major role in the reduction of blood
pressure. Potentially, a third application lies in inhibiting an enzyme that can
degrade enkephalin. Such inhibitors may be approaching clinical trial as
painkillers.

8.2 Carbonic anhydrase

An enzyme exists in a large number of tissues including kidney, erythrocyte
and pancreas catalysing the hydration of carbon dioxide to form carbonic
acid, which immediately ionizes to form a bicarbonate ion:

$$CO_2 + H_2O \rightleftharpoons H_2CO_3 \rightleftharpoons HCO_3^- + H^+$$

The enzyme can also catalyse the hydration of aldehydes and ketones, such as pyruvate, but it is unlikely that any of these reactions have physiological significance. It may seem unnecessary for an enzyme to carry out this reaction — estimates of the ration of enzyme-catalysed to uncatalysed rate (catalytic rate enhancement) range from 8000 (Maren, 1984) to 10^5 (Coleman, 1984) depending on the source of the enzyme. Nevertheless, the ubiquity and high level of the enzyme in, for example, the kidney suggest that the rate enhancement of this reaction plays a very important role in the metabolism of the cell.

Carbonic anhydrase normally exists in at least two forms (or isoenzymes) designated CA1 and CA2, which are distinguished mainly by a considerable difference in catalytic rate with CA2 being the most active. The isoenzymes also differ to some extent in primary sequence but not in overall length of some 260 amino acid residues, resulting in a molecular weight of approximately 30 kDa. There are 105 differences between human erythrocyte CA1 and CA2 which represents 59·6% homology — the similarity is particularly close about the active site as would be expected. Other species show greater homology between the two forms (Deutsch, 1984).

As shown by X-ray crystallographic data, carbonic anhydrase contains one zinc atom per active site coordinated to 3 nitrogen atoms from imidazole side-chains of histidines residues 94, 96 and 109, and to one oxygen atom from a water molecule in an approximately tetrahedral conformation. The reaction mechanism has been the subject of much discussion over the years (reviewed in Coleman, 1984).

The zinc atom based in a hydrophobic pocket acts to lower the pK_a of the water molecule and binds a coordinated hydroxide ion although the pH of the surrounding medium may be neutral. A buffer molecule accepts the proton, indeed the overall reaction is buffer catalyzed at neutral pH in that the reaction rate is found to be faster in the presence of buffer than in unbuffered solution. The hydroxide ion behaves as a nucleophile, by attacking carbon dioxide to form bicarbonate ion. A sigmoid pH profile is obtained for both hydration and dehydration reactions but in opposing senses, which may be a consequence of the requirement for a base such as the hydroxide ion in the active site with a pK_a in the region of 7 as an intermediate for both reactions (Coleman, 1984). A 5-coordinated zinc atom (with a second water molecule attached via its oxygen atom to the zinc) has been shown to have theoretical advantages for the catalysis in that the second water molecule is ready to slip into the place of the first water molecule as soon as reaction has occurred, but it is at present difficult to be sure how many water molecules are at the active site (Cook and Allen, 1984).

The detailed interaction of the sulphonamides, used originally as diuretics, with carbonic anhydrase has been a subject of considerable interest. The sulphonamide group has to be unsubstituted for activity, with the amide nitrogen atom binding to the zinc atom at the fourth coordination site displacing the water molecule. One of the two sulphonamide oxygen atoms links with the fifth coordination site of the zinc (see Vedani and Meyer, 1984).

Kinetic studies have shown that the sulphonamides are competitive inhibitors of the dehydration of carbonic acid as well as the hydration of carbon dioxide, as one would expect from the principle of microscopic reversibility. In this situation, competitive kinetics have shown that the inhibitor binds at the substrate binding site (Cook and Allen, 1984).

Carbonic anhydrase is normally present in tissue in great excess, as indeed are a number of other enzymes, and over 99% of the enzyme has to be inhibited before physiological effects became apparent. The importance of this high concentration relates to bicarbonate ion transport across epithelial cell membranes where sodium is the counter-ion, in that a large excess of enzyme is required to act on the substrate.

One example of this occurs in the proximal tubule of the kidney where the majority of bicarbonate ion that has previously been filtered out in the glomerulus is reabsorbed (Kokko, 1984) (see Figure 8.1). In a coupled transport process, sodium ion is reabsorbed while hydrogen ion is secreted into the filtrate passing through the lumen of the proximal tubule (see Figure 8.2a). Hydrogen and bicarbonate ions form carbonic acid which disproportionates to carbon dioxide and water, catalyzed by carbonic anhydrase in the luminal membrane (Lucci *et al.*, 1983). Carbon dioxide diffuses across the luminal membrane of the proximal tubule cell and the cytoplasmic enzyme reconverts it into bicarbonate ion which is transported across the basement membrane of the cell into the peritubular fluid that drains into the blood. The proton that is also formed in the reaction is re-circulated across the luminal membrane in exchange for sodium (DuBose *et al.*, 1981; Lucci *et al.*, 1983). The overall result is the net reabsorption of sodium and bicarbonate ion.

If insufficient hydrogen ion is secreted to react with all the tubular bicarbonate, then the urine will contain larger amounts of bicarbonate than usual, and consequently will have a higher pH. The content of sodium ion will also increase (Kokko, 1984) and in order to maintain the osmotic pressure water will be secreted together with the ions (see Figure 8.2a). This situation occurs as the consequence of carbonic anhydrase inhibition because the supply of protons is greatly reduced. It appears that at least two-thirds of the carbon dioxide reabsorption is mediated by carbonic anhydrase because, when the enzyme is completely inhibited, the proportion of bicarbonate reabsorbed falls to about 37% of its control value (Maren, 1984). Carbonic anhydrase inhibitors were originally developed for use as diuretics but their action is only weak because their major effect is in the proximal tubule (Kokko, 1984). As a consequence the greater volume of fluid reaching the distal tubule produces a compensatory reabsorptive effect there which markedly reduces the effect of the drug.

Another family of diuretics, the widely-used thiazides, were originally developed out of studies on the sulphonamides. Although the thiazides are weak inhibitors of carbonic anhydrase, they act primarily at the early portion of the distal tubule and, since carbonic anhydrase is mainly located in the

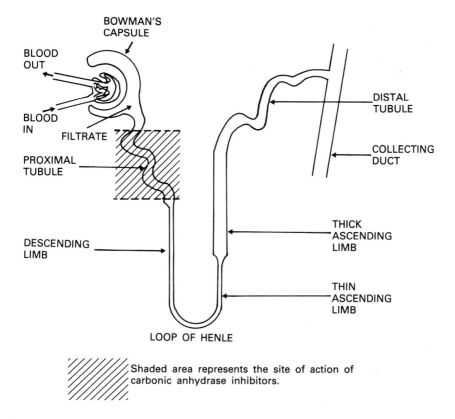

BOWMAN'S
CAPSULE

BLOOD
OUT

BLOOD
IN

FILTRATE

PROXIMAL
TUBULE

DESCENDING
LIMB

DISTAL
TUBULE

COLLECTING
DUCT

THICK
ASCENDING
LIMB

THIN
ASCENDING
LIMB

LOOP OF HENLE

Shaded area represents the site of action of
carbonic anhydrase inhibitors.

Low molecular weight substances are ultrafiltered from the blood through a capillary
system in the glomerulus. The filtrate passes into Bowman's capsule and then through the
proximal tubule, the loop of Henle and the distal tubule. It finally reaches a collecting duct
that drains several nephrons.

Figure 8.1 Site of action of carbonic anhydrase inhibitors.

proximal tubule, inhibition of this enzyme is unlikely, therefore, to play much part in the mode of action of the thiazide diuretics. Lant (1985) concludes that, at present, no satisfactory biochemical mechanism has been proposed to explain the mode of action of these drugs.

One use of carbonic anhydrase inhibitors at present is to increase the pH of the urine. This becomes necessary when, for example, drugs that cause tumours to diminish in size (oncolytic) release large amounts of the breakdown product of purines, namely uric acid. At normal pH this could begin to precipitate and so a carbonic anhydrase inhibitor is used, often in conjunction with sodium bicarbonate, to keep the acid in the form of the more soluble sodium salt.

Carbonic anhydrase inhibitors are also used for the treatment of glaucoma, where the object is to reduce the volume of liquid in the aqueous humor of

(a) Proximal tubule of the kidney.

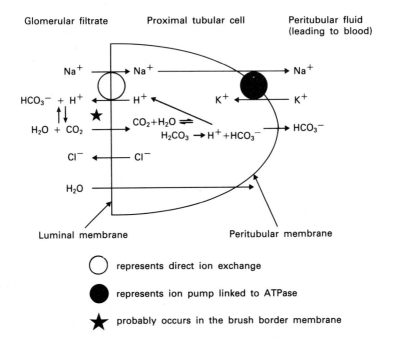

represents direct ion exchange

represents ion pump linked to ATPase

probably occurs in the brush border membrane

(b) Ciliary epithelium in the eye.

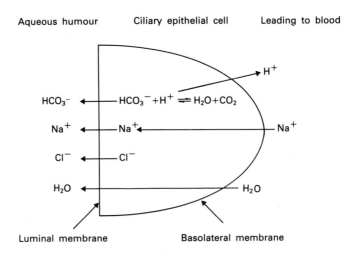

Figure 8.2 Ion transport in kidney and eye.

the eye where this has become excessive and is putting a great deal of pressure on the eyeball. The situation in the eye differs from that in the kidney in that the protons are secreted across the basolateral membrane into the blood and not into the lumen or aqueous humor. Bicarbonate ion is secreted into the aqueous humor, however, and inhibition of carbonic anhydrase reduces the level of the ion in the humor by about 50% thus reducing the concomitant fluid outflow (Maren, 1984). The details of ion transport in the eye are less well understood, however, than in the proximal tubule cell of the kidney (see Figure 8.2b).

Carbonic anhydrase inhibitors are particularly useful for the treatment of glaucoma since they reduce the pressure behind the iris, and assist in opening the angle between the iris and the cornea thus allowing the aqueous humor to leak into the anterior chamber and be reabsorbed. Drugs are used in the acute condition in order to manage the attack while the patient is prepared for surgery — normally a necessity because a physical obstruction has usually induced the condition in the first place. The main intention of acute treatment is to reduce liquid volume, and thus intra-ocular pressure, so preventing damage to the optic nerve (AMA Drug Evaluations, 1983). The drugs are less useful in treating chronic primary open angle glaucoma, which slowly progresses as a consequence of a gradual closing of the channel through which the aqueous humor circulates until it may result in irreversible loss of vision.

Acetazolamide was originally used to treat glaucoma but its use has largely been superseded by the more recently synthesized analogue, methazolamide, as the key requirement is to obtain a sufficient concentration of drug in the ciliary epithelium of the anterior chamber of the eye. Methazolamide is more lipophilic and penetrates this tissue more readily (AMA Drug Evaluations, 1983).

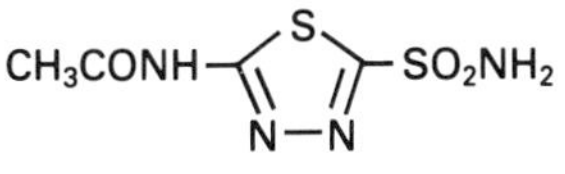

Acetazolamide 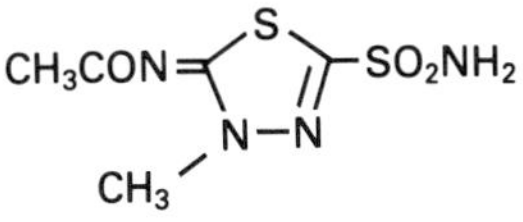   Methazolamide

8.3 Angiotensin converting enzyme (ACE)

The second zinc metalloenzyme to be discussed in this chapter is the metallo-peptidase, angiotensin-converting enzyme (ACE), the search for inhibitors of which has turned out to be both a very interesting intellectual study and also of great importance medically. The search was initiated because ACE converts the biologically inactive decapeptide, angiotensin I, to the highly active pressor (blood pressure raising) octapeptide, angiotensin II, by removing the C-terminal dipeptide His-Leu (Figure 8.3). The enzyme will also cleave the nonapeptide, bradykinin, which is a depressor agent, to inactive metabolites.

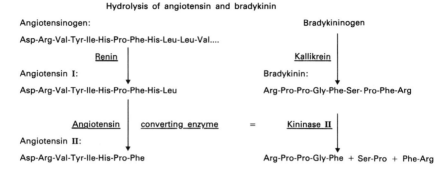

Angiotensinogen, the precursor protein, is cleaved at one specific site by renin, secreted by the kidney, to yield biologically inactive angiotensin I. This is converted by angiotensin-converting enzyme to the active octapeptide angiotensin II.

Precursor bradykininogen is converted to the active nonapeptide bradykinin by kallikrein and thence to the inactive pentapeptide by angiotensin-converting enzyme.

Figure 8.3 Hydrolysis of angiotensin and bradykinin.

ACE is still known in some quarters as kininase 2, and it plays a pivotal role in the control of blood pressure.

In general, the enzyme prefers substrates with hydrophobic amino acids in the ante-penultimate position. It will cleave a variety of substrates, both peptide and ester, provided that the C-terminal amino acid has a free car- boxyl group, and proline is not present in the penultimate position. Thus angiotensin II is not hydrolyzed further by ACE, whereas bradykinin is attacked in three places (Figure 8.3), and so the enzyme may be regarded as not particularly specific (Ondetti and Cushman, 1984).

Angiotensin II is one of the most active biological agents known. It raises blood pressure by constricting blood vessels and, in addition, it causes aldos- terone to be released from the adrenal cortex. The latter acts on the kidney to retain sodium ion and, therefore water too, thereby raising blood volume and pressure. The complex system for maintaining blood volume and pressure initially depends on the action of the endopeptidase renin. When blood volume or sodium ion concentration in the plasma falls below a certain critical value, renin is released from the kidney and acts on an α-globulin known as angiotensinogen — a glycoprotein of molecular weight 60 kDa — to yield angiotensin I. Figure 8.4 outlines the interactions of the renin–angiotensin– aldosterone system.

It was clear in the 1970s from this knowledge that an inhibitor of ACE might be expected to have anti-hypertensive properties, and a study began to design agents that would bind to the active site. Although ACE had not been isolated at all at that time, advantage was taken of the much greater under- standing of the active site of another zinc metallopeptidase, namely carboxy- peptidase, an enzyme which is elaborated by the pancreas and cleaves the C- terminal amino acid from peptides and proteins.

Interactions of the renin-angiotensin-aldosterone axis

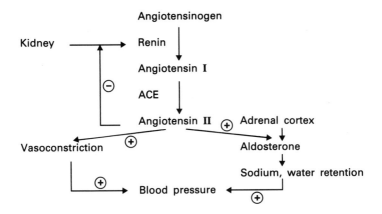

Figure 8.4 Interactions of the renin–angiotensin–aldosterone axis.

The X-ray structure of carboxypeptidase A had been known for some time (refined in 1981 by Rees *et al.*), and it showed that the zinc atom was held in position by three amino acid residues; His-69, Glu-72 and His-196. With two bonds from imidazole nitrogens and two from the carboxylate oxygens of the glutamate, the fifth ligand in the natural state is the oxygen of a water molecule. This is replaced in the enzyme/substrate complex by the oxygen of the carbonyl group of the peptide link to be broken, and the latter is thereby polarized by the action of the zinc atom acting as a Lewis acid. Glutamate-270 then functions as a base or nucleophile to attack the carbon of the carbonyl group to yield a transient acyl-enzyme intermediate. Tyrosine-248 donates a proton to the departing amino group. The free carboxylate of the substrate is linked by attraction between a negative and positive ion (salt bridge) to Arg-145.

The nature of the group that, in effect, breaks up the enzyme–substrate complex by attacking the carbonyl carbon is likely to be a hydroxide ion, which may be activated by Glu-248 or possibly the zinc (Lipscomb, 1980). It would be intellectually satisfying if the zinc atom played a role in carboxypeptidase mechanism similar to that in carbonic anhydrase, in lowering the pK_a of the water molecule and allowing a hydroxide ion to attack a carbon of a carbonyl group. There is some support for this idea in a low-temperature study of carboxypeptidase hydrolysing an ester substrate, where an ionizing group is identified as an enzyme-bound water molecule (Makinen *et al.*, 1979). Lipscomb (1980) points out, however, that changes in temperature may change the rate-limiting step in the reaction mechanism and so the situation is at present unresolved. For a diagrammatic representation of a possible enzyme mechanism see Figure 8.5.

Although the information regarding the active site of angiotensin-converting enzyme is still fairly sparse, it is clear that similar groups are involved in the catalysis, as shown by group-specific inactivating agents backed up by the use of ligands to protect against inactivation (Brunning, 1983; reviewed in Ondetti and Cushman, 1984). One difference between the two enzymes is the requirement for chloride ion demonstrated by ACE. Chloride ion has no effect on the rate of modification of arginyl, carboxyl and tyrosyl residues by these reagents, but it does affect the rate of reaction of pyridoxal phosphate with the enzyme, which suggests that lysine, which forms a Schiff's base with the aldehyde group of pyridoxal, may bind the halide ion. Activation by chloride ion is seen with most substrates and at all pHs, and is characterized by a lowering of K_m but no change in K_{cat} (Ondetti and Cushman, 1984). Recent studies have defined the active site glutamic acid residue in bovine lung ACE in the sequence: H_2N-Phe-Thr-Glu-Leu-Ala-Ser-Glu. This

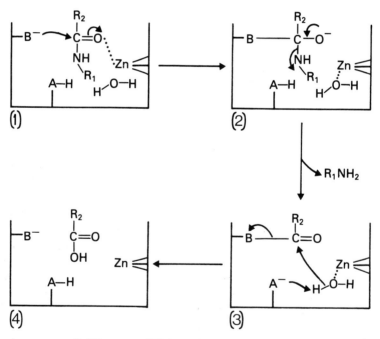

1. The base group B (Glutamate 270 in carboxypeptidase) attacks the carbon of the carbonyl group already distorted by activation by the zinc.
2. An acyl-enzyme intermediate is formed, and then rearranges to release the amine R_1NH_2 taking the proton from the acid group A (Tyrosine 248).
3. The hydroxyl of the water molecule activated by the zinc attacks the carbonyl group to yield the acid R_2COOH, while acid group A receives the proton.
4. The acid R_2COOH is released and the active site resumes its original format.

Figure 8.5 Putative Zn-metallopeptidase mechanism.

sequence shows considerable homology with bovine carboxypeptidase A, but little with the bacterial zinc endopeptidase, thermolysin (Harris and Wilson, 1985).

8.3.1 Angiotensin converting enzyme inhibitors

The story of the development of ACE inhibitors that are marketed today is a triumph of logical and imaginative planning, combining features of both substrates and inhibitors. As mentioned above, the information about the active site of ACE was strictly limited. Nevertheless, the potential similarities between ACE and carboxypeptidase A were exploited to the full. Early studies with inhibitory peptides showed that having proline or an aromatic amino acid in the C-terminal position gave the most effective inhibitors, but proline inhibitors gave the better results *in vivo*; Val–Trp is some 150 times more active *in vitro* than Ala–Pro but no more active *in vivo* (Petrillo and Ondetti, 1982).

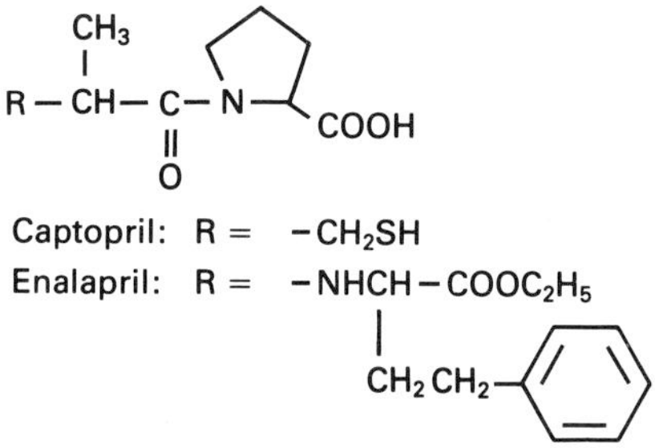

$$
\begin{array}{c}
CH_3 \\
| \\
R-CH-C-N \\
\quad \quad \| \\
\quad \quad O
\end{array}
\quad COOH
$$

Captopril: R = $-CH_2SH$
Enalapril: R = $-NHCH-COOC_2H_5$
$\quad \quad \quad \quad \quad | $
$\quad \quad \quad \quad CH_2\,CH_2-\langle \text{phenyl} \rangle$

Succinylproline could be regarded as a good starting point for inhibitor development with four sites interacting with the protein; the carboxylate group of the proline, the proline ring itself, the carbonyl of the amide link and the carboxylate of the succinate all binding to the zinc atom (I_{50} of 3.3×10^{-4} M). A methyl group placed on the carbon atom next to the carbonyl group of the amide improved the binding markedly. Replacement of the carboxylate by a sulphydryl gave a further enormous improvement, to produce captopril with an I_{50} of 2.3×10^{-8} M. Enalaprilic acid was derived by three further changes; the addition of further binding affinity by a phenethyl group (possibly interacting with the binding site for the phenylalanine side-chain characteristic of angiotensin I), together with a restoration of the carboxyl group in place of sulphydryl, and the introduction of a secondary amino group. These changes lowered the I_{50} to 1×10^{-9} M (reviewed in Ondetti and Cushman, 1984; Mackaness, 1985); see Figure 8.6.

Detailed kinetic studies with different substrates and inhibitors, excluding captopril and enalaprilic acid, have revealed a complex pattern of inter-actions. It appears that an inhibitor can bind to more than one form of the enzyme and, in addition, there are parallel reaction pathways. Moreover, both captopril and enalaprilic acid are slow and tight binding inhibitors (see

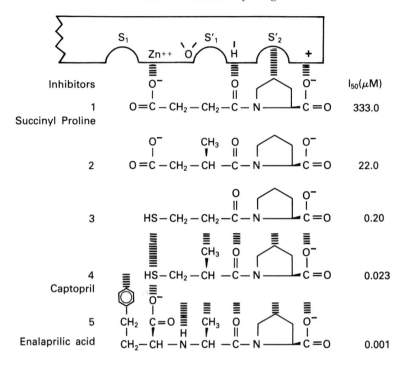

Figure 8.6 The development of ACE inhibitors.

Reproduced from *J. Cardiovasc. Pharmacol.* **7** (Suppl 1) S31 (1985), by permission of the Copyright Owner.

Morrison, 1982). As a consequence, sufficient time must be allowed for complete binding to place, and, in addition, allowance must be made for the depletion of active enzyme by drug binding as both drug and enzyme concentrations are in the nanomolar range. K_i values of 3.3×10^{-10} M and 5.0×10^{-11} M have been calculated for captopril and enaprilic acid respectively (Shapiro and Riordan, 1984). A two-step inhibition is identified characterized by a fast step followed by a slow second one which is probably an isomerization of the complex. The overall type of inhibition is judged to be competitive.

8.3.2 Pharmacology of ACE inhibitors

Captopril and enalapril have made a major contribution to our understanding of the mechanism for maintaining blood pressure as well as to the treatment of hypertension. These drugs have a powerful effect in reducing blood pressure. Whether this is entirely due to the blockade of angiotensin II production or whether inhibition of bradykinin degradation may also play a part is not yet resolved although positive evidence for the latter mechanism is lacking (see Ondetti and Cushman, 1984). Enaprilic acid is administered as the ethyl ester, enalapril, as the acid is poorly absorbed from the gastro-intestinal tract. Subsequent hydrolysis by esterases yields the active agent.

In the clinic, captopril and enalapril are extremely effective at lowering the blood pressure in hypertensive patients, particularly those in whom the renovascular system is the originator of the condition and the renin–angioten-sin–aldosterone axis is involved. Elevated aldosterone and angiotensin II levels in the plasma lead to increased sodium and water retention and so to increased blood pressure. Treatment with ACE inhibitors lowers these levels to normal. The efficacy of the drugs extends to essential hypertension, i.e. of unknown origin, whether mild, moderate or severe, but in these cases a diuretic is usually prescribed, in addition, to lower sodium levels and blood volume and thereby assist the action of the ACE inhibitor (Brunner *et al.*, 1983; Ferguson *et al.*, 1984).

Since these drugs were designed to combat angiotensin II action, we might expect a correlation between initial plasma renin concentration and the blood pressure lowering effects, and this is indeed found to be the case for captopril (see Ferguson *et al.*, 1984) and enalapril (see Brenner *et al.*, 1983). Further-more, the levels of angiotensin II prior to treatment correlate with falls in blood pressure as a consequence of treatment (MacGregor *et al.*, 1983). Plasma renin rises as a consequence of captopril treatment — not unexpec-tedly in view of the positive feedback effect of lowered sodium levels and blood volume on the secretion of renin. Plasma renin may not be the only target, however, because both renin and ACE have been found in the blood vessel walls, where renin levels are elevated as a consequence of captopril treatment (Unger *et al.*, 1983). In addition, the renin-angiotensin system has been found in brain, and it is possible that ACE inhibitors that can cross the blood–brain barrier will have a further antihypertensive effect (Unger *et al.*, 1983).

In principle, we would expect that plasma renin would be an indicator of blood pressure, since elevated blood pressure usually acts to lower plasma renin levels. This does not always seem to occur, however, since a number of patients with high blood pressure have normal, instead of lowered, renin levels (noted by Laragh, reported in Petrillo and Ondetti, 1982). ACE inhibitors are found to be effective even when renin levels are normal, as in these cases.

There are two other effects of these drugs that have emerged as a conse-quence of their use. One is the interaction of angiotensin II with the transmis-sion of adrenergic impulses, as the peptide stimulates the release of noradrenaline from pre-synaptic vesicles and inhibits the re-uptake of the neurotransmitter. Consequently, angiotensin II potentiates the vasoconstric-tor response to sympathetic nerve stimulation as well as to exogenously applied noradrenaline. Clearly, the inhibition of angiotensin II production will limit this effect, and supplement the action of ACE inhibitors as vasodila-tors (reviewed in Unger *et al.*, 1983).

Another very interesting effect that seems to be associated with ACE inhibitor treatment is euphoria (Zubenko and Nixon, 1984). The reason for this is not entirely certain, but it may be due to the inhibition of the hydrolysis of the enkephalin precursor, [Met]-enkephalin-Arg^6-Phe^7 which is probably carried out by ACE in the brain (Normal *et al.*, 1985). It is already clear that enkephalinase, which cleaves [Met]-enkephalin at the Gly^3-Phe^4 bond, is not inhibited by captopril, and so it is intriguing to speculate that other opioid peptides are responsible for the euphoria; particularly as [Met]-enkephalin-Arg^6-Phe^7 is derived physiologically from the proenkephalin A precursor protein. Opioid peptides are discussed in section 9.7 in Chapter 9.

8.4 *Endopeptidase 24.11*

The treatment of pain in an effective but non-addictive manner remains one of the most important pharmacological goals of the future. The discovery of the opioid peptides as the endogenous ligands for the morphine receptor suggested that prolongation of their action could lead to a new treatment of pain, as an alternative to the use of morphine (see Chapter 9). One way of doing this is to inhibit peptidases that hydrolyse enkephalins. This approach has led to two structures which may soon reach clinical trial as analgesics or painkillers; retro-thiorphan and L-phe-[N]-L-phe-β-ala.

One enzyme that can hydrolyse enkephalins is endopeptidase 24.11 (it has been given this curious shorthand as an abbreviation of the Enzyme Commission number 3.4.24.11). The enzyme has been isolated from kidney microvilli and shown to be identical with an enzyme from pig brain (Matsas *et al.*, 1983). The enzyme should not, strictly speaking, be referred to as enkephalinase because it will hydrolyze various peptide substrates (Hersh, 1986). Nevertheless, in the central nervous system its main role does appear to be termination of the action of enkephalins by cleaving [Met]- and [Leu]-enkephalin to yield Tyr-Gly-Gly and either Phe-Leu or Phe-Met — acting in fact as a dipeptidylcarboxypeptidase.

Endopeptidase 24.11 is a zinc-containing enzyme and bears a considerable resemblance to the family of zinc metallopeptidases. Approaches to inhibition based on the strategy that originally led to the angiotensin-converting enzyme inhibitors has yielded two compounds that are analogous to captopril and enalapril respectively, but which show considerable selectivity for endopeptidase 24.11. Retrothiorphan has a K_i for ACE of $>1\times10^{-5}$ M and 6×10^{-9} M for endopeptidase 24.11, while L-phe-[N]-L-phe-β-ala is also very selective with corresponding figures $>1 \times 10^{-4}$ M and 2×10^{-8} M (Mumford *et al.*, 1982).

These compounds give rise to pain relief in animal tests, which is reversed by naloxone, the characteristic opioid antagonist, and thus the inhibition appears to potentiate the body's own pain-relieving system. It is hoped that use of the drugs will not give rise to addiction (Chipkin, 1986).

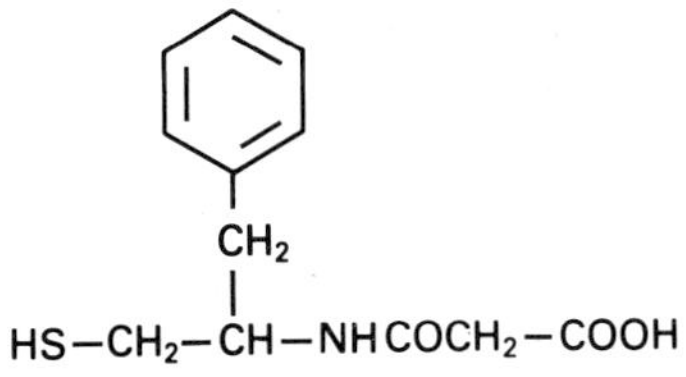

$$HS-CH_2-CH-NHCOCH_2-COOH$$

with CH_2 attached to a phenyl ring above the CH.

Retrothiorphan

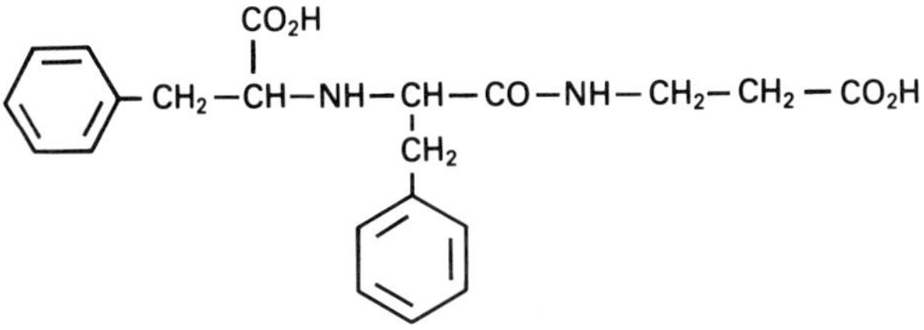

$$CH_2-CH-NH-CH-CO-NH-CH_2-CH_2-CO_2H$$

with CO_2H above the first CH (attached to a phenyl ring), and CH_2 below the second CH attached to a phenyl ring.

L-phe-[N]-L-phe-β-ala

References

AMA Drug Evaluations (1983). AMA Drug Division 5th Ed. (Philadelphia: W.B. Saunders), Ch. 19.

Brunner, H.R., Turini, G.A., Waeber, B., Nussberger, J. and Biollaz, J. (1983). *Clin. Exper. Hypertens.*, **A5**, 1355–1366.

Brunning, P. (1983). *Clin. Exper. Hypertens.*, **A5**, 1263–1275.

Chipkin, R.E. (1986). *Drugs of the Future*, **11**, 593–606.

Coleman, J.E. (1984). *Ann. N. Y. Acad. Sci.*, **429**, 26–48.

Cook, C.M. and Allen, L.C. (1984). *Ann. N. Y. Acad. Sci.*, **429**, 84–88.

Deutsch, H.F. (1984). *Ann. N. Y. Acad. Sci.*, **429**, 183–194.

DuBose, T.D., Pucacco, L.R. and Carter, N.W. (1981). *Am. J. Physiol.*, **240**, F138–146.

Ferguson, R.K., Vlasses, P.H. and Rotmensch, H.H. (1984). *Am. J. Med.*, **77**, 690–698.

Harris, R.B. and Wilson, I.B. (1985). *J. Biol. Chem.*, **260**, 2208–2211.

Hersch, L.B. (1986). *Trends in Pharmacol. Sci.*, **7**, 346–347.

Kokko, J.P. (1984). *Am. J. Med.*, **77** (5A), 11–17.

Lant, A. (1985). *Drugs*, **29**, 57–87.

Lipscomb, W.N. (1980). *Proc. Nat. Acad. Sci. (U.S.A.)*, **77**, 3875–3878.

Lucci, M.S., Tinker, J.P., Weiner, I.M. and DuBose, T.D. (1983). *Am. J. Physiol.*, **237**, F443–449.

MacGregor, G.A., Markandu, N.D., Smith, S.J., Sagnella, G.A. and Morton, J.J. (1983). *Clin. Exper. Hypertens.*, **A5**, 1367–1380.

Mackaness, G.B. (1985). *J. Cardiovasc. Pharmacol.*, **7**, Suppl. 1, S30–34.

Makinen, M.W., Kuo, L.C., Dymovski, J.J. and Jaffer, S. (1979). *J. Biol. Chem.*, **254**, 356–366.

Maren, H. (1984). *Ann. N.Y. Acad. Sci.*, **429**, 568–579.

Matsas, R., Fulcher, I.S., Kenny, A.J. and Turner, A.J. (1983). *Proc. Nat. Acad. Sci. (U.S.A.)*, **80**, 3111–3115.

Morrison, J.F. (1982). *Trends in Biochem. Sci.*, **7**, 102–105.

Mumford, R.A., Zimmerman, M., ten Broeke, J., Joshua, H. and Bothrock, J.W. (1982). *Clin. Exper. Hypertens.*, **A5**, 1362–1380.

Norman, J.A., Autry, W.L. and Barbaz, B.S. (1985). *Molec. Pharmacol.*, **28**, 521–526.

Ondetti, M.A. and Cushman, D.W. (1984). *CRC Crit. Rev. Biochem.*, **16**, 381–411.

Petrillo, E.W. and Ondetti, M.A. (1982). *Medicinal Res. Rev.*, **2**, 1–41.

Rees, D.C., Lewis, M., Honzatko, R.B., Lipscomb, W.N. and Hardman, K.D. (1981). *Proc. Nat. Acad. Sci. (U.S.A.)*, **78**, 3408–3412.

Shapiro, R. and Riordan, J.F. (1984). *Biochemistry*, **23**, 5225–5233.

Unger, T., Ganten, D. and Lang, R.E. (1983). *Clin. Exper. Hypertens.*, **A5**, 1333–1354.

Vallee, B.L. and Galdes, A. (1984). *Adv. Enzymol.*, **56**, 281–430.

Vedani, A. and Meyer, E.F. (1984). *J. Pharm. Sci.*, **73**, 352–358.

Zubenko, G.S. and Nixon, R.A. (1984). *Am. J. Psychiat.*, **141**, 110–111.

Chapter 9

Neurotransmitter action and metabolism

Table 9.1. Drugs and their targets discussed in Chapter 9

Section	Receptor/ Enzyme	Type	Drug	Use in
9.2	Adrenergic	α_2	Clonidine	Hypertension
		α_1	Prazosin	Hypertension
		α_1	Phentolamine	Congestive heart failure
		β	Propranolol	Arrhythmia
		β_1	Metoprolol	Hypertension
		β_1	Atenolol	Hypertension
		β_2	Salbutamol	Asthma
		α,β	Labetalol	Hypertension
9.2	Dopamine	D_1	Bromocryptine	Parkinsonism
		D_1	L-Dopa + carbidopa	Parkinsonism
		D_1	Fenoldapam	Hypertension
		D_2	Haloperidol	Schizophrenia
		D_2	Pimozide	Schizophrenia
9.4	Serotonin	5-HT_1	Ketanserin	Hypertension
		5-HT_2	Mianserin	Depression
		$5\text{-HT}_{1,2}$	Methysergide	Migraine
9.5	Serotonin/ Noradrenaline uptake		Imipramine	Depression
			Iprindole	Depression
			Maprotiline	Depression
			Zimelidine	Depression
9.6	Monoamine oxidase	MAOA	Clorgyline	Depression
		MAOA	Toloxatone	Depression
		MAOA	Cimoxatone	Depression
9.7	Cholinergic Muscarinic	M_1	Pilocarpine	Glaucoma
		M_1	Pirenzepin	Ulcers
		$M_{1,2}$	Atropine	Pre-anaesthetic medication
		$M_{1,2}$	Trihexyphenidyl	Parkinsonism
	Acetylcholin-esterase		Pyridostigmine	Myasthenia gravis
			Neostigmine	Myasthenia gravis
9.8	4-Amino-butyrate	GABA_A	Chlordiazepoxide	Hypnotic/sedative/ muscle relaxant
		GABA_A	Diazepam	
		GABA_A	Avermectin	Helminth infection
		GABA_B	Baclofen	Spasticity

Table 9.1 contd.

Section	Receptor/ Enzyme	Type	Drug	Use in
9.9	Opiate	$\mu(\kappa,\sigma)$	Morphine	Pain
		$\mu(\kappa,\sigma)$	Pentazocine	Pain
		$\mu(\kappa,\sigma)$	Loperamide	Diarrhoea
9.10	Histamine	H_1	Mepyramine	Allergic reactions
		H_1	Chlorpheniramine	Hay fever
		H_2	Cimetidine	Ulcers
		H_2	Ranitidine	Ulcers

Legend to Table 9.1

For some drugs there are additional receptors to which the drug may bind and which may contribute either to its mode of action and/or to its toxicity. Chlorpromazine, for example, is known to bind to several other types of receptor, possibly because of its affinity for membranes, and these effects are likely to contribute to its toxicity.

Furthermore, some of these drugs have other uses, as for example, neostigmine for the reversal of neuromuscular blockade in surgery.

9.1 Introduction

In this chapter we consider the drugs that have been developed deliberately, or found in hindsight (serendipitously), to modulate neurotransmitter action or metabolism. In order to understand more about the way in which drugs can interfere with the transmission of nerve impulses we need to know how impulses are transmitted along a nerve.

The resting nerve cell or neurone maintains a potential difference between the cytoplasm and extracellular fluid of approximately 75 mV (cytoplasm negative). This is achieved largely by the functioning of a membrane-bound sodium pump acting to export positive sodium ions from the cell using the energy of hydrolysis of ATP. The membrane is practically impermeable to sodium ions at this potential, thus a tenfold concentration gradient is maintained between the nerve cell cytoplasm and the extracellular fluid.

An impulse may be generated by a variety of means such as heat, mechanical deformation, specific chemicals or neurotransmitters etc., depending on the type of neurone in question, and it produces a reduction in the membrane potential difference to approximately 60 mV. When this happens, 'gates' or channels in the membrane open at one end of the neurone, which are specific for the active transport of sodium ions, causing the sodium cation to rush in and thereby neutralize the negative charge inside the membrane (depolarization). The neurone remains in this state of depolarization, usually for 1 to 3 milliseconds in man, until sufficient potassium ion has left the cell through specific channels to repolarize the membrane. This condition is also known as refractory because no further impulses can pass until the resting potential of 75 mV is re-established. The intracellular concentrations of potassium and sodium ion are then restored by the action of the Na^+/K^+ adenosine triphosphatase which uses the energy of hydrolysis of ATP to drive an exchange of sodium inside for potassium outside the cell. This enzyme is discussed further in Chapter 10.

The impulse or action potential passes down the nerve cell until it reaches the end of the cell body opposite another nerve cell at a junction called a synapse. The cell from which the action potential arrives at the synapse is known as the pre-synaptic cell. Communication with the second (post-synaptic) cell is performed by a chemical agent known as a neurotransmitter, specific for that synapse. The chemical is stored in vesicles immediately before the synapse and the release of these is triggered by the arrival of the action potential. The contents of the vesicles are extruded into the gap between the two nerve cells, known as the synaptic cleft, and the neurotransmitter passes across to bind to specific protein targets on the post-synaptic cell membrane, known as receptors. This complex formation may elicit a response of either an excitatory nature, i.e. a continuing impulse (accompanied by depolarization) or an inhibitory response, i.e. to prevent an impulse arising in the responding cell. The latter condition is accompanied by *hyper*-polarization up to -100 to -120 mV, usually by the inward transport of anions such as chloride ion, and this renders the cell insensitive to stimulation.

Clearly, the action of the neurotransmitter needs to be terminated as soon as it is completed, and hydrolytic enzymes accomplish this action; acetylcholine is hydrolysed by acetylcholinesterase in the synaptic cleft, while the transmitters that contain a primary amine group (aminergic) such as noradrenalin, dopamine and serotonin are taken up into the pre-synaptic cell and oxidatively deaminated to aldehydes by monoamine oxidase. Catecholamine transmitters may also be taken up outside the neurone where catechol O-methyltransferase renders the amine inactive by catalysing the transfer of a methyl group to the *meta*-hydroxyl on the catechol ring.

For a substance to be confirmed as a neurotransmitter, two major actions have to be seen: firstly the agent must be released from the appropriate nerve ending by an impulse, and secondly the action of the compound when applied to post-synaptic membranes must mimic the action that occurs when the nerve is stimulated in the normal way.

Furthermore, a number of other factors may be taken into account. The compound has to be present in the appropriate nerve terminals in sufficient quantity to act as a transmitter. This means that the enzymes for its biosynthetic pathway must occur in those terminals. Substances that interfere with the binding of the natural ligand to its receptor in subcellular preparations should have a similar action on the natural transmitter released in the normal way. In addition, there is usually a re-uptake mechanism and/or a metabolic enzyme for removal of the transmitter from the synaptic cleft. These conditions have been met for all of the transmitters covered in this chapter, with the possible exception of histamine. The similarities of drugs binding to histamine receptor compared with ligand binding to the other receptors discussed in this chapter suggest its inclusion here.

It should be noted that receptor binding requires the correct stereochemical form of an agonist. In the case of the catecholamines it is the naturally occurring L-form that is active while the D-isomers are at least ten times less active. Mere binding of an agent to a membrane preparation that contains a receptor does not guarantee that a pharmacological response will ensue. Parallel studies on isolated tissues or whole animals should also be carried out to confirm the receptor occupancy as a valid pharmacological result.

9.1.1 Signal transduction and receptor occupancy

The chain of molecular events that connect receptor occupancy with cell response is not yet fully understood. Nevertheless it is clear that there are two principal methods of transducing the extracellular signal to activate intracellular pathways. One involves the activation of the membrane-bound enzyme adenylate cyclase which catalyses the conversion of ATP to cyclic adenosine $3',5'$-monophosphate (cAMP) — this in turn activates a specific protein kinase which is crucial in regulating the activity of many enzymes by controlling their phosphorylation state. The second pathway involves hydrolysis of phospholipids in the membrane to yield prostaglandins and leukotrienes, as well as release of calcium from intracellular stores, leading to a wide variety of physiological effects. Usually a particular type of receptor will only activate one of other of these pathways, which are now described in more detail.

At rest it is proposed that the receptor protein, located on the outer surface of the cell, is complexed with a guanine nucleotide binding protein, thus preventing it making any contact with the catalytic unit of the enzyme, which is situated on the inner surface of the membrane. The binding of an agonist to the receptor on the extracellular surface of the cell causes the complex to dissociate, thus allowing GTP to bind to the regulatory protein. This protein moves across the membrane bilayer to bind to the enzyme and thereby modulates its activity. There are two sorts of guanine nucleotide-binding protein with which we are concerned here, one of which, designated N_S activates adenylate cyclase, while the other, designated N_I, inhibits the enzyme.

The guanine nucleotide regulatory protein is a trimer consisting of an α-subunit of molecular weight 45 kDa for N_S and 41 kDa for N_I; the β-subunit of 35 kDa and the γ of 10 kDa do not differ between N_I and N_S. When GTP binds to the α-subunit it is activated (a process which involves dissociation of the trimer) either to stimulate or inhibit adenylate cyclase. For every catalytic turnover of the enzyme one GTP molecule is hydrolysed to GDP which dissociates from the binding protein thereby rendering the latter inactive. GTP has to bind again before activity is restored (Gilman, 1984; Brodde, 1986).

The α-subunit is the target for cholera toxin and studies of the interaction have revealed many details of adenylate cyclase activation. This toxin acts as an enzyme catalysing the transfer of the adenosine-diphosphoribosyl moiety from nicotinamide adenine dinucleotide to an arginine residue on the α-subunit of the protein. Usually the binding of the β-subunit to the α would inhibit the action of the latter. ADP-ribosylation of the α-subunit decreases its affinity for the β-subunit but not for adenylate cyclase and so the enzyme is activated — permanently in this case since GTP hydrolysis cannot take place (Kahn and Gilman, 1984). Cyclic AMP levels are thereby raised sufficiently high to pump ions and water out of the gastric cell and severe dehydration results (since the organism that produces the toxin, *Vibrio cholerae*, is found in the gut the main effects of the toxin are on the gastric cell). A number of compounds containing the guanidine group, designed to bear some resemblance to the side-chain of arginine, have been found to inhibit the enzyme activity of cholera toxin (Coulson *et al.*, 1984).

When adenylate cyclase is activated, the rate of breakdown of adenosine triphosphate to adenosine 3′,5′-cyclic monophosphate and pyrophosphate is raised from a very low, almost undetectable level in some cases, and the cyclic nucleotide is able to carry out various reactions — the most important of which is the activation of protein kinases. These exist as inactive tetramers composed of two catalytic and two regulatory subunits. Cyclic AMP binds to the regulatory subunits causing them to dissociate from the catalytic dimer, leaving the latter free to express its activity in phosphorylating a number of crucial proteins. These include triglyceride lipase — that catalyses the initial and rate-limiting step in lipolysis — and the phosphorylase kinase which activates phosphorylase, again by phosphorylation, to hydrolyse glycogen to glucose (Rodbell, 1980; Gilman, 1984).

The process as a type of cascade allows a considerable amplification of the original hormone signal through the catalytic activity of adenylate cyclase and the protein kinases. The adenylate cyclase system is outlined in Figure 9.1, while the receptors discussed in this chapter are listed by the signal transduction pathway that they activate in Table 9.2.

The second major method of signal amplification that the target cell uses is via hydrolysis of the phospholipid phosphatidylinositol that is found primarily in the plasma membrane of the cell. When membrane receptors are activated through ligand binding, a chain of events is set in motion whereby phospholipase C cleaves the phospholipid to release, on the one hand, arachidonic acid which is further metabolized to prostaglandins, thromboxanes and leukotrienes (see Chapter 7). Recent studies suggest that the receptor and phospholipase C may be connected by a guanine nucleotide regulatory protein in a similar fashion to the cAMP pathway (reviewed in Taylor and Merritt, 1986).

The other products of phospholipid hydrolysis are diacylglycerol and either inositol 1-phosphate, the 1,4-diphosphate or the 1,4,5-triphosphate. Diacylglycerol activates protein kinase C in the presence of calcium ion and phospholipid to phosphorylate a number of proteins (see Figure 9.2). The triphosphate acts intracellularly to induce the release of calcium from the endoplasmic reticulum in order to produce a transient rise in the cytoplasmic concentration of the cation.

Cell membrane

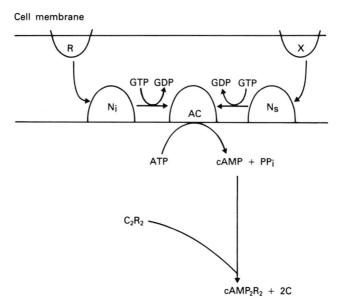

R and X represent receptors for hormones and neurotransmitters
N_i represents the inhibitory guanine nucleotide binding protein
N_S represents the stimulatory guanine nucleotide binding protein
AC is adenylate cyclase catalytic subunit
cAMP is cyclic AMP, PP_i is pyrophosphate
C_2R_2 is the tetramer of protein kinase

The binding of a hormone or neurotransmitter to receptors R or X causes release of N_i or N_S, thus allowing it to interact with adenylate cyclase. For every turnover of the enzyme, one molecule of GTP is hydrolysed to GDP. The enzyme catalyses the hydrolysis of ATP to cAMP and PP_i. cAMP binds to the regulatory subunit of protein kinase thus allowing the catalytic subunit to phosphorylate various proteins with crucial rate-limiting roles.

Figure 9.1 Adenylate cyclase modulation by receptor occupancy.

This transient rise in intracellular calcium ion level is sufficient to sustain the release of aldosterone by angiotensin II for example (reviewed in Rasmussen, 1986), but a more prolonged charge is required for smooth muscle contraction and for the release of catecholamines from the adrenal medulla by nicotine or by potassium ion. This is obtained by a cycling process the details of which are not fully understood but it is proposed that protein kinase C is involved in the latter case because polymyxin B, an inhibitor, also inhibits the release of catecholamine while phorbol diester, an activator, reverses this inhibition (Wakade *et al.*, 1986). Indeed, activation by phorbol diester is usually a clear indication that protein kinase C and phosphatidyli-nositol hydrolysis are involved in a secretory process.

In the case of excitatory cells such as those neurones from which acetylcho-line is released and also cardiac cells, the depolarization of the membrane is

Table 9.2 Association of ligands with signal transduction pathways

1. Adenylate cyclase

Receptor subtype		GTP binding protein
Serotonin	5-HT$_1$	N$_S$
Muscarinic	M$_2$	N$_S$
GABA	B	N$_i$
Opiate	μ	N$_i$
Histamine	H$_2$	N$_i$
Dopamine	D$_1$	N$_S$
	D$_2$	N$_i$
Adrenergic	α_2	N$_i$
	$\beta_{1,2}$	N$_S$

2. Inositol phospholipid

Serotonin	5-HT$_2$	
Muscarinic	M$_1$	
Nicotinic		
Histamine	H$_1$	
Adrenergic	α_1	

N$_i$ represents the inhibitory guanine nucleotide regulatory protein leading to inhibition of adenylate cyclase.
N$_S$ represents the stimulatory GTP binding protein.

All the various receptors discussed in this chapter are listed, in order of the discussion, with respect to the signal transduction pathway that they activate. The only exception is GABA$_A$ which does not appear to operate through either of these pathways and, moreover, unlike the receptors above, contains a chloride channel as part of the receptor.

sufficient to open voltage-sensitive channels, so that as the voltage across the membrane falls calcium is drawn into the cell (reviewed in Rasmussen, 1986). Calcium ion activates a number of proteins either directly or as a complex with the specific binding protein, calmodulin. The latter complex acts frequently through protein kinases or may activate other proteins directly (such as cyclic nucleotide phosphodiesterase) thereby interacting with the cyclic nucleotide system.

9.1.2 Drug development

Drugs can interfere with the process of synaptic transmission in various ways. Occasionally, a false substrate can be provided for the biosynthetic pathway of the neurotransmitter, resulting in a false neurotransmitter with altered properties which competes with the natural agent. The mechanisms for removal of used neurotransmitter (both re-uptake and metabolism) have been found useful as targets for the rational development of drugs. Probably the major effort in drug development, however, has gone into the development of agents which act like the neurotransmitter (agonists) and those which antagonize its action.

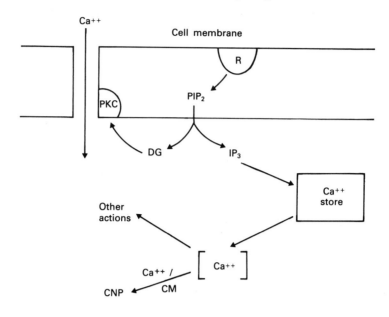

R is the receptor
PIP$_2$ is phosphatidylinositol 4,5-di-phosphate
IP$_3$ is inositol 1,4,5-triphosphate
DG represents diacylglycerol
PKC represents protein kinase C
CNP is cyclic nucleotide phosphodiesterase

Binding of a ligand to the receptor induces the hydrolysis of PIP$_2$ to IP$_3$ and DG. IP$_3$ causes calcium to be released from intracellular stores such as mitochondrion and, in muscle, sarcoplasmic reticulum. DG activates PKC to phosphorylate various proteins and possibly to open a voltage dependent channel to allow calcium influx into the cell. Calcium ion may activate certain processes directly or via its complex with calmodulin, e.g. CNP, one form of which is calcium-dependent, and often used to assay calmodulin.

Figure 9.2 Activation of inositol phospholipid pathway by receptor occupancy.

It is a feature of pharmacology in particular, and possibly science in general, that initial concepts are simple but become progressively more complex as more information is obtained. Neuropharmacology is no exception — for every neurotransmitter discussed in this chapter, at least two kinds of receptor, with different effects, have been identified, although drugs for use as agonists or antagonists at some of the sub-types have not been devised yet!

For many of the receptors discussed in this chapter the endogenous ligand has been known for some time, as in the case of the receptors acted on by the catecholamines, but for others, e.g. the opiate receptors, drug binding studies suggested that there were tight binding sites before a natural ligand was discovered. The opiate peptides, such as enkephalin and endorphin, were subsequently found to fit the bill. At the present time, the diazepam binding site, although part of, and modulatory of, the 4- (or γ-)aminobutyric acid

(GABA) receptor is clearly separate from the latter, and may also have an unknown endogenous ligand.

9.2 Adrenergic receptors

Adrenergic receptors occupy a privileged place in receptor pharmacology, in that the first separation of hormonal action into distinct activities was proposed on the basis of excitatory and inhibitory action of the catecholamines (named because of the 1,2-dihydroxybenzene or catechol moiety that they contain) on smooth muscle (Ahlquist, 1948). These activities were termed α and β, and from this arose the concept of a specific receptor for each type of physiological activity.

The three catecholamines [noradrenaline, adrenaline and isopropylnoradrenaline (isoprenaline); see Figure 9.3 for the biosynthesis of the two former agents], differ markedly in their activities on the two receptor types. Noradrenaline and adrenaline are the most active on α receptors, while isoprenaline is the more potent agonist on β-receptors. In cardiac muscle the predominant receptor is the β-type, which in this tissue governs the increase

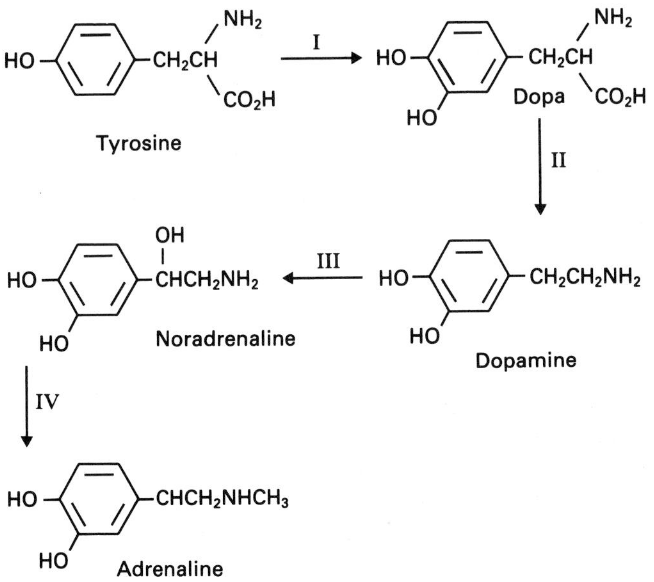

I: Tyrosine hydroxylase, tetrahydrobiopterin
II: Aromatic amino acid decarboxylase, pyridoxal phosphate
III: Dopamine β-hydroxylase, ascorbate
IV: Phenylethanolamine N-methyltransferase, S-adenosylmethionine
(Enzymes are noted together with their prosthetic groups)

Figure 9.3 Pathway of catecholamine biosynthesis.

in rate and force of contraction of the heart, known respectively as the positive chronotropic and inotropic effects.

Subsequent studies with specific agonists and antagonists have shown that both α and β receptors can be divided into subtypes: α_1, α_2, β_1 and β_2. The inhibition of radioligand binding in various tissues suggests that both β_1 and β_2 co-exist in a variety of tissues (reviewed in Minneman and Molinoff, 1980). The type of activity governed by the various receptors varies, so that for example in smooth muscle the α_2-receptor usually governs stimulatory actions while β_2-receptor occupancy is inhibitory. The use of agents that interact selectively with the different receptor sub-types has markedly improved our understanding of the way these systems work, while at the same time proving immensely useful in the therapy of high blood pressure, asthma and other disorders.

9.2.1 α-Adrenergic receptors

The detailed study of the molecular pharmacology of the α-receptor has lagged behind that of the β-receptor, possibly because selective pharmacological modulators of the α-receptors have only been discovered more recently. Noradrenaline and adrenaline are agonists, by definition, at the α-receptor while agents such as phentolamine are antagonists. Although the catecholamines do not distinguish between the α-receptors, prazosin is a selective antagonist at the α_1 site while the alkaloid yohimbine is antagonistic at the α_2 site. Clonidine, on the other hand, is an α_2 agonist.

Structures of α-receptor; agonists and antagonists

Phentolamine

Prazosin

Yohimbine

Clonidine

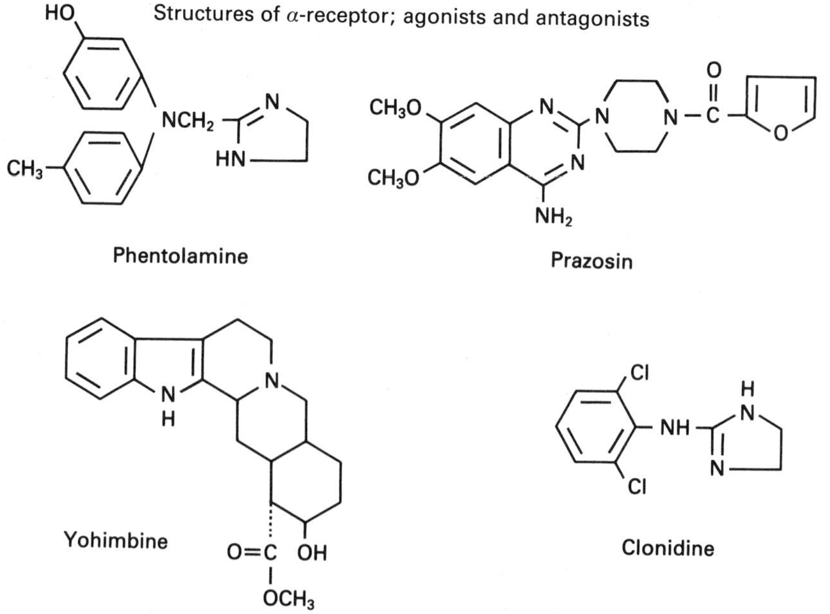

Yohimbine, prazosin and phentolamine are antagonists. Clonidine is an agonist.

The anatomical distribution of the receptor has been worked out. The α_1-receptors generally occur post-synaptically where they mediate smooth muscle contraction. α_2-receptors can appear both pre- and post-synaptically and in the former position they control the constriction of vascular smooth muscle. A pre-synaptic receptor can control the release of the neurotransmitter while both subtypes are apparently required for the initiation of post-synaptic excitatory activity (Van Zwieten, 1986; Brodde, 1986). Post-synaptic α_2-receptors are found in several tissues including brain and parotid gland.

The α_2-receptor appears to be linked to adenylate cyclase in an inhibitory fashion in sharp contrast to the stimulatory action of β-adrenergic agonists. This effect occurs through an inhibitory regulatory protein that binds guanine nucleotides (see Lefkowitz and Caron, 1986, for a review) and the complex reduces the affinity of the α_2-receptor for agonists but not antagonists (see Limbird *et al.*, 1982).

The presence of calcium ions is essential for the expression of α_2 agonism, whether it be vasoconstriction that is involved (Lew and Angus, 1985) or inhibition of water outflow in enterotoxin induced diarrhoea (Newsome *et al.*, 1984). Sodium ion also appears to regulate the receptor probably at a different locus from that of the GTP-binding protein (Limbird *et al.*, 1982).

9.2.1.1 α_2-Agonists

The binding affinity of clonidine at the α_2-receptor is high. Whether the receptor is in the intact membrane of the human platelet or solubilized in the presence of digitonin, the binding constant lies between 4 and 9×10^{-9} M (Limbird *et al.*, 1982). A similar binding constant was obtained for the rat brain α_2-receptor in membrane-bound and detergent solubilized form (Matsui *et al.*, 1984). Since GTP reduced the binding of clonidine to the solubilized receptor in the latter preparation, it is probable that the receptor/GTP binding protein complex remained intact during the solubilization procedure. Sodium ion also inhibited clonidine binding.

Clonidine is used to reduce blood pressure often in conjunction with another drug such as a thiazide diuretic (Rudd and Blaschke, 1980) or a β-adrenergic receptor blocker (Vauholder *et al.*, 1985). The advantage of the latter combination is that clonidine tends to induce a rapid fall in blood pressure while the reduction is much slower with β-blockers, so that an immediate but also longer lasting effect can be obtained with the drug combination.

9.2.1.2 α-Adrenergic antagonists

Prazosin, a selective α_1 antagonist, is used to reduce blood pressure because of its action as a vasodilator. It appears to be moderately selective as a blocking agent at post-synaptic receptors. Prazosin thereby inhibits the vaso-constriction produced by noradrenaline that is released from the sympathetic nerve endings, while the drug's lack of activity on α_2-receptors allows the neurohormone to exert negative feedback control of its own release, thus reducing the cardiac stimulation that follows non-selective α-adrenergic blockade (Rudd and Blaschke, 1985).

Prazosin binds very tightly to α_1-receptors with a binding constant of between 1 and 6×10^{-10} M to membrane-bound receptors, and appears to be some four orders of magnitude more selective for α_1- than α_2-receptors (see McPherson and Summers, 1982 and references therein).

Phentolamine is a non-selective antagonist at α-receptors. Its structure is related to that of clonidine (see structures) but the molecular changes of an additional benzene ring and different substituents on it have turned a specific agonist into a non-specific antagonist. In addition, it has activity at acetyl-choline and histamine receptors. Phentolamine is sometimes used in cases where excessive adrenergic activity has produced constriction in the blood vessels, such as congestive heart failure.

9.2.2 β-Adrenergic receptors

The β-receptor has been identified and characterized in many tissues by the use of radiolabelled β-blockers. The link between ligand binding and acti-vation of adenylate cyclase has been confirmed by various workers including Lefkovitz (1975) and Minneman *et al.* (1979). The β-adrenergic receptor has been purified from frog erythrocyte membranes by a combination of affinity chromatography with a β-adrenergic blocking drug followed by high perform-ance liquid chromatography (Shorr *et al.*, 1982). The molecular weight of the glycoprotein receptor is 64 kDa, and it has been shown to contain binding sites for both ligand and guanine nucleotide binding protein (Lefkowitz and Caron, 1986). The distinction between β_1 and β_2 subtypes has been firmly underlined in the sense that adenylate cyclase, activated by isoprenaline, is antagonized by propranolol in all tissues but by the selective β-blocker, practolol, more effectively in heart than in lung (Lefkovitz, 1975).

The biochemical events that occur as a result of receptor agonism and antagonism have been worked out in great detail for the β-receptor. The receptor is coupled to adenylate cyclase via a guanine nucleotide-regulatory protein (see Figure 9.1). The hormonal signal is thus transmitted to cyclic adenosine $3',5'$-monophosphate (cAMP) which is an intracellular 'second messenger', and the metabolic actions that are associated with hormone action immediately follow, such as stimulation of glycogenolysis, gluconeo-genesis and lipolysis etc. (see section 9.1.1).

The β-receptor appears to exist in states of both low and high affinity. The binding of GTP to the guanine nucleotide regulatory protein converts the high-affinity receptors to low affinity. Agonists induce the formation of a quaternary complex involving the receptor, GTP-binding protein (N_S) and GDP; GDP is lost from the binding protein and GTP takes its place; the receptor is converted to its low affinity state and GTP/N_S proceeds to bind to the catalytic subunit of adenylate cyclase, thereby activating the enzyme and hydrolysing GTP to GDP in the process (Brodde, 1986).

One interesting feature of agonist activity is its transient nature. Frequently, it is found that the pharmacological activity of the catecholamines in particular rapidly falls; the phenomenon is known as desensitization. This occurs as a result of receptor phosphorylation by cAMP-activated protein kinase (Lefkowitz and Caron, 1986). Two or three serine residues are the targets, and the agonist isoprenaline causes a two to three-fold increase in the rate of phosphorylation, which is completely blocked by antagonists.

9.2.2.1 β-Adrenergic antagonists

The structure of isoprenaline underwent a number of modifications in order to devise a compound that would block the β activity of the catecholamines and be pharmacologically valuable as an antihypertensive. These modifications initially concentrated on extending the side-chain so that first three and then four atoms were interposed between the nitrogen atom and the benzene ring. Thus β-antagonists were derived from β-agonists. The β-receptor is, therefore, the target for a number of drugs that are either agonists or antagonists. Propranolol, the first effective antagonist or β-blocker, is still used to lower blood pressure and in the management of cardiac arrhythmias (see Chapter 10 for a discussion on arrhythmia). The drug is, however, non-selective in that it blocks the β-response in heart, lung and a variety of other tissues non-specifically (Lefkovitz, 1975), and this can lead to severe broncho-constriction in the lung — highly undesirable in an asthma patient, where propranolol should not be used.

A search for more selective β-blockers has resulted in the development of, for example, metoprolol and atenolol which are more active on cardiac receptors (termed β_1) although they still inhibit lung β_2-receptors at higher doses. For example, metoprolol is 10 to 20 times more selective for β_1 while atenolol is about three times more potent. Propranolol, on the other hand,

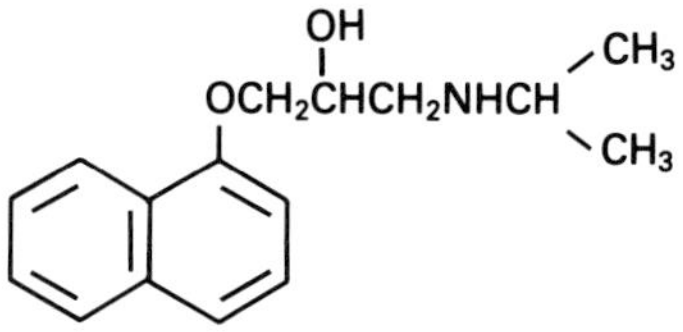

Propranolol

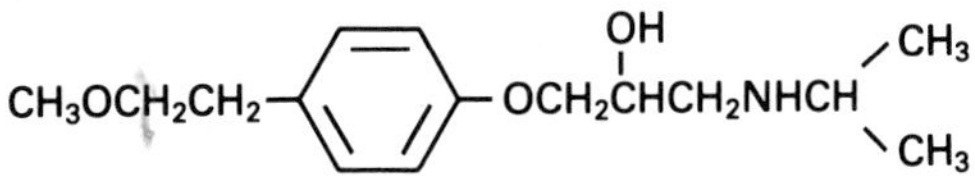

Metoprolol

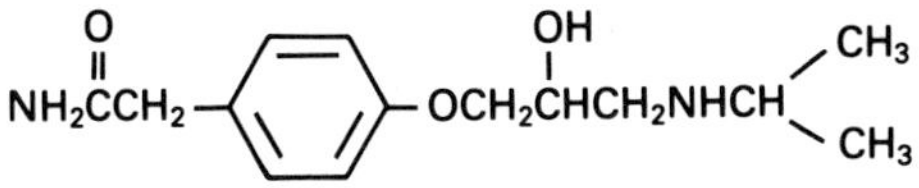

Atenolol

shows no such selectivity. There is a problem in that the concentrations of drug required to have an effect on membrane preparations are much higher than those in intact tissue. This has been rationalized as being a consequence of some distortion of the normal receptor/enzyme coupling mechanism in the broken membrane fragments (Minneman *et al.*, 1979).

9.2.2.2 β-Adrenergic agonists

Isoprenaline is the prototype of β-receptor agonists and found a use in earlier years as a treatment for asthma. It suffered, however, from a lack of specificity since it activated receptors in the cardiovascular system and gave rise to unpleasant side-effects, notably palpitations and occasionally cardiac arrhythmias (uneven beating of the heart) through its positive inotropic effect. In addition, the drug is rapidly inactivated by catechol O-methyltransferase and thus when given had a very short half-life. A longer-acting, more selective agent was required.

Accordingly, salbutamol was synthesized specifically for this purpose (Cullum *et al.*, 1969), and was shown to be much more specific for lung than for cardiac receptors. As a consequence *in vivo* the drug showed lower activity in raising heart rate than in raising bronchial muscle tone in cats (Apperley *et al.*, 1976). *In vitro* the dose for half-maximal effect (ED_{50}) in contracting the guinea pig trachea was in the region of 3×10^{-9} M while ED_{50} for contraction of the right atrium was 10^{-7} M (Buckner and Abel, 1974).

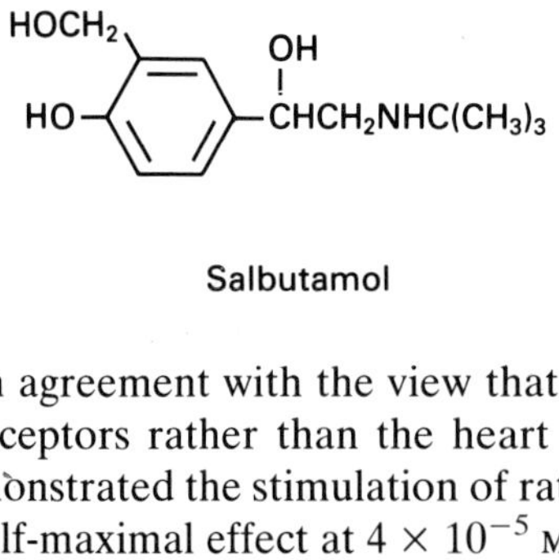

Salbutamol 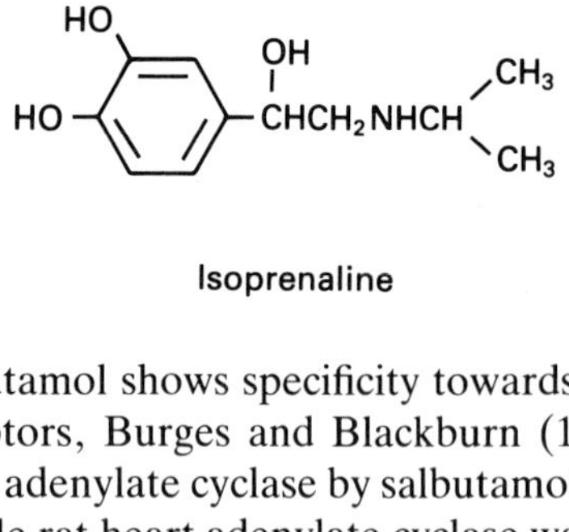   Isoprenaline

In agreement with the view that salbutamol shows specificity towards lung β-receptors rather than the heart receptors, Burges and Blackburn (1972), demonstrated the stimulation of rat lung adenylate cyclase by salbutamol with a half-maximal effect at 4×10^{-5} M, while rat heart adenylate cyclase was not stimulated by concentrations as high as 10^{-3} M. The link with raised intracellular cAMP levels was demonstrated by Mitchell *et al.* (1979) who used lung parenchyma from guinea pigs. Although low concentrations (3×10^{-8} M) of salbutamol failed to raise cAMP levels, the combined action of salbutamol and theophylline (3×10^{-4} M) did, and the result was explained by the inhibition of cyclic nucleotide phosphodiesterase by the latter. At the same time anaphylactic contraction and histamine release were also inhibited.

The elevation of cAMP levels has been associated with a reduction in allergic histamine release by Bourne *et al.* (1972). A β-agonist would therefore be expected to have a therapeutic value in asthma, reinforced by the evidence from more recent work that has indicated that the leukotrienes provide the major effect in the development of the condition (Church and Warner, 1985) and β-agonists can block leukotriene release from mast cells.

9.2.3 Mixed α- and β-antagonist

Although it might be imagined that the α and β receptors would only respond to the catecholamines, a drug called labetalol has blocking activity at both. The drug is about one-tenth as potent as phentolamine in blocking α-receptors and a third as effective as propranolol at the β-receptor, probably at the β_2-receptors (Sybertz *et al.*, 1981). It will be seen that the structure resembles noradrenaline and salbutamol, with the nitrogen atom only two carbon atoms away from the ring and a hydroxyl group on the α-carbon. It is not surprising therefore that labetalol has some intrinsic sympathomimetic activity largely on the β_2-receptors as exemplified by vasodilatory activity that can be blocked by propranolol (Baum and Sybertz, 1983). Labetalol is a potent antihypertensive agent and has been used to treat essential hypertension.

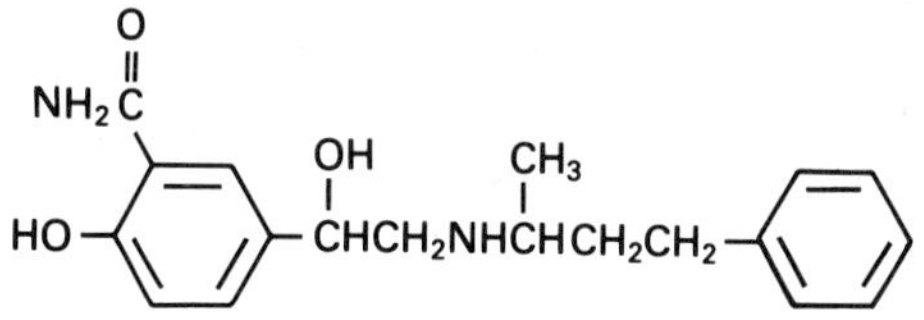

$$NH_2C(=O)-\text{(HO-ring)}-CH(OH)CH_2NHCHCH_3CH_2CH_2-\text{(ring)}$$

Labetalol

9.3 Dopamine receptors

Dopamine receptors are found in the periphery and in the central nervous system where they mediate inhibitory actions on other neurotransmitter pathways. They are usually hyperpolarizing, although there are indications that occasionally depolarization is the major effect i.e. activation occurs (Brown and Arbuthnott, 1983). Dopamine receptors may be classified on an anatomical basis into three types:
   (a) cell-body receptors that control dopaminergic neuronal activity,
   (b) pre-synaptic receptors (autoreceptors) that were originally thought to control the synthesis and release of dopamine, but the evidence now is against this view,
   (c) post-synaptic receptors that inhibit the special response characteristic of the particular pathway, as for example, acetylcholinergic and glutamate activated pathways (Brown and Arbuthnott, 1983).
   More recently, another type of classification of the brain receptors has become possible by the use of binding studies with radiolabelled ligand, coupled with studies of adenylate cyclase which is activated by the binding of

dopamine to its receptors. At least two main types of receptor have been identified by this means: those that are sensitive to sub-nanomolar concentrations of dopamine and where coupled adenylate cyclase is activated (D_1 receptors) and those (D_2) where nanomolar levels of the neurotransmitter are sufficient for activation (Seeman *et al.*, 1985), and adenylate cyclase is either inhibited or no effect is detected. The reason for this variability with respect to D_2 receptors lies in the fact that it has been difficult to demonstrate effects on adenylate cyclase where both types of receptor are present, but it appears that when D_1 receptors are blocked by a specific antagonist (SCH 23390) it can be shown that D_2 receptors inhibit the enzyme (Cooper *et al.*, 1986). With respect to the earlier anatomical classification, D_1 and D_2 receptors can be both pre- and post-synaptic.

This separation between D_1 and D_2 receptors has been amplified by the realization that guanine triphosphate modulates the binding of agonists, but not apparently antagonists, to D_1 receptors (Seeman 1982; Martres *et al.*, 1984). Moreover, GTP (and sodium ion) is required for the inhibition of adenylate cyclase by dopamine at D_2 receptors (Cooper *et al.*, 1986).

Spiroperidol is a selective antagonist at D_2 receptors. Recent studies have resolved a problem with spiroperidol in that its dissociation constant when measured directly with radiolabelled ligand was $5 \cdot 2 \times 10^{-11}$ M but when measured against ^{3}H-dopamine was $9 \cdot 1 \times 10^{-10}$ M. As noted above, dopamine itself has a greater affinity for the D_1 than D_2 receptors so that it would preferentially label the former, and confuse the issue. When, however, D_1 receptors were preferentially blocked by a specific antagonist, SCH 23390, the binding constant was found to be $3 \cdot 7 \times 10^{-11}$ M by either direct binding or by antagonism (Seeman and Grigoriadis, 1985). This may seem a relatively trivial point but it illustrates the difficulties faced in receptor pharmacology when specific antagonists are not available to bind to interfering receptors.

The D_1 receptor has been solubilized from rat striatal membranes and shown to bind the specific antagonist, SCH 23982, with a binding constant of $1 \cdot 8 \times 10^{-9}$ M (Sidhu and Fishman, 1986) — SCH 23892 is the active R-enantiomer of SCH 23390 (see structures). This compound is the first specific antagonist to be discovered for the D_1 receptor (Iorio *et al.*, 1983), although not in use as a drug. Indeed, the discovery of this specific antagonist has greatly increased our understanding of the D_1 receptor and its pharmacological correlates (see Kebabian *et al.*, 1986). SCH 23390 binds to the D_1 receptor with a binding constant of between 0·6 and 4×10^{-9} M. It antagonizes spiroperidol binding to the D_2 receptor, on the other hand, with a much weaker binding (K_d is $8 \cdot 8 \times 10^{-7}$ M (Kebabian *et al.*, 1986), so there is a considerable degree of selectivity. SCH 23390 also blocks dopamine stimulated adenylate cyclase activity with a K_i of 1×10^{-8} M (Iorio *et al.*, 1983).

Dopamine receptors in the periphery, rather confusingly, are classified as DA_1 and DA_2. Vascular dopamine receptors, DA_1, mediate vasodilatation in kidney and mesentery; DA_2 are situated on sympathetic nerve terminals where they inhibit the release of transmitter, i.e. decrease sympathetic tone.

There are considerable similarities between DA_1 and D_1 receptors, particularly with regard to antagonism by SCH 23390 which has similar inhibitory effects on both receptors. Similarly, DA_2 and D_2 receptors show resemblances (Kebabian *et al.*, 1986). It seems possible that this separate classification may soon be amended.

The behavioural correlates of dopamine action are believed to be exemplified by the effects of dosage with apomorphine, a dopamine agonist (dopamine does not pass the blood–brain barrier), namely a stereotypic type of behaviour in rats that involves persistent biting, gnawing and licking, and hypothermia in mice etc. These activities can all be antagonized by ligands at both types of receptor. A further action of dopamine is to inhibit the production of prolactin which appears to be under the control of D_2 receptors since it is only antagonized by the neuroleptic drugs and not by SCH 23390 (Iorio *et al.*, 1983).

9.3.1 Dopamine agonists

One of the most widely used dopamine agonists is bromocryptine, one of the ergot alkaloids derived from the growth of the fungus *Claviceps purpurea* on damp rye. The fungus forms a purple growth — the ergot — around the grain from which the ergot alkaloids are obtained. In the Middle Ages ingestion of infected rye gave rise to a number of severe reactions including gangrene, abortion and, after high doses, death. A variety of pharmacologically active agents have been isolated and identified from the ergot, including the related structures lysergic acid diethylamide and bromocryptine.

The latter binds to dopamine receptors in pituitary and brain with affinities in the region of 10^{-9} M as measured by competition with the binding of ^3H-spiroperidol (Creese *et al.*, 1977). Consistent with the belief that these receptors are D_2 is the fact that bromocryptine does not stimulate adenylate cyclase activity, although no inhibitory effect on the enzyme has been reported.

Dopamine and acetylcholine exert balancing inhibitory and excitatory control over voluntary movements. When the dopamine neurones in a certain portion of the brain known as the substantia nigra serving the basal ganglia fail to inhibit these movements, Parkinsonism is the consequence. The brain can suffer the loss of many of these dopamine neurones without any apparent changes in behaviour, but when the loss reaches about 80%, the symptoms of Parkinsonism appear, namely involuntary movements or tremors, tics, rigidity, a gradual loss of motor function and eventually, in the late stages of the disease, the inability to stand upright. Furthermore, the main psychic correlate is apathy (Bianchine, 1985). The substantia nigra governs the transmission of motor impulses that arise elsewhere in the brain, and contains a large number of pigmented neurones with several synaptic junctions with other neurones. Dopamine acts here as an inhibitory neurotransmitter.

The reason for this loss of neurones and the dopamine deficiency is not known but a toxin present as an impurity in some forms of heroin (1-methyl-4-phenyl-1,2,3,6-tetrahydropyridine: MPTP) can give rise to symptoms in animals indistinguishable from Parkinsonism in man. A similar destruction of dopaminergic neurones in the substantia nigra seems to occur (see section 9.6 where monoamine oxidase inhibitors are discussed). Some dopamine antagonists, specifically the drugs used to treat schizophrenia, can also give rise to symptoms of Parkinsonism by blocking the post-synaptic dopamine receptors in the basal ganglia. This condition is reversible, however, since it disappears when the drug is withdrawn (Borison *et al.*, 1973; Baldessarini, 1979).

The dopamine deficiency in Parkinsonism can be treated in a number of ways. Dopamine can be supplied to the remaining neurones or dopamine agonists can be used which bypass the damaged neurones and act on the post-synaptic receptors. Alternatively, the excitatory transmission governed by acetylcholine can be suppressed in order to achieve a more appropriate balance of inhibition and excitation (Pearce, 1984; Calne, 1984). The latter approach is discussed in the section on acetylcholine receptors in this chapter (9.7).

Dopamine cannot be supplied directly to the brain since systemically administered catecholamines do not cross the blood–brain barrier. The precursor of dopamine, L-dopa (L-*di*hydro*x*yphenyl*a*lanine), can however be administered as a pro-drug (see Figure 9.3 for the biosynthetic pathway). L-Dopa is decarboxylated by aromatic amino acid decarboxylase to yield dopamine. Pyridoxal phosphate is the cofactor. Unfortunately, much of the amino acid is decarboxylated in the liver and only about 5% is available to cross the blood–brain barrier. Accordingly, a decarboxylase inhibitor, which itself cannot cross into the brain is usually co-administered with L-dopa. The most useful is carbidopa, which acts by forming a Schiff's base between the terminal amino moiety of its phenylhydrazine group and the aldehyde of pyridoxal. Although daily dosage with L-dopa is often effective for a number of years it gradually loses its efficacy and must be supplemented in various ways.

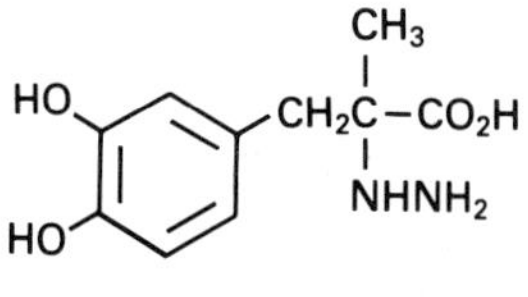

Carbidopa

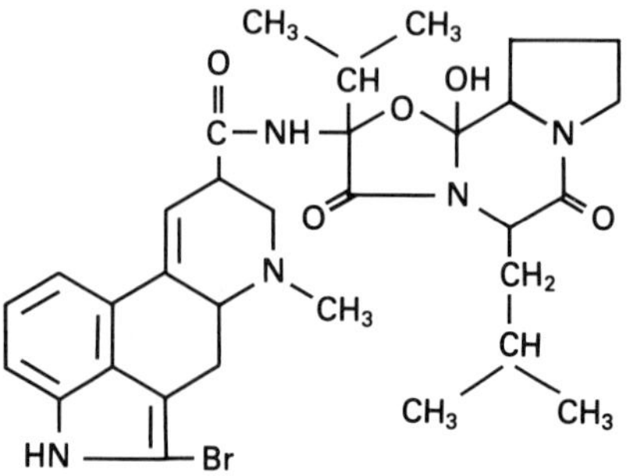

Bromocryptine

Bromocryptine can be used in conjunction with L-dopa in order that the dose of the latter may be reduced (see Lieberman and Goldstein, 1985, for a review on the use of bromocryptine in Parkinsonism). The use of an additional drug is also helpful in reducing the side-effects of L-dopa treatment, namely psychosis, hallucinations, nausea and respiratory distress.

Another condition where bromocryptine has found a medical role is in the treatment of infertility due to the hypersecretion of prolactin (Weinstein *et al.*, 1981). Prolactin secretion is suppressed by dopamine and bromocryptine can also inhibit this secretion at the level of the pituitary; this is consistent with the view noted above that bromocryptine acts at D_2 receptors. High levels of prolactin will inhibit the release of gonadotrophins in humans and, consequently, will suppress ovulation. This effect is clearly of great biological value when nursing mothers are feeding their children because it blocks the possibility of another pregnancy, but in cases of inappropriate prolactin release infertility is the result. Bromocryptine can also be used to treat the symptoms of excessive secretion of prolactin by a pituitary tumour. In general, therefore, bromocryptine is widely used in situations where an agent that acts like dopamine is required.

DA_1 agonists also have pharmacological uses as antihypertensive agents. Essential hypertension is characterized by constriction of the blood vessels of the kidney. There are receptors for dopamine in these blood vessels that cause dilatation. Fenoldopam, an analogue of SKF 38393 is a relatively selective agonist at these receptors producing hypotension, an effect that is antagonized by SCH 23390 (Sengupta and Lokhandwala, 1985). The drug is effective at reducing blood pressure in man (Harvey *et al.*, 1986), and is undergoing clinical trial. Structurally it is unusual in being a derivative of catechol, and this may be responsible for its short duration of action, through relatively rapid metabolism in the liver (Harvey *et al.*, 1986).

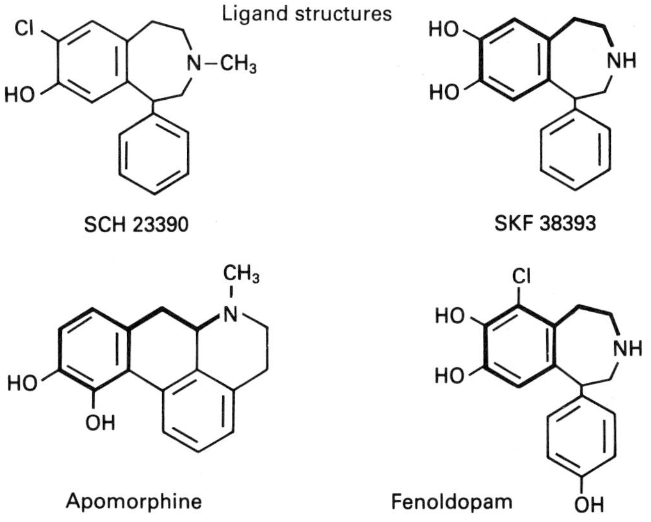

Ligand structures

SCH 23390

SKF 38393

Apomorphine

Fenoldopam

Some of these structures contain the framework of the dopamine structure, drawn in heavy type.

9.3.2 Dopamine antagonists

A large number of drugs that are used for the treatment of psychosis, notably schizophrenia, are antagonists of dopamine at the D_2 receptor. Schizophrenia is a complex disorder over the diagnosis of which there has been considerable confusion in the past, and so an outline is presented here. Recent attempts to categorize the diagnosis more accurately have apparently led to greater treatment success (Manschrek, 1981). The Diagnostic and Statistical Manual of the American Psychiatric Association (DSM-III) has laid down a number of criteria which must be satisfied before schizophrenia can be diagnosed. These include:

(a) A period of at least 6 months in which social withdrawal, impaired self-awareness, and markedly peculiar behaviour predominate. This is known as the 'prodromal phase'.
(b) An active phase with delusions and hallucinations possibly associated with incoherent and fragmented thinking together with flatness of emotional expression.
(c) A phase similar to the prodromal phase in which the symptoms subside (residual phase).
(d) At least 6 months continuous signs of the condition in addition to (a).
(e) The onset must occur under 45 years of age, and not be due to any organic disorder.

The course of the disease varies widely but the active phase may be interspersed with residual phases over a number of years. The active phase can be treated and the signs greatly ameliorated by antipsychotic drugs, and in this respect the treatment has been greatly improved. The residual phase, social withdrawal and emotional aspects of the condition are much more difficult to treat. The antipsychotic agents ameliorate the condition but do not effect a cure although remission can occur (Manschrek, 1981).

A large number of antipsychotic agents have been developed for the treatment of schizophrenia. The phenothiazine group of drugs, of which the first was chlorpromazine, interact with both types of dopamine receptor (and a number of other receptors as well). In fact, these drugs are extremely non-selective in their widespread biological actions. The more selective agents such as the butyrophenones (e.g. haloperidol) and diphenylbutylpiperidines (e.g. pimozide) interact with D_2 receptors more specifically, although not totally selectively, and the relative binding affinities of the ligands correlate with their anti-psychotic potency in man (Snyder, 1981). The term neuroleptic is often used to describe these drugs, mainly to distinguish them from the other types of psychoactive drugs that depress the central nervous system such as general anaesthetics, sedatives, etc.

Other characteristic behaviour of the neuroleptics stems from their binding to dopamine receptors in that they reduce the firing of dopaminergic neurones (Bunney *et al.*, 1973), reduce the rates of both biosynthesis of dopamine

Phenothiazines

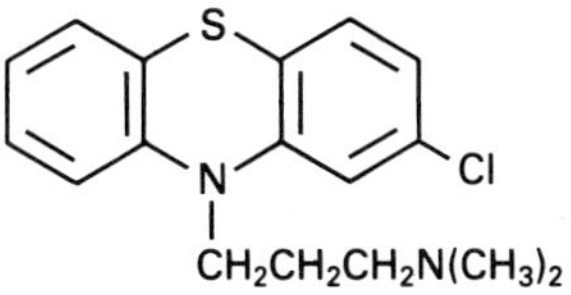

Chlorpromazine

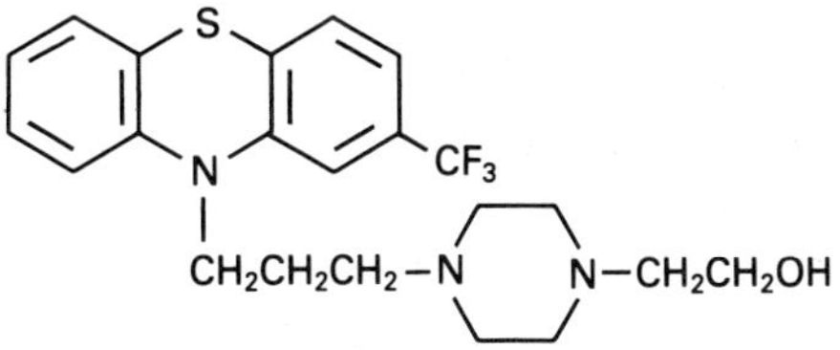

Fluphenazine

Diphenylbutylpiperidine

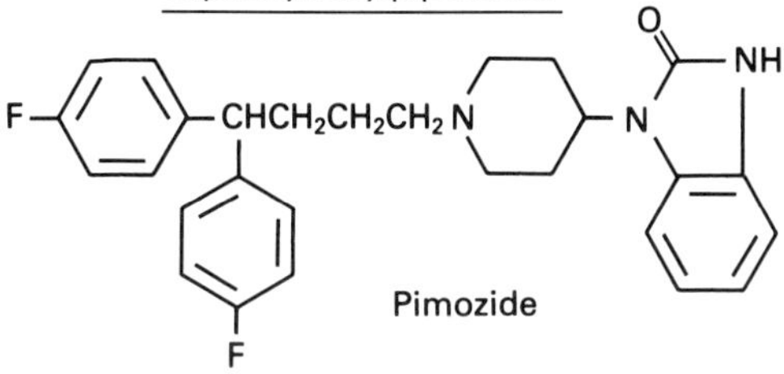

Pimozide

Butyrophenones

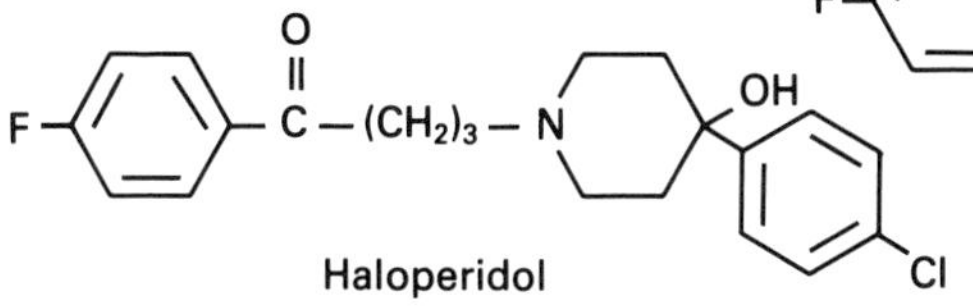

Haloperidol

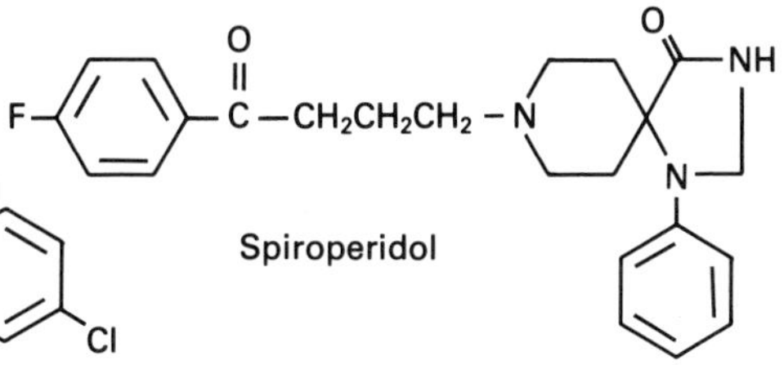

Spiroperidol

from tyrosine and of its metabolism to 3-methoxytyramine, homovanillic and dihydroxyphenylacetic acids (Figure 9.4) (Anden and Stock, 1973).

It is probably incorrect, however, to assume that dopaminergic systems are overactive in schizophrenia — on the other hand a reduction in dopaminergic activity is certainly beneficial to the patient. The effect is reversed on withdrawal of the drug and so they are at best a palliative for the condition. An associated feature of the action of these drugs is the slow onset of clinical effect in man compared with the rapid onset of their pharmacological action *in vitro*. This suggests that, perhaps like the tricyclic antidepressants (see Section 9.5), secondary, indirect changes produced by the drug administration may make a major contribution to their clinical efficacy. In addition, the use of these drugs increases the number of D_2 receptors in the brain as shown by measurement of dopamine binding to post-mortem material (Snyder, 1981) — an interesting result of the brain trying to nullify the effect of the drug.

Although in general the discussion of drug side-effects is not intended to be a major feature of this book, the neuroleptic drugs show at least three types of

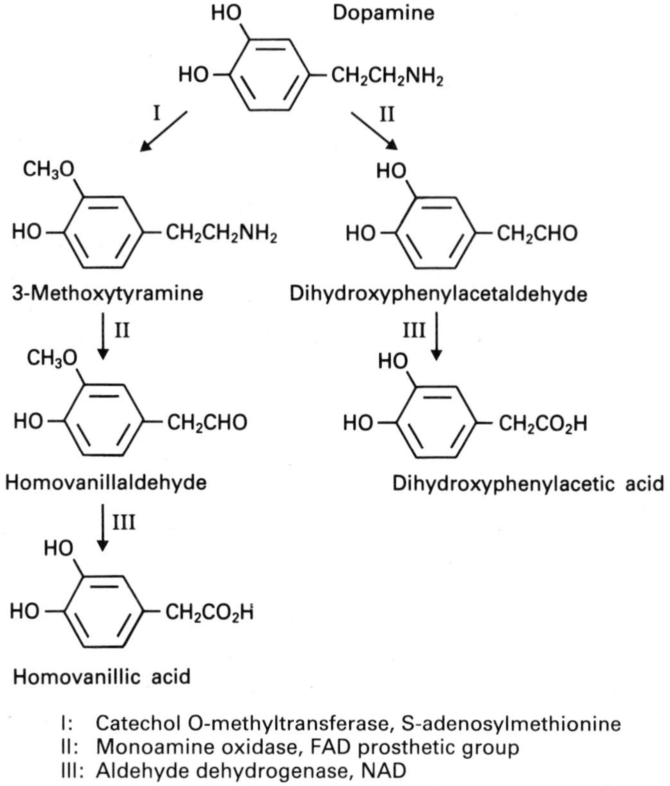
HO
Dopamine
HO
CH2CH2NH2
I
II
CH3O
HO
CH2CH2NH2
3-Methoxytyramine
HO
HO
CH2CHO
Dihydroxyphenylacetaldehyde
II
CH3O
HO
CH2CHO
Homovanillaldehyde
III
HO
HO
CH2CO2H
Dihydroxyphenylacetic acid
III
HO
HO
CH2CO2H
Homovanillic acid
I: Catechol O-methyltransferase, S-adenosylmethionine
II: Monoamine oxidase, FAD prosthetic group
III: Aldehyde dehydrogenase, NAD

Figure 9.4 Dopamine metabolism.

side-effect that are a consequence of dopamine receptor blockade in other parts of the brain. On the one hand, the therapeutic effects of the neuroleptic drugs result from their blockade of receptors in the projections to the limbic system that regulates emotion in the brain. Parkinsonism, however, can result as a consequence of post-synaptic blockade of the basal ganglia in the corpus striatum which are served by nerve fibres from the substantia nigra (see Snyder, 1986). Neuroleptic treatment for schizophrenia can, therefore, cause some similar effects. Tardive dyskinesia, which is characteristic of Parkinsonism, and prolactin release may also result as side-effects of neuroleptic treatment. The former is characterized by a variety of movement disorders such as:

(a) constant chewing movements with the tongue intermittently darting out of the mouth (fly-catcher tongue),
(b) body rocking while sitting,
(c) marching on the spot.

These symptoms take time to develop, hence the term tardive. They are induced by receptor supersensitivity caused by prolonged drug treatment, in other words the receptors that are not blocked by the drug are supersensitive to dopamine (Borison *et al.*, 1983).

Prolactin release from the pituitary is inhibited by dopamine acting at $\bar{D}_2$ receptors. Dopamine blockade has the effect, therefore, of increasing prolactin levels in the plasma leading to breast engorgement and a persistent or recurrent discharge of milk from the breast (galactorrhoea).

The hope that clarification of the mode of action of the neuroleptic drugs might lead to a greater understanding of the aetiology of schizophrenia has not yet been realized. Hyperactivity of the dopaminergic receptor is not a feature of the condition. Various attempts to demonstrate the formation *in vivo* of hallucinogenic by-products, e.g. by methylation of the neurotransmitter amines to form such compounds as bufotenin or mescaline, have failed. Nevertheless, we desperately need a much greater understanding of a disease which is very complex and difficult to treat.

9.4 Serotonin (5-hydroxytryptamine–5-HT) receptors

The effects of serotonin are widespread and vary between stimulation and inhibition on cardiovascular, gastrointestinal and respiratory systems. This variation is partly due to the reflex nature of serotonin effects which depend on a number of variables such as route and speed of injection, anaesthetic state etc. Repeated tests tend to give effects which get smaller and smaller (known as tachyphylaxis) (Douglas, 1985). It is probably easier to consider the various effects under the separate classes of serotonin receptors.

The direct correlation of radioligand binding studies with changes expressed on an organ or tissue basis has always been a fundamental goal in receptor pharmacology. Nowhere is this more essential than for serotonin or 5-HT receptors, and some of the criteria required for specific agonists and antagonists still cannot be satisfied at the present time. Nonetheless, a tripartite classification into 5-HT_1-like, 5-HT_2 and 5-HT_3 has been proposed (Bradley *et al.*, 1986).

The 5-HT_1-like receptors control the pre-synaptic release of serotonin from nerve terminals (see Figure 9.5 for serotonin biosynthesis), and contraction of smooth muscle both intra-cranial and some vascular (Peroutka, 1984; Bradley *et al.*, 1986). There is a degree of heterogeneity and no selective antagonist at this site. Methysergide is a potent antagonist in the micromolar range, but ketanserin, a strong antagonist of 5-HT_2 receptors is ineffective, while serotonin binds in the nanomolar range. Adenylate cyclase is linked to the 5-HT_1 receptor and is activated and inhibited by these levels of agonists and antagonists in synaptic membranes from rat brain. Guanine nucleotides inhibit the binding of serotonin but do not affect antagonist potencies.

5-HT_2 receptors mediate the contraction of gastrointestinal smooth muscle, as well as platelet aggregation and other inflammatory responses (Leysen *et al.*, 1984; Bradley *et al.*, 1986). Ketanserin is a powerful and specific antagonist of serotonin binding whilst methysergide also inhibits in the nanomolar range. This receptor is coupled to the hydrolysis of inositol phospholipid and protein kinase C activation in the rat cerebral cortex (Kendall and Nahorski,

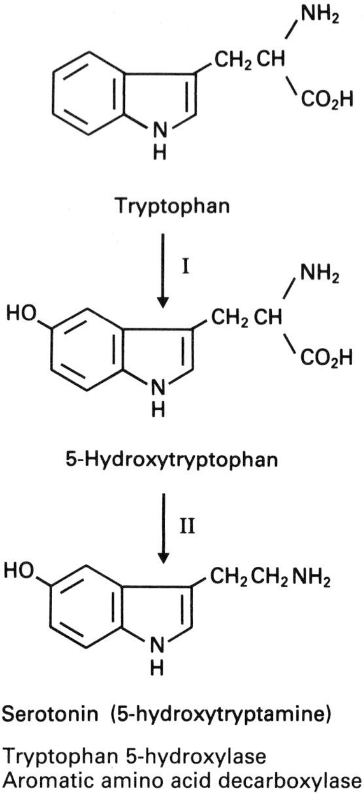
NH2
CH2 CH
CO2H
N
H
Tryptophan
I
HO
5-Hydroxytryptophan
II
CH2 CH2 NH2
Serotonin (5-hydroxytryptamine)
I: Tryptophan 5-hydroxylase
II: Aromatic amino acid decarboxylase

Figure 9.5 Serotonin biosynthesis

1985), and in the rat aorta where phosphoinositide turnover is modulated by phorbol ester (Roth *et al.*, 1986). The calcium influx mediated by these changes (see section 9.1) is responsible for the contraction of smooth muscle.

Methysergide exhibits a binding constant of between 1 and 16×10^{-9} M to various parts of rat brain, as measured by antagonism of ^{3}H-ketanserin binding (Leysin *et al.*, 1984). Methysergide is used prophylactically for the treatment of attacks of migraine, but it is of no value once an attack has already started. The connection of 5-HT$_2$ receptors with migraine is, however, tenuous and the mechanism of action of the drug is not completely understood (Speight and Avery, 1972). Clearly, one of the major aims of studying serotonin receptors is to develop a more effective drug for the treatment of migraine.

Mianserin, a clinically effective anti-depressant, is a potent antagonist at 5-HT$_2$ receptors. It also increases brain noradrenaline turnover, possibly as a result of antagonism of pre-synaptic α_2-adrenergic receptors (i.e. Charney *et al.*, 1981), and reduces cAMP production induced by noradrenaline in rat brain slices, which appears to be a feature of prolonged administration of most, if not all, anti-depressants (see Section 9.3). Ketanserin also is able to reduce this production of cAMP by noradrenaline in suspensions of synaptic membranes after three weeks' dosing of rats at 10 mg/kg. There is a possibility, however, that ketanserin and mianserin binding sites differ, but both are coupled to the post-synaptic 5-HT$_2$ receptor complex (Gandolfi *et al.*, 1985).

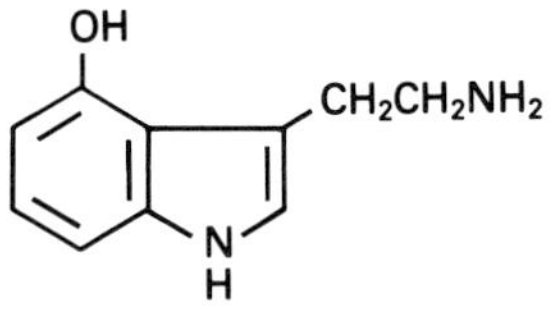

Serotonin

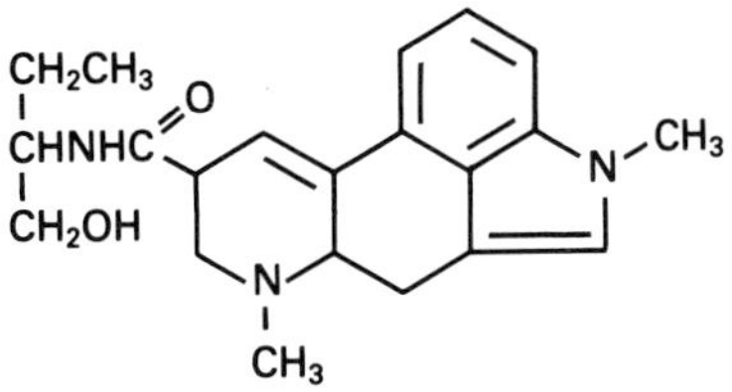

Methysergide

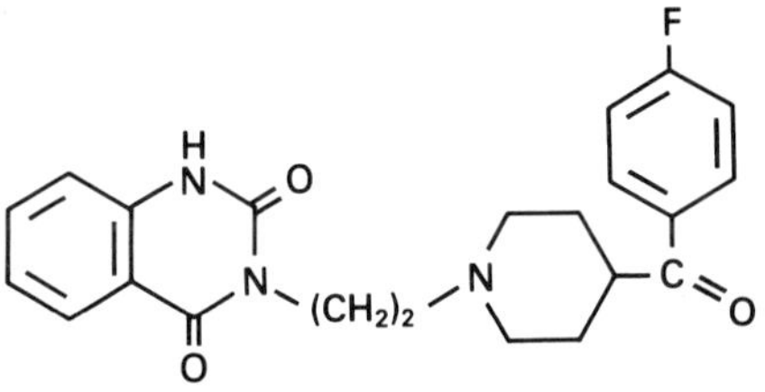

Ketanserin

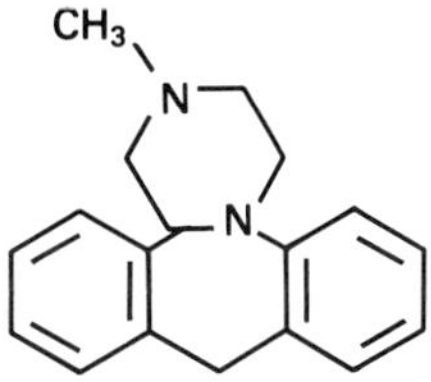

Mianserin

In contrast to mianserin, ketanserin has been found of value in the treatment of mild to moderately severe essential hypertension, although this effect may be a result of α_1-adrenergic receptor antagonism. It is too early to tell if the use of the drug will become widespread in this condition (Andren *et al.*, 1983). One might expect that if depression results from a reduced activity of serotonin nerves (Goodwin and Post, 1983), the development of a serotonin receptor agonist would be a good approach, but this does not seem to have been addressed. Indeed, the effect of mianserin as a 5-HT$_2$ receptor antagonist would seem to argue the opposite.

The third type of receptor is a more recent classification forced by the discovery of antagonists to serotonin that had no effect on the types of receptor noted above. Serotonin mediates the release of noradrenaline from sympathetic nerves innervating the rabbit heart and cocaine antagonizes this action, albeit weakly; indoleamine-like compounds with a tropane side-chain were therefore synthesized (for example ICS 205-903) which have I_{50} in the region of 10^{-8} to 10^{-11} M. 5-HT$_3$ receptors appear to modulate the depolarization of neurones in the periphery (Bradley *et al.*, 1986). This receptor subtype may also be heterogeneous (Richardson *et al.*, 1985).

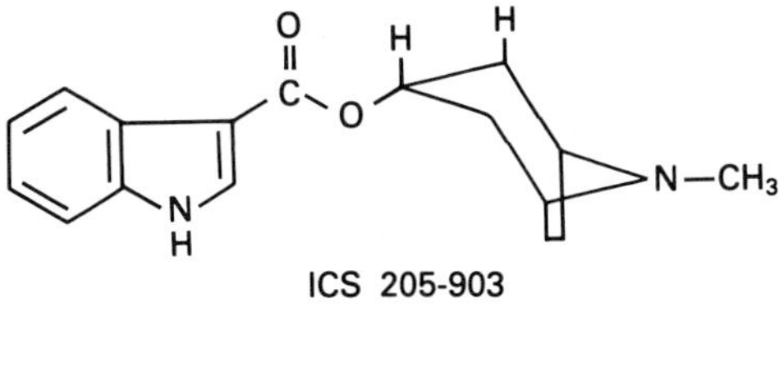

ICS 205-903

9.5 *Serotonin and noradrenaline re-uptake mechanisms and depression*

As noted in the introduction, there are uptake systems that efficiently remove the amine neurotransmitters from the synaptic cleft in order to terminate the signal. These mechanisms are believed to be related in the case of serotonin to re-uptake mechanisms in the blood platelet, and the latter are often used as models of the neuronal system even though platelets and synapses differ markedly in other ways.

Depression is a condition that can be treated by different types of drug, notably the monoamine oxidase inhibitors and the tricyclics in addition to electroconvulsive therapy. The aetiology of the condition, in spite of intensive research over the last 20 years, still defies a full understanding. A reduction in activity in either serotonin or noradrenaline transmission has been variously canvassed, and the question is still open at the present time.

The inhibition of monoamine oxidase A, the isoenzyme that primarily catalyses the oxidation of noradrenaline and serotonin, in principle raises the brain levels of both monoamines. It does not, however, answer the question as to which, if either, monoamine is functionally inadequate in depression. Indeed, there is a further question which needs to be answered; although depression can be treated by drugs that promote monoaminergic transmission, is this a necessary and sufficient explanation of their antidepressive activity, particularly in view of the long time interval between potentiation of monoamine activity and the onset of clinical relief of symptoms? This interval may be of the order of days or even weeks, and has led to a shift in research emphasis from acute effects to a study of slower adaptive changes induced by chronic antidepressive therapy.

The discussion was initiated in 1967 by Coppen who suggested that noradrenaline levels were associated with drive and energy, and serotonin with mood. This view may owe a lot to the age-old connection between the catecholamines and 'fight or flight' that is a classic feature of school biology textbooks. There are difficulties here, in that MAO inhibitors' effect in animal tests that measure activity (tetrabenazine-induced sedation, for example) have been shown to correlate with raised levels of serotonin rather than with noradrenaline (Christmas *et al.*, 1972). Nevertheless, there are always problems in extrapolating from the results of animal tests to effects in man in any aspect of drug development, and particularly so in the area of depression because of the difficulty in designing tests that measure mood.

The debate has been thrown into sharp relief by studies on a group of drugs known as the tricyclic antidepressants, of which the best known is probably imipramine. These compounds bear a structural resemblance to the phenothiazines since they have three aromatic rings fused in a linear fashion as well as a side-chain containing an amine group, but they do not affect dopamine receptors. They do, however, interfere with the re-uptake of serotonin and noradrenaline into the presynaptic nerve terminal, and in some cases have a variety of effects on receptors for other neurotransmitters. One view is that tertiary amines tend to favour the inhibition of serotonin uptake while secondary amines affect noradrenaline re-uptake (Carlsson, 1984). This difference appears to correlate with the clinical picture, at least in the case of zimelidine where the tertiary amine is a more effective agent (Ogren *et al.*, 1981).

On the other hand, there are other agents which are effective antidepressants and inhibit the re-uptake of noradrenaline selectively. One such compound is maprotiline which has a secondary amine side-chain attached to a tetracyclic structure, not greatly different from the classical tricyclics (Pinder *et al.*, 1977).

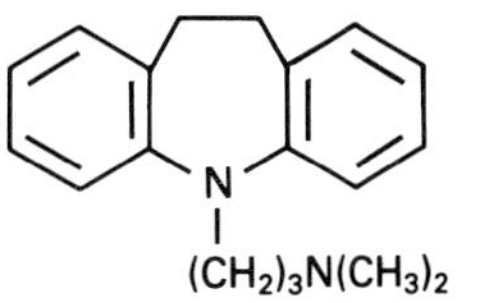

Imipramine

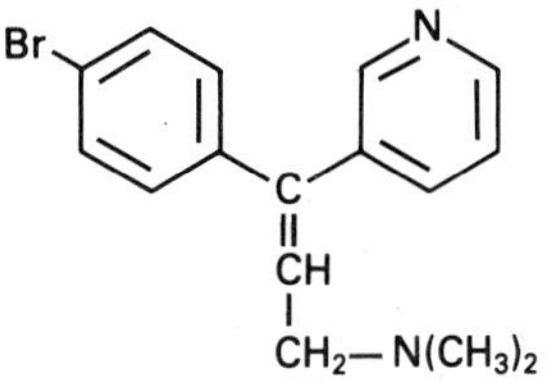

Zimelidine

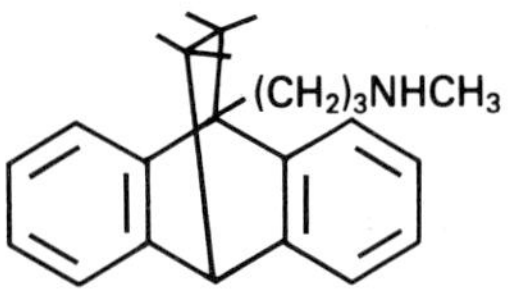

Maprotiline

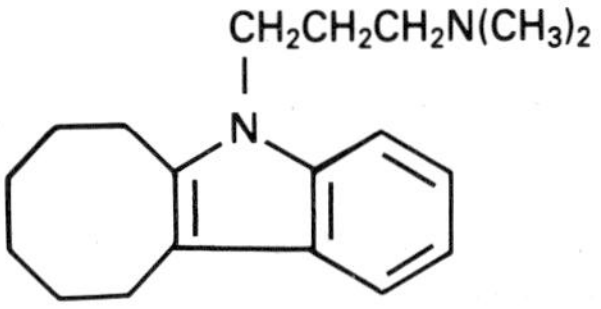

Iprindole

Another approach has been to investigate the brain levels of serotonin, and its metabolite 5-hydroxyindoleacetic acid (to indicate turnover), in the brains of patients who suffered from depression, suicide etc. (reviewed in Goodwin and Post, 1983). Clear indications have been found of a lowered level of serotonin *activity* at least, and in some cases a lowered *level* of the amine. In addition, the level of imipramine binding sites falls in cases of depression; these sites are located on the serotonin neurone near the serotonin receptor and may be allosteric effectors of the uptake mechanism.

A number of attempts have been made to rationalize these viewpoints (Costa *et al.*, 1983; Charney *et al.*, 1981). In outline, prolonged antidepressive therapy increases the sensitivity of post-synaptic α_1-adrenergic (see Section 9.2) and serotonin receptors and decreases the sensitivity of pre-synaptic receptors (the post-synaptic receptor participate in the neurotransmission of the nerve impulse, whereas the pre-synaptic receptors regulate neurotransmitter release and re-uptake).

The net effect of these changes is enhanced neurotransmission, and a correction of a subnormally functioning receptor/transmitter relationship.

Whether this involves serotonin or noradrenaline obviously depends on the drug involved, but it is possible that there are interactions between the two systems which may lead to a similar result after long-term administration. Indeed, Costa *et al.* (1983) suggest that there is a direct controlling role for a serotonin receptor over the noradrenaline neurone in order to maintain the firing of the latter. Alternatively, there may be at least two major subtypes of depressive illness that result from inadequate transmission at either aminergic nerve.

In sharp contrast to this view, it has been found that almost all antidepressants, including those like iprindole (which has no effect on the re-uptake system for either monoamine) do share a common action in desensitizing the adenylate cyclase system to noradrenaline activation. This effect requires chronic administration (Charney *et al.*, 1981; Sugrue, 1981). Even electroconvulsive therapy and chronic administration of monoamine oxidase inhibitors produce this effect.

As to the nature of the adrenergic receptor involved, it seems likely that it is not only the β-receptor, since long-term treatment with mianserin will ultimately reduce the stimulation of adenylate cyclase by noradrenaline but not by isoprenaline (Charney *et al.*, 1981). This is another characteristic of the drug besides its ability to block $5\text{-}HT_2$ receptors. Other evidence suggests that the α-receptor is not involved, particularly as it is usually negatively coupled to adenylate cyclase. Alternatively, it could be argued that this chronic desensitization (down-regulation) is independent of any acute blocking effects on a particular adrenergic receptor, and occurs by another means.

The importance of this effect to the aetiology of depression is not at all clear at present. It may be fortuitous or it may indicate that a hyperactive noradrenaline neuronal system is a feature of depression — a view diametrically opposed to the earlier hypothesis that the condition derived from a deficiency in monoamine level or function (Sugrue, 1981). To discover which of these hypotheses is right is clearly of paramount importance in understanding the aetiology of the condition, and in the development of new and more effective drugs.

9.6 Monoamine oxidase

The importance of monoamine oxidase (MAO) lies in its function of destroying excess neurotransmitter after action. Clearly, any interference with MAO activity will prolong the action of the transmitter, and such interference has been found of therapeutic value in the treatment of depression.

The initial discovery came after two drugs, isoniazid and iproniazid, were put on the market in 1951 for the treatment of tuberculosis (iproniazid is the isopropyl analogue of isoniazid; the latter is still used for the therapy of tuberculosis). Eventually, it was found that the tubercular patients became euphoric (i.e. highly elated) on iproniazid. Depressed patients subsequently were also found to respond. Iproniazid was eventually withdrawn from the market in the United States because of liver toxicity, although it is still available in Britain, and a number of other compounds were developed in its wake. One characteristic they all have in common is the irreversible inhibition of MAO.

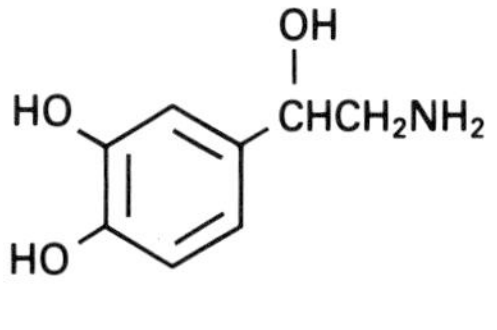

Noradrenaline

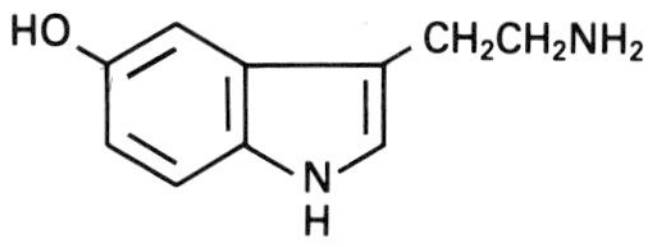

Serotonin

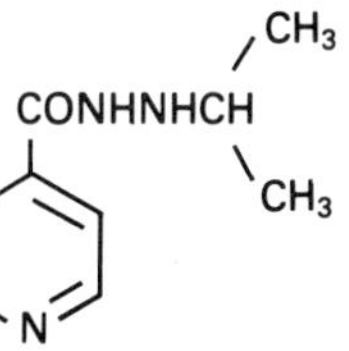

Iproniazid 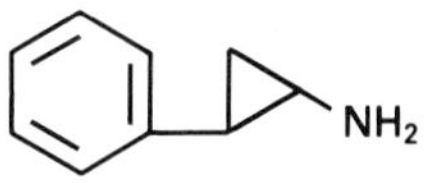   Tranylcypromine

Monoamine oxidase is a flavoprotein, present in the outer mitochondrial membrane, which catalyses the oxidation of the primary amine group of neurotransmitters (for example, noradrenaline, dopamine and serotonin) to an aldehyde; hydrogen peroxide is formed in the process (see Figures 9.6 and 9.7). Serotonin is thereby oxidized to 5-hydroxyindoleacetaldehyde and noradrenaline to 3,4-dihydroxymandelaldehyde. Subsequent conversions yield either acid or alcohol, and in the case of the catecholamines, methylation of one of the catechol hydroxyl groups takes place primarily in the liver by catechol O-methyltransferase (COMT). COMT will also catalyse the methylation of the original monoamines (the methyl group is obtained from S-adenosylmethionine).

A detailed study of monoamine oxidase was carried out using a number of acetylenic inhibitors. Two isoenzymes were identified (called A and B by Johnston, 1968). MAO-A was primarily responsible for the oxidation of serotonin and noradrenaline, and was inhibited specifically by clorgyline, whereas MAO-B was responsible for the oxidation of phenylethylamine (benzylamine, although not occurring physiologically, is often used as the substrate *in vitro*). Deprenyl was subsequently found to be specific for MAOB (Knoll and Magyar, 1972). Other naturally occurring monoamines, such as dopamine, tryptamine and tyramine are substrates for both isoenzymes. Most tissues contain both isoenzymes but there are cases where only one isoenzyme

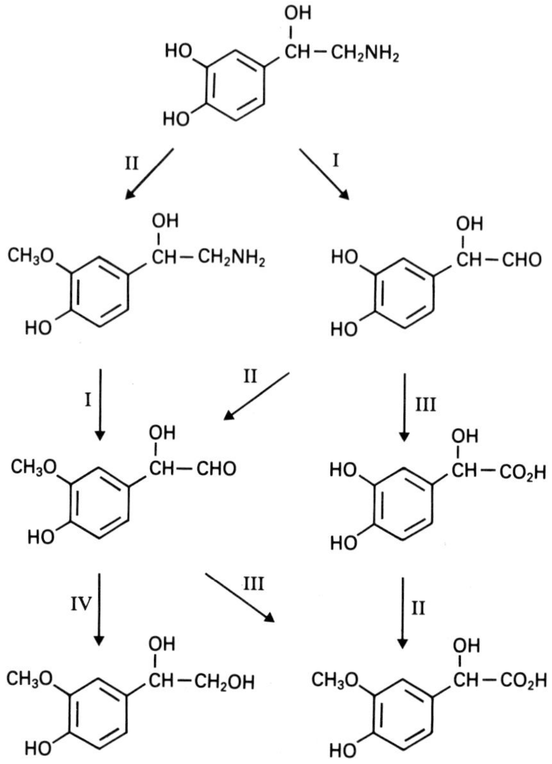

I: Monoamine oxidase
II: Catechol O-methyltransferase
III: Aldehyde dehydrogenase
IV: Aldehyde reductase
VMA: Vanillylmandelic acid
MOPEG: 3-Methoxy-4-hydroxyphenylethyleneglycol

Figure 9.6 Noradrenaline metabolism.

is present that is able to catalyse, slowly, the oxidation of substrates of the other isoenzyme.

The mechanism underlying this specificity has been addressed by Fowler *et al.* (1982) who showed that, since the structures formed with the selective acetylenic inhibitors are similar, the selectivity must derive from a kinetic difference. The inhibitors are K_{cat} or suicide inhibitors (see Chapter 1) and the reaction can be represented as follows:

$$E + I \underset{k_{-1}}{\overset{k_{+1}}{=}} EI \overset{k_2}{\to} EI^* \qquad (K_i = k_{+1}/k_{-1})$$

where E and I represent the free enzyme and inhibitor respectively, EI represents the non-covalently bound inhibitor/enzyme complex and EI* is the

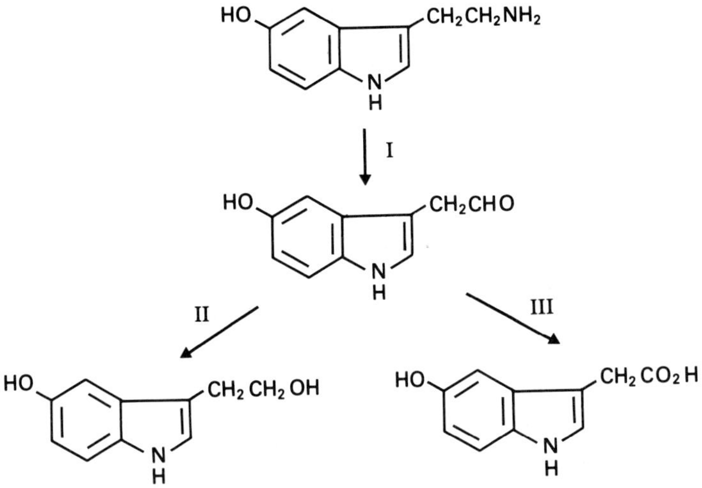

5-HTP

5-HIAA

I: Monoamine oxidase
II: Aldehyde reductase
III: Aldehyde dehydrogenase
5-HTP: 5-Hydroxytryptophol
5-HIAA: 5-Hydroxyindoleacetic acid

Figure 9.7 Serotonin metabolism.

covalent complex. In the case of clorgyline the K_i for MAO-A form is $5{\cdot}4 \times 10^{-8}$ M and $5{\cdot}8 \times 10^{-5}$ M for MAO-B. This difference in the rate of formation of the reversible complex is sufficiently large to account for almost all the specificity although k_2 for the inactivation of MAO-A is greater by one order of magnitude than k_2 for MAO-B. Deprenyl, on the other hand, owes much of its selectivity for MAO-B to a much higher k_2 for that form, i.e. the rate of formation of the irreversible adduct is much faster.

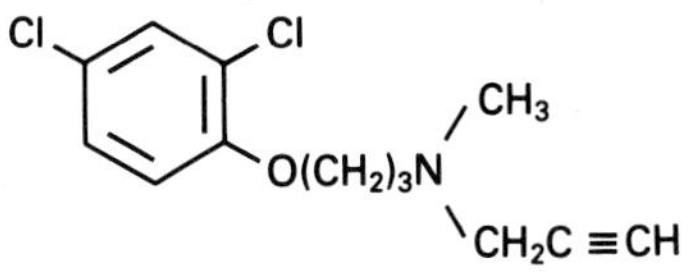

Clorgyline

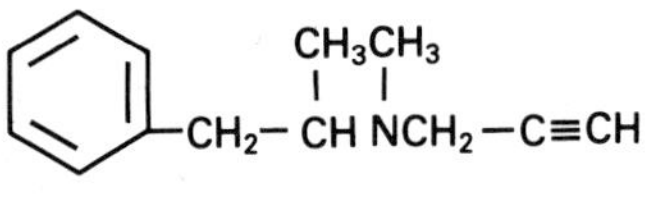

Deprenyl

A detailed study of the interaction of acetylenic irreversible inhibitors with MAO has shown that the flavin portion is the target for the inhibitor. The inhibitor initially may be oxidized to an allenic grouping that subsequently reacts irreversibly with the N^5 position of the isoalloxazine ring (Maycock *et al.*, 1976) to give a structure of the type:

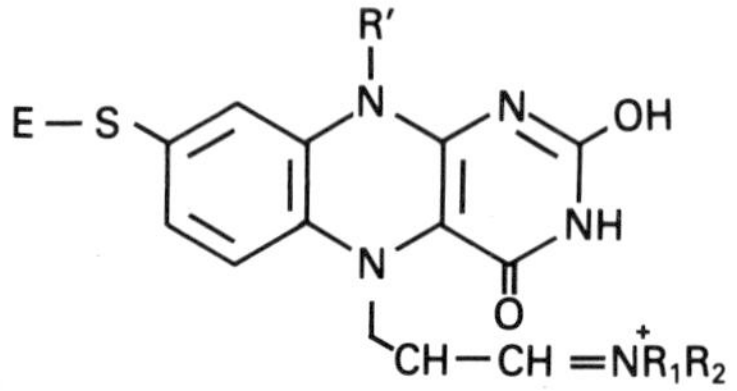

R₁, R₂ represent alkyl groups
R' the ribityl side-chain and
E the enzyme

(the flavin is linked covalently to the enzyme through a sulphur atom). In contrast, the cyclopropylamine type of inhibitor, tranylcypromine for example, although also a suicide substrate, yields a product that can be attacked by a nucleophilic sulphhydryl group and no binding at all occurs to the flavin (Silverman, 1983):

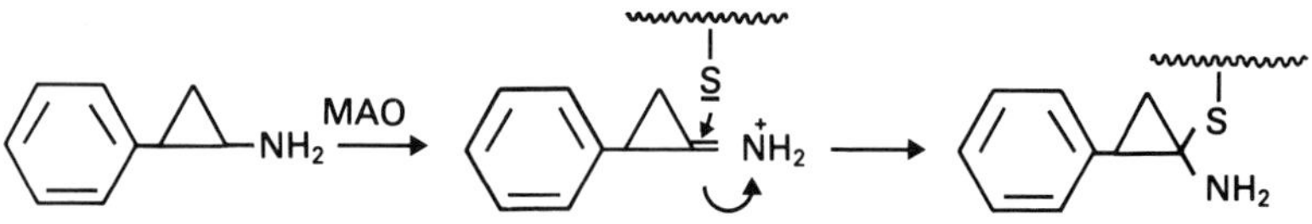

This adduct can be broken down more easily at neutral pH after the protein has been denatured.

The precise difference between the isoenzymes A and B remains to be elucidated. It used to be thought that the forms might differ by virtue of different lipid environments surrounding the same protein. Indeed, the nature of the active site peptide that is attached to the complex between pargyline (another irreversible monoamine oxidase inhibitor) and the flavin, does not differ between the two (Yu, 1981). Recently, a new procedure for the separation of the isoenzymes from human brain, has revealed that they are distinct proteins and not the same protein with a different lipid content (Pearce and Roth, 1984). This is in agreement with the genetic evidence which implies that the A and B isoenzymes are coded for by different gene loci (Cawthon and Breakfield, 1979).

Monoamine oxidase is also found in many other tissues, and one of its principal tasks in the gut is to detoxify monoamines absorbed in the diet as, for example, tyramine from cheese, pickled herrings, chianti etc., or dopamine in broad beans. These amines will otherwise reach the circulation and are likely to produce a hypertensive crisis by releasing noradrenaline from nerve endings (see Section 9.2). Clearly, any drug that irreversibly inhibits MAO non-specifically, and is given on a daily basis, will result in a greatly reduced level of the enzyme, since it takes almost three weeks to regain its original level by re-synthesis after one single dose. This was found to be the case with the earlier non-specific irreversible inhibitors such as tranylcypromine, and use of these agents was eventually restricted to a hospital environment where close supervision of the patient could be maintained. Even the use of selective but irreversible inhibitors could cause hypertensive crises in patients brought on by eating cheese.

More recently, a new class of reversible competitive inhibitor has been described that shows specificity for MAO-A, but does not apparently potentiate tyramine's ability to raise blood pressure (reviewed in Benedetti and Dostert, 1985). Toloxatone is a competitive inhibitor with an I_{50} for MAO-A of 1.4×10^{-6} M (5-HT as substrate) and 2.0×10^{-4} M for MAO-B (β-phenethylamine as substrate — Kan *et al.*, 1978). This inhibition was sufficient to cause a modest rise of 37% in 5-HT levels after a rather high oral toloxatone dose of 100 mg/kg. Noradrenaline and dopamine levels were also raised but to a lesser extent (Keane *et al.*, 1979). Indeed, these effects may be too small to explain antidepressive action via monoamine oxidase inhibition. Another related structure, cimoxatone, is a competitive inhibitor (Kan and Benedetti, 1981).

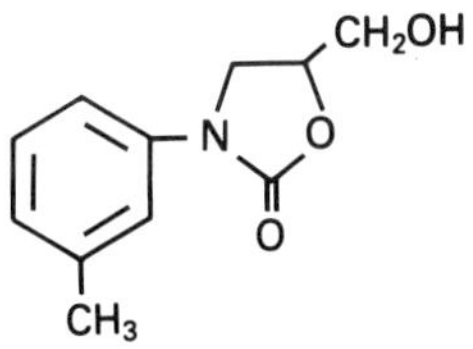

Toloxatone

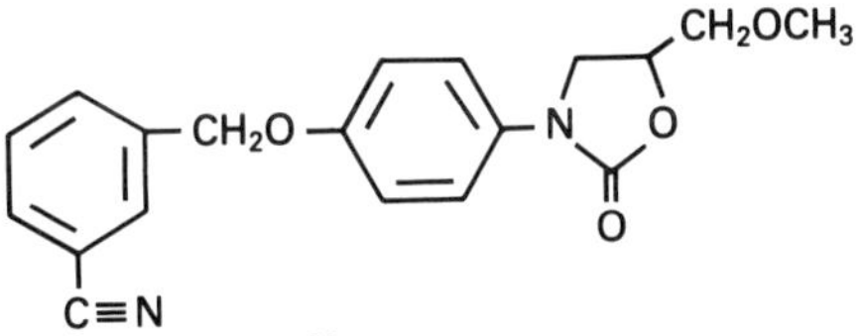

Cimoxatone

9.6.1 Monoamine oxidase and Parkinsonism

An interesting recent development involving monoamine oxidase is the discovery that it may be involved in the generation of some types of Parkinsonism. Various heroin addicts injected synthetic heroin contaminated with 1-methyl-4-phenyltetrahydropyridine (MPTP). They subsequently developed irreversible symptoms characteristic of Parkinsonism (slowness of movement, tremor and rigidity) normally confined to those over 55 years of age. Parkinsonism develops as a consequence of the destruction of the dopamine neurones in the nigro-striatal region of the brain and at least 80% of these must be inactivated before the disease manifests — hence the connection with old age. There is now considerable concern that other chemicals in the environment may also be responsible for the development of Parkinsonism.

MPTP itself does not produce the effect. The compound is oxidized by MAO-B, surprisingly fast for a tertiary amine, via a two electron oxidation to the dihydropyridine (MPDP) which subsequently disproportionates into MPTP and the aromatic *N*-methyl-4-phenylpyridinium ion (MPP$^+$ — see Figure 9.8). This MPP$^+$ is taken up by the catecholamine uptake system into the nigro-striatal region and probably reacts with the surrounding tissue. This transformation also takes place in other parts of the brain and it is not clear

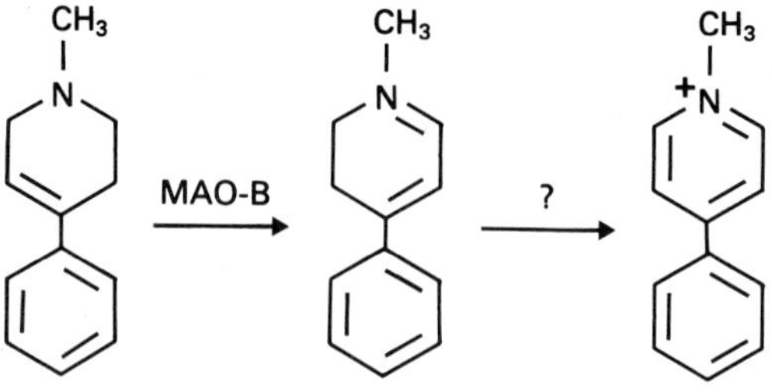

MPTP: 1-Methyl-4-phenyl-1,2,3,6-tetrahydropyridine
MPDP: 1-Methyl-4-phenyl-2,3-dihydropyridine
MPP^+: 1-Methyl-4-phenylpyridinium ion

MPTP is oxidized to MPP^+ in a two step process with the dihydropyridine as an intermediate. MAO-B carries out the first step while the second may occur without the intervention of an enzyme. Toxic products may arise from MPDP or MPP^+, or either of these two may be responsible for the damage to the nigro-striatal region of the brain.

Figure 9.8 The oxidation of MPTP by MAO-B.

why the nigro-striatal area should either be particularly sensitive to the toxin, or should concentrate it so effectively.

MPTP also binds to receptor-like high-affinity sites, which appear to be monoamine oxidase, in a number of brain regions (reviewed in Langston, 1985), presumably because it is both substrate and irreversible inhibitor of the enzyme (Singer *et al.*, 1985). It appears that a small proportion of MPTP will react covalently with the enzyme while the majority is converted to MPDP. MAO-B is associated with the uptake mechanism for dopamine (Melamed *et al.*, 1985).

9.7 Acetylcholine action

Acetylcholine is probably the most widely found neurotransmitter in the human body as it serves large areas of the central and peripheral nervous systems, and mediates a variety of actions both inhibitory and excitatory. In the autonomic nervous system, i.e. the system that governs involuntary functions such as heart rate, digestion etc., acetylcholine is the predominant neurotransmitter. Acetylcholine is, therefore, responsible for carrying messages to a variety of organs including heart, blood vessels, glands and smooth muscles.

$$CH_3CO_2CH_2CH_2\overset{+}{N}(CH_3)_3$$

Acetylcholine

Two major sub-divisions of acetylcholine receptors have been defined on the basis of agonist and antagonist action, namely nicotinic and muscarinic. The former are characterized by the action of nicotine as an agonist, and of tubocurarine as an antagonist, and are located in autonomic ganglia (i.e. where two neurones or sets of neurones interact), skeletal muscle and some of the synapses in the central nervous system. The nicotinic receptors on ganglia and skeletal neuromuscular junctions differ, however, in some respects; tubocurarine blocks both but its action is much more marked on the latter, and there are other more selective antagonists.

Muscarinic receptors, on the other hand, are found in smooth muscle, cardiac muscle and glands, and are the predominant form of the acetylcholine receptor in the central nervous system. Muscarine and pilocarpine are agonists; atropine and hyoscine antagonists.

9.7.1 Muscarinic receptor

There appear to be various different types of muscarinic receptor. Among the agonists, pilocarpine is able to distinguish between neuronal (i.e. brain and ganglionic) receptors (classified as M_1) and those in heart and guinea pig ileum (M_2), (ED_{50} for ganglion receptors is 4.3×10^{-7} M and for guinea pig ileum receptors is 1.5×10^{-5} M — Caulfield and Stubley, 1982). Birdsall *et al.* (1978) studied the competition between agonists and [^{3}H]propylbenzilylcholine for the muscarinic receptors in rat brain synaptosomes, and explained their results in terms of the existence of two sets of binding sites with different affinities for the agonists. Furthermore, the binding of carbamylcholine, for example, takes place over 5 orders of magnitude of concentration to the receptors in part of rat brain and it has been suggested that the agonist distinguishes between three sets of receptors, although the binding must obey the Law of Mass Action at each individual receptor (see Birdsall and Hulme, 1983).

With respect to antagonists, a sharp differentiation is shown by pirenzepin which favours binding to M_1 receptors from sympathetic ganglia by 50 times compared with M_2 receptors in the stomach ($K_d = 2 \times 10^{-8}$ M *vs* 10^{-6} M). The binding was measured indirectly by competition with the antagonist ^{3}H-N-methylhyoscine. In contrast, gallamine, which is used as a neuromuscular blocking agent by virtue of its action at nicotinic receptors on the skeletal neuromuscular junction, has a greater affinity for heart muscarinic receptors than for those in other sites. Apparently, the action of the drug is non-competitive, and gallamine binds to an allosteric site on the receptor protein, thereby modulating the binding of agonists and antagonists to the main site — possibly in a fashion similar to the benzodiazepins at the 4-aminobutyric acid site (see Section 9.7) (reviewed in Birdsall and Hulme, 1983).

A number of agents can influence the binding of ligands to the muscarinic receptor. Thus GTP enhances the binding of antagonists but lowers the binding affinity for agonists, probably by converting the various receptors into the one form with lowest affinity. This tends to support the view that the receptor subtypes are different by virtue of the different environment of the receptor protein rather than by intrinsic differences of the protein themselves. In sharp contrast, GTP converts sites of low affinity for antagonists into ones of higher affinity, but only under conditions when inorganic ions are absent (Ehlert *et al.*, 1981). The relevance of this finding to the normal physiological state is uncertain since we might expect inorganic ions to be present in quantity at all times. The relationship between receptor occupancy and pharmacological response has been explored to some extent by Hulme *et al.* (1978), who were able to show a good correlation between binding affinity to rat brain synaptosomes (measured by competition with [³H]propylbenzilyl-choline) and antagonism of smooth muscle contraction for 20 agents.

The modulation by GTP of muscarinic receptor binding immediately suggests the involvement of GTP-binding proteins. In this context, carbamyl-choline has been shown to inhibit GTP-stimulated adenylate cyclase activity in the heart (see Sokolovsky *et al.*, 1983). It is the M_2 type of receptor that is involved in this case.

On the other hand, cGMP levels are frequently raised by muscarinic antagonists, probably through activation of the M_1 receptor. The connection of response and stimulus is indirect since no acetylcholine-stimulation of guanylate cyclase has been found. This enzyme is found largely in the soluble phase, not bound to the membrane where it could couple directly to the receptor. There appears to be a definite link, however, between receptor occupancy, the opening of calcium channels and activation of guanylate cyclase (see Richelson and El-Fakahary, 1981) since:

(a) calcium channel blockers can antagonize receptor-mediated cGMP synthesis

(b) calcium ions (and calcium ionophores) can stimulate cGMP synthesis in various systems

(c) muscarinic receptor stimulation increases calcium influx.

Acetylcholine receptor occupancy by agonists stimulates the production of phosphatidic acid in rat cerebral cortex (as a consequence of phospholipase C hydrolysis of phosphatidylinositol phosphates to yield diacylglycerol). Pirenzepin antagonizes this action with a binding constant of about 2.5×10^{-8} M very close to the figure for binding to M_1 receptors (Smith and Yamamura, 1985). The efficiency of coupling of receptors to inositol phosphate hydrolysis, however, may vary across different parts of the brain (Fisher and Bartus, 1985).

The muscarinic receptor from brain has been isolated and shown to be a single polypeptide chain of molecular weight 86 kDa (Sokolowsky *et al.*, 1983) or 80 kDa (Birdsall *et al.*, 1979). In the latter case, a receptor protein similar to that found in brain was also detected with extracts of smooth muscle. The heterogeneity of the receptor may be a consequence of differing proteins, as shown by the recent cloning of the pig brain and heart receptors (Fukuda *et al.*, 1987).

The receptor used to be regarded as exclusively post-synaptic but recently it has become clear that there are pre-synaptic receptors as well (particularly in the central nervous system), sometimes referred to as auto-receptors. The pre-synaptic receptors appear to govern acetylcholine release by a feedback mechanism, as agonists inhibit and antagonists potentiate the release of acetylcholine (Sokolowsky *et al.*, 1983). In the guinea pig enteric nervous system the receptors are of the M_2 type. In this instance the release of noradrenaline is also inhibited.

9.7.1.1 Muscarinic antagonists

Pirenzepin has found a role in the treatment of gastric and duodenal ulcers, and appeared to be as effective as cimetidine in one large multi-centre trial (Giorgio-Conciato *et al.*, 1982). Some side-effects were noted, characteristic of antimuscarinic drugs and due to blockade of the parasympathetic nervous system, notably dry mouth and blurred vision, but they were of low incidence and severity. The lack of central nervous system effects suggests that although the drug has a great affinity for neuronal receptors, it does not cross the blood–brain barrier because it is too hydrophilic. The implication is that the selectivity of pirenzepin arises from a high affinity for the muscarinic receptors in the parietal cells of the gastric mucosa, albeit not measured, which are responsible for acid secretion (Hammer *et al.*, 1980) together with its inability to cross the blood–brain barrier.

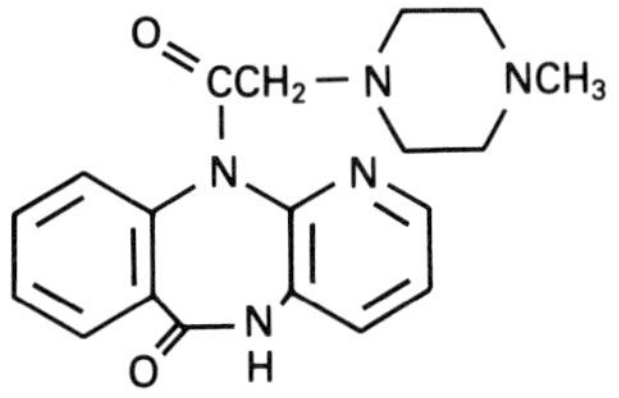

Pirenzepin

Atropine is a powerful muscarinic antagonist that does not distinguish between the receptor subtypes. Binding to guinea pig brain homogenates gives a binding constant of 2 to 3×10^{-10} M, as measured by antagonism of the binding of another antagonist, ^{3}H-quinuclidinyl benzilate (Fisher and Bartus, 1985). Atropine is useful to counteract the effects that stem from the stimulation of the vagus nerve during anaesthesia, and so is usually part of operation pre-medication.

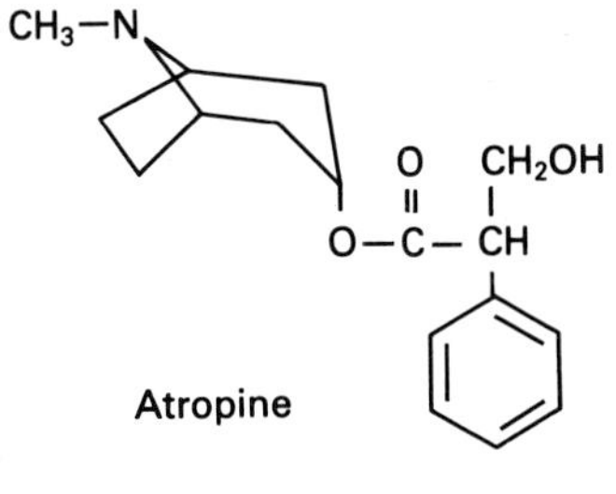

Atropine

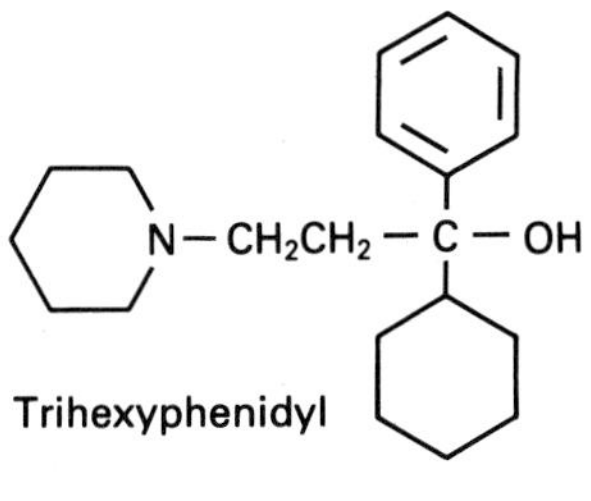

Trihexyphenidyl

Muscarinic antagonists, as for example trihexyphenidyl, are useful in the treatment of mild to moderate Parkinsonism. Although this disease results from a deficit of dopamine neurones which act in an inhibitory fashion, acetylcholine neurones which are excitatory activate the same area of the brain (the nigro-striatal). Accordingly, some degree of amelioration may be achieved by damping down the acetylcholinergic activity to restore the excitatory/inhibitory balance (Calne, 1978). Trihexyphenidyl antagonizes the action of acetylcholine in the guinea pig ileum with a K_1 in the region of 2×10^{-9} M (Barlow *et al.*, 1972).

9.7.1.2 *Muscarinic agonists*

One type of glaucoma is characterized by a build-up of water in the aqueous humor of the eye, a condition which is particularly prevalent in the elderly. The iris and the cornea are normally separated by a channel through which the aqueous humor can filter and be dispersed by absorption into the nearby blood vessels. In later life the iris can be pushed forwards, possibly because of inflammation or increasing size of the lens, thus closing the channel. The aqueous humor can no longer filter out and pressure on the eyeball increases, with particular danger of destruction of the optic nerve fibre leading to blindness.

Pilocarpine is one of the drugs recommended for treatment of this condition in the United States (AMA Drug Evaluations, 1983). The drug acts as a miotic, i.e. it narrows the pupils by causing them to contract, thus drawing the iris away from the cornea and allowing the fluid to drain away. This action is mediated by muscarinic receptors.

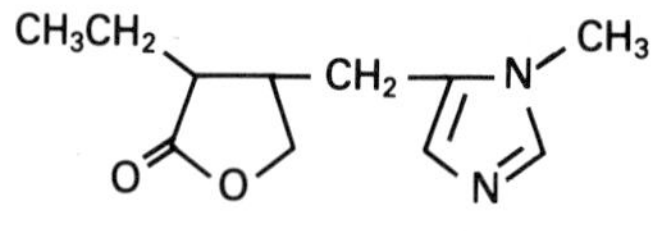

Pilocarpine 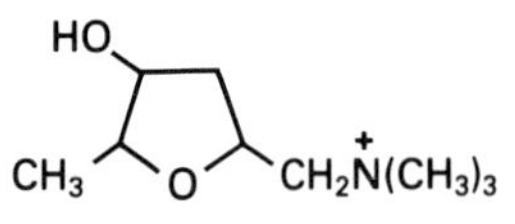   Muscarine

9.7.2 Acetylcholinesterase

Closely associated with cholinergic synapses is the enzyme acetylcholinesterase, which rapidly hydrolyses the neurotransmitter and is essential to ensure that the signal is terminated as soon as it has passed. Recent studies have suggested that the enzyme is attached to the lipid environment of the membrane via a link with inositol phosphate and diacylglycerol because phospholipase C will liberate the enzyme from its normal membrane-bound state (reviewed in Low *et al.*, 1986).

Acetylcholinesterase hydrolyses the ester link of the neurotransmitter to yield acetate and choline with a very high turnover number. The enzyme has two well defined parts in its binding site for substrates; one anionic which binds the cationic head of the substrate (and of inhibitors), and an esteratic site at which hydrolysis takes place with the formation of an acyl-enzyme intermediate. There is apparently another anionic binding site which can be occupied by bis-quaternary ligands (see the extensive review by Rosenberg, 1975). A charge-relay system similar to that defined for trypsin, chymotrypsin and elastase with aspartate carboxyl, histidine imidazole and serine hydroxyl is proposed at the active site of acetylcholinesterase. The serine is in a sequence -Gly-Asp-Ser-Gly-, but it is not clear whether the aspartate residue is the one forming part of the charge relay system.

Aspartate   Histidine   Serine

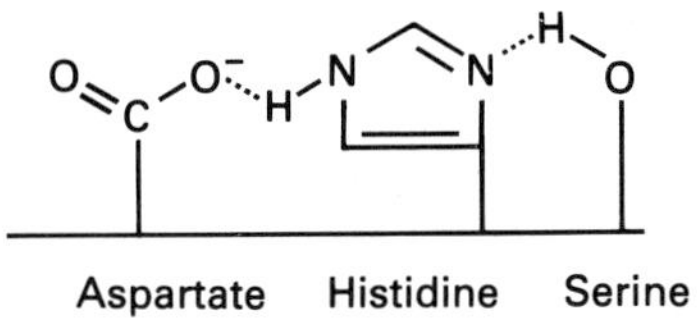

As with the proteases, the serine hydroxyl is activated by the charge relay system sufficiently to be a strong nucleophile thus forming a transient acyl-enzyme intermediate, which is easier to detect with acetylcholinesterase as the complex dissociates more slowly.

A family of compounds with structures related to acetylcholine and including a cationic head group and carbamate ester, inhibit acetylcholinesterase by acting as false substrates, but with very low turnover numbers, so that the true substrate is unable to be hydrolysed. One of the best known is pyridostigmine which inhibits acetylcholinesterase by binding to the anionic site with its cationic pyridine nitrogen, while the carbamate ester is available to be hydrolysed by the esteratic site. The tetrahedral complex formed on hydrolysis, a dimethylcarbamoyl-enzyme, resembles the transition state of the normal reaction but is extremely slow to hydrolyse and prevents the hydrolysis of the normal substrate. The drug is therefore a type of suicide inhibitor. The I_{50} for the reaction is $1 \cdot 6 \times 10^{-6}$ M (Wilson *et al.*, 1961) which is effectively k_1/k_2 in the following scheme:

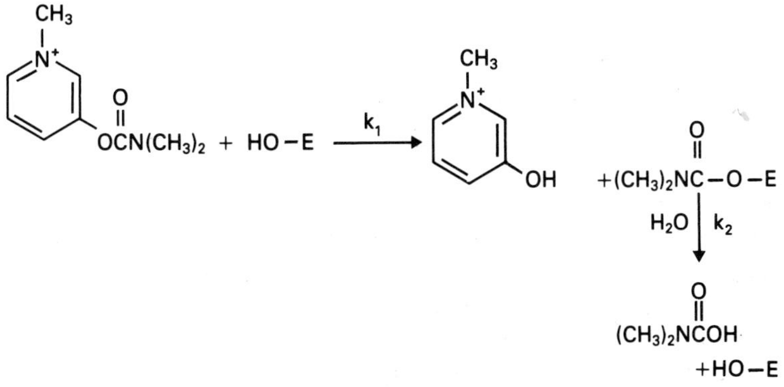

Inhibitors of acetylcholinesterase would be expected to potentiate the action of acetylcholine and this has been made use of in the development of the organophosphate insecticides. In medicine, less toxic inhibitors have been used in conditions where the receptors are damaged or diminished in number, as in myasthenia gravis. The damage is caused by auto-antibodies to the acetylcholine receptor on the post-junctional side of the neuromuscular junction, and is a genuine autoimmune disease (Scadding and Havard, 1981). The plasma antibody level does not always parallel the severity of the disease, but this may be because the method of measuring antibodies by radioimmunoassay does not necessarily detect all the proteins that have widely differing deleterious effects.

There are various possibilities as to how the auto-antibodies act to develop the condition. The simplest is that the post-synaptic receptor is blocked by the auto-antibody which acts as a kind of antagonist. This may not be the whole story, however, as receptors are subsequently cross-linked as in the case of anaphylactic reactions on the mast cell surface (see Chapter 10), and the rate of receptor degradation is increased. In addition, a degree of membrane lysis occurs, and a considerable amount of debris is found in the synaptic cleft. The net result in morphological terms is a widening of the synaptic cleft and elongation of the synaptic membrane (Scadding and Havard, 1981) so that the stimulatory signal cannot be transmitted from nerve to muscle.

Myasthenia gravis is most obviously recognized by muscle weakness which characteristically starts around the eyes and face and then works its way down via the limb girdle, outer limbs and trunk in that order. There is a marked fatigability of skeletal muscle, in that if a motor nerve of a patient is stimulated, synaptic transmission rapidly falls away, unlike the results obtained with a normal subject. Pyridostigmine and neostigmine are widely used for this condition today. They compensate for the deficit of acetylcholine receptors by slowing down the rate of degradation of acetylcholine and so increase the duration of its activity.

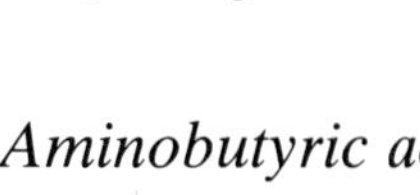

Pyridostigmine   Neostigmine

9.8 *4-Aminobutyric acid receptor*

4-Aminobutyrate (γ-aminobutyrate; GABA) is a neurotransmitter that acts at inhibitory synapses often pre-synaptically, i.e. the inhibitory nerve terminal secretes GABA on to a neighbouring excitatory nerve terminal and thereby reduces the output of the excitatory transmitter (often acetylcholine).

By this means specific neuronal pathways converging on another neurone can be eliminated without influencing the efficacy of others. The relevant biosynthetic pathway is found in GABA neurones, as found with other neurotransmitters; GABA neurones can synthesize GABA from glutamate with the help of glutamate decarboxylase (see Figure 9.9).

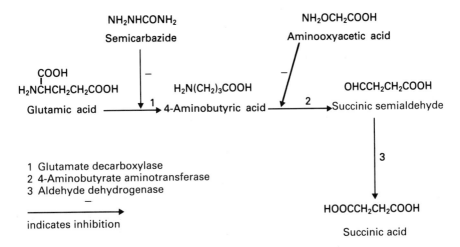

Figure 9.9 The pathways of GABA synthesis and breakdown.

Like the other receptors discussed in this chapter that are subdivided into various types, GABA receptors are divided into 2 major types. GABA$_A$ receptors contain a channel for chloride ion in the receptor itself, whereas some of the other receptors discussed in this chapter may indirectly cause an ion channel such as for calcium to be opened. The binding of agonists thus allows negative ions to pass into the cell and increases the negative membrane potential (hyperpolarization) so rendering it unable to transmit the nerve impulse or action potential. The consequence of this action is to induce muscle relaxation, to reduce anxiety and to block convulsions. Sedation may also be a consequence (Sanger, 1985).

H$_2$NCH$_2$CH$_2$CH$_2$CO$_2$H

4-Aminobutyrate

Bicuculline

In addition, there are at least three types of binding site on the $GABA_A$ receptor protein. One site that binds the neurotransmitter itself and muscimol, a close structural analogue of GABA as an agonist while bicuculline is an antagonist. Agents that bind to the second class of binding site include the benzodiazepines of which the best known are probably the anxiolytic (tranquillizer) drugs diazepam and chlordiazepoxide (these are generic names; the drugs may be more familiar under the tradenames of Valium and Librium). Picrotoxin antagonizes GABA action by binding to a third site near the chloride channel.

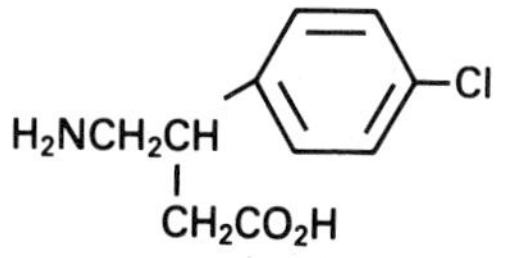

Baclofen 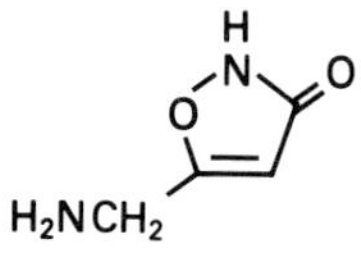   Muscimol

Recent studies have suggested that $GABA_B$ receptors are not linked to chloride channels since baclofen, an agonist with a structural resemblance to GABA, does not influence the rate of chloride passage across brain membranes, unlike the $GABA_A$ agonists (Allan and Harris, 1986). $GABA_B$ receptors do not bind diazepins and are linked in an inhibitory fashion to adenylate cyclase — modulated as usual by guanine nucleotides (Wojcik and Neff, 1984) — see Table 9.2. The function of these receptors is not yet known. Bowery (1984) has reviewed the present views on these two types of receptor.

9.8.1 Benzodiazepines

A well known family of psychoactive drugs used as hypnotics, sedatives and tranquillizers, the benzodiazepines, that were originally marketed in the late 1960s mainly for the treatment of anxiety, were discovered subsequently to bind very tightly to specific sites in the brain (Speth *et al.*, 1978). A considerable body of evidence linked this site with the $GABA_A$ receptor and, not for the first time, the experimental use of a drug advanced the state of scientific knowledge, having initially been of therapeutic value to the population at large (Sanger, 1985).

Benzodiazepines act in such a way as to simulate the effects of exogenously administered GABA, but if GABA levels are depleted there is no such effect. This occurs, for example, when semicarbazide is co-administered; glutamate decarboxylase, an enzyme with pyridoxal phosphate as a prosthetic group, is inhibited through the formation of a Schiff's base and GABA levels are lowered. Conversely, benzodiazepine effects are potentiated by aminooxyacetic acid, an inhibitor of GABA oxidation to succinic semialdehyde by GABA aminotransferase. It is clear that benzodiazepines do not act directly

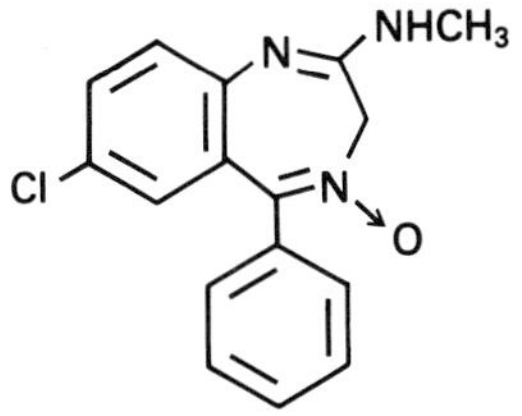

CH₃ (structure)

Flunitrazepam $-NO_2$ F
Diazepam $-Cl$ H

Chlordiazepoxide

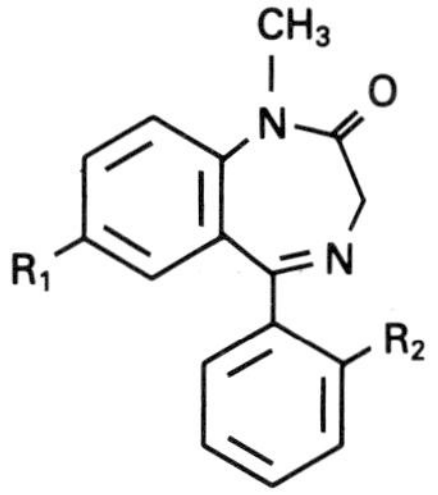

on GABA$_A$ receptors, but rather bind nearby, probably on the same macro-molecule, and enhance the binding of the neurotransmitter (Harvey, 1985).

The best characterized benzodiazepine ligand is flunitrazepam which binds to homogenates of human brain with a binding constant in the range $1–3 \times 10^{-9}$ M (Speth *et al.*, 1978). This was measured both by kinetic analysis of the rates of association and dissociation, and also by Scatchard analysis. Inhibition of flunitrazepam binding by a number of benzodiazepines correlated well with a number of pharmacological tests (notably inhibition of convulsions induced by pentylenetetrazole in mice, and muscle relaxation in cats) and also with the dose used to treat anxiety in man. Braestrup *et al.* (1977) using ^{3}H-diazepam binding to a membrane preparation from human brain obtained similar results. They also showed that GABA receptors varied in density between different parts of the brain and this density paralleled the distribution of benzodiazepine receptors.

Conversely, the binding of GABA and GABA agonists such as muscimol potentiate the binding of benzodiazepines to their receptors, an action which can be reversed by the GABA$_A$ antagonist bicuculline (Braestrup *et al.*, 1980). The receptor complex has been isolated from calf brain and shown to retain the potent binding of flunitrazepam, GABA and muscimol. In addition chloride ion stimulates drug binding. The receptor complex appears to be a single macromolecule through which allosteric effects may be transmitted, and which also contains a chloride channel (Gavish and Snyder, 1981), and a muscimol binding site (Martini *et al.*, 1983).

More recent studies, however, have cast doubt on this simple picture (reviewed in Sanger, 1985). If rats are trained to respond to food and drink but responding, in addition, is linked to an aversive stimulus such as an electric shock, benzodiazepines greatly increase what would otherwise be a low rate of respondance. Furthermore, the magnitude of the drug effect correlates highly with the clinical potency in treating anxiety. Not all agonists increase the rate of respondance when given systemically, but do if injected into the appropriate part of the brain. These differences may arise as a consequence of non-uniform distribution of agents in the brain, or it may be that some of the anxiolytic actions of benzodiazepines are not involved with GABA — unlike the muscle relaxant and anti-convulsant actions (Sanger, 1985).

9.8.2 Avermectin

GABA acts as a neurotransmitter in a number of other organisms besides mammals, including arthropods and nematodes, which are themselves pathogenic for man. Conspicuous among these infections is one caused by a nematode worm, *Onchocerca volvulus*, a native of West and Central Africa, which infects man to produce the disease known as river blindness or onchocerciasis. The worm is injected into man by a bite from a gnat of the genus *Simulium* and migrates in the surface tissue where it is found in nodules particularly around the joints. It may also migrate to the eye where it usually causes blindness by attacking the optic nerve. The only drugs that have been used are suramin which is extremely toxic, and diethylcarbamazine which is less toxic but is liable to kill the adult worms swiftly and precipitate an extreme allergic reaction by virtue of the large amount of nematode protein rapidly entering the circulation. This reaction can cause more damage than the nematode itself, particularly in the eye.

Recently, however, a major improvement has been effected by the use of ivermectin. This is a member of the avermectin family of compounds, first obtained from the fermentation of *Streptomyces avermitilis* (Egerton et al., 1979). The avermectins which possess a sixteen-membered lactone ring with a spiroacetal system involving carbon atoms 17 to 25 and a disaccharide substituent at the C13 position (see structures for details). The naturally occurring avermectins fall into four major (A_{1a}, A_{2a}, B_{1a} and B_{2a}) and four minor (A_{1b}, A_{2b}, B_{1b} and B_{2b}) series (Campbell et al., 1983). Ivermectin is synthesized from avermectin B_1 by reduction of the 22,23 double bond (see structure).

When given in a single oral dose to patients in Mali in a double blind trial, avermectin B_{1a} killed the larval and adult nematodes slowly, without producing an allergic reaction, and its effects lasted for a number of months. The eye lesions healed. Furthermore, the skin count of organisms dropped markedly. Unlike with diethylcarbamazine treatment, the fertility of the remaining worms was greatly reduced, and those that were fertile produced abnormal offspring (Lariviere et al., 1985).

Ivermectin was originally introduced to the veterinary market for the treatment of internal parasites (endoparasitic nematodes) of mammals of economic importance such as sheep, cattle and horses and of external parasites (ectoparasites), namely arthropods such as ticks and mites. Lobsters are also sensitive to the drug! The common link between these organisms is their possession of neuronal synapses at which GABA acts as a transmitter (see Campbell et al., 1983; Wright, 1987 for reviews on avermectin and its biological action).

Most of the experimental studies have been carried out with avermectin

B_{1a}, referred to as avermectin below. In *Ascaris suum*, a relatively large nematode that infects pigs, there are at least two sites of drug action: one is a block between a nerve cell (known as an interneurone) that forms a synapse with an excitatory motor neurone on the dorsal side of the nematode and leads to depolarization of the muscle membrane. The second site is at the neuromuscular junction of an inhibitory motor neurone on the ventral side and muscle, the activation of which leads to hyperpolarization. At the former site the action of 6×10^{-6} M avermectin B_{1a} in reducing the depolarization can be mimicked by the agonists muscimol and piperazine, and reversed by the antagonist picrotoxin, which indicates that the interaction is with the $GABA_A$ receptor (Kass *et al.*, 1984). On the other hand, avermectin's action in reducing hyperpolarization at the second site does not respond to picrotoxin reversal and is presumably nothing to do with $GABA_A$ receptors.

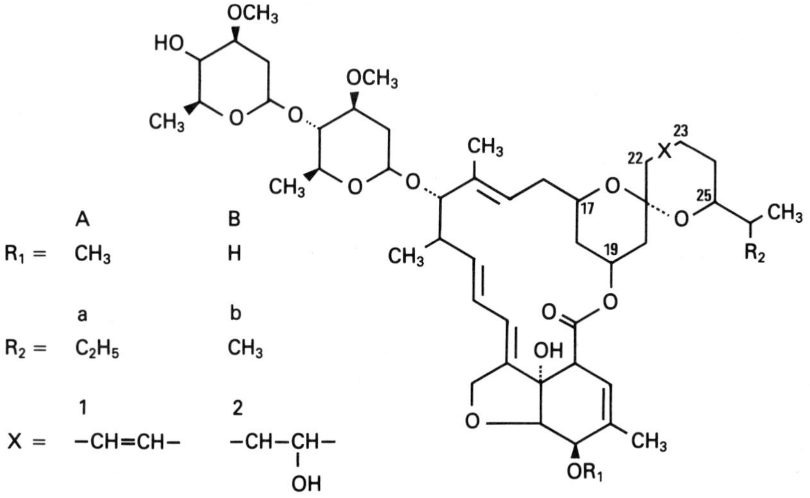

	A	B
$R_1 =$	CH_3	H

	a	b
$R_2 =$	C_2H_5	CH_3

	1	2
$X =$	$-CH=CH-$	$-CH-CH-$
		$\quad\quad\;\; |$
		$\quad\quad\; OH$

Avermectins

$$R = -\overset{\displaystyle OH}{\underset{\displaystyle CH_3}{C}}-CH_3 \quad : \text{Picrotin}$$

$$R = -\overset{\displaystyle CH_3}{\underset{\displaystyle CH_2}{C}} \quad : \text{Picrotoxinin}$$

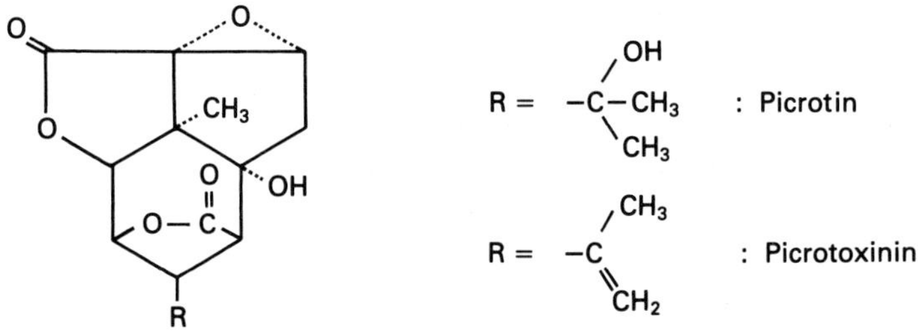

Picrotoxin (1 Mole Picrotin: 1 Mole Picrotoxinin)

In sharp contrast, at 1×10^{-7} M concentration the drug appears to antagonize the action of GABA in reducing the length of time that ion channels are open, and the probability of their opening — measured by using microelectrodes implanted into the nematode (Martin, 1987). The apparent contradiction between these findings remains to be resolved, although the use of a low physiological level of the drug may be crucial.

The arthropod target for avermectin is also a neuromuscular junction as shown by experiments in lobsters where the action of the drug was synergistic with GABA. The neuromuscular junction is known to be controlled by one excitatory axon with glutamate as the transmitter and one inhibitory axon with GABA as the transmitter. Avermectin reduced both excitatory and inhibitory post-synaptic potentials, probably by opening chloride channels and hyperpolarizing the membrane. The response was restored by picrotoxin. It appears likely that avermectin acts as a GABA agonist either by potentiating the binding of GABA to its receptor (see Campbell *et al.*, 1983) and/or by releasing GABA pre-synaptically. The net effect of drug on both nematode and arthropod is muscular paralysis.

In vitro studies indicate that avermectin B_{1a} can stimulate the high affinity binding of GABA to rat brain membranes by increasing the number of available binding sites, rather than the binding constant, at an I_{50} of 7×10^{-6} M. This effect is chloride ion dependent and is antagonized by picrotoxin and bicuculline (Pong and Wang, 1982). Avermectin cannot reach these sites *in vivo* as it does not cross the blood–brain barrier. Avermectin also enhances the binding of diazepam to rat brain membranes in the micromolar region, and potentiates the pharmacological activity of the tranquillizer, notably a decrease in motor activity and muscle power (Williams and Yarbrough, 1979).

More recent studies in locust leg muscle (the exterior tibiae) have shown that there are at least two other chloride channels which can be modulated by avermectin besides that connected to the GABA receptor: one is probably linked to glutamate receptors and the other is not linked to any known receptor (Duce and Scott, 1985). The relative importance of these three types of ion channel with respect to the action of avermectin is as yet unknown.

In conclusion, the relatively simple picture of avermectin acting to potentiate GABA requires some qualification. Clearly, in some cases this view is correct, but in other situations avermectin may act as an antagonist, and in yet others may have actions which do not relate to GABA action but may still be connected with chloride channel opening.

9.8.3 Baclofen

Baclofen is a close structural analogue of GABA in medical use for the treatment of spasticity (Bianchine, 1985). The drug is believed to exert its effects by depressing synaptic transmission in the spinal cord. It acts at $GABA_B$ sites that are insensitive to bicuculline, and apparently do not increase chloride ion conductance. Hyperpolarization of cell membranes appears to be the consequence of baclofen's action — possibly by increasing potassium conductance, i.e. the cation flows out of the cell and thus raises the potential difference between membrane and extracellular fluid (Newberry and Nicoll, 1984).

9.9 Opiate receptors

The opiates are a class of naturally occurring opium alkaloids of which the best known is probably morphine. Opioids are chemically synthesized agents such as meperidine, methadone and naloxone which mimic or antagonize the pharmacological actions of morphine. Both opiates and opioids bind to receptors designated as opioid.

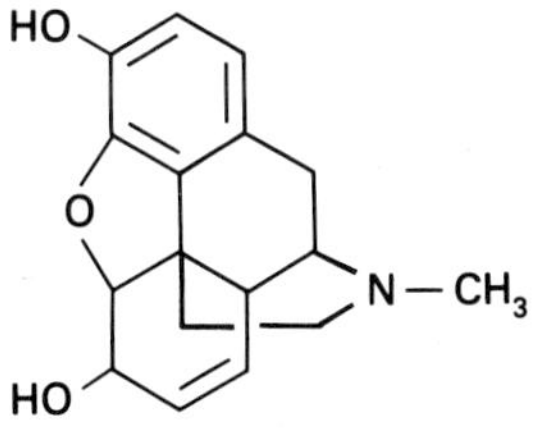

Morphine

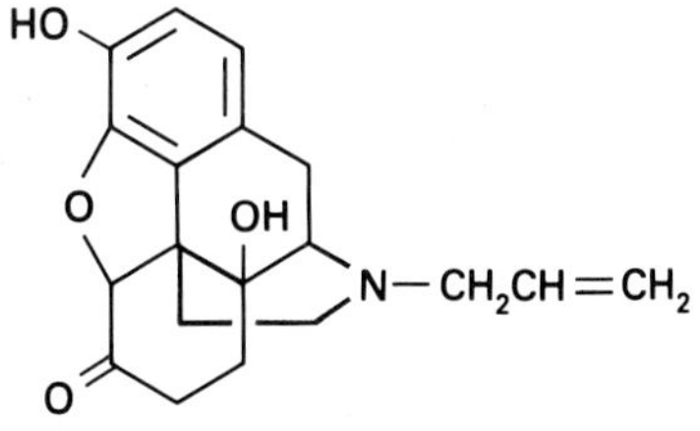

Naloxone

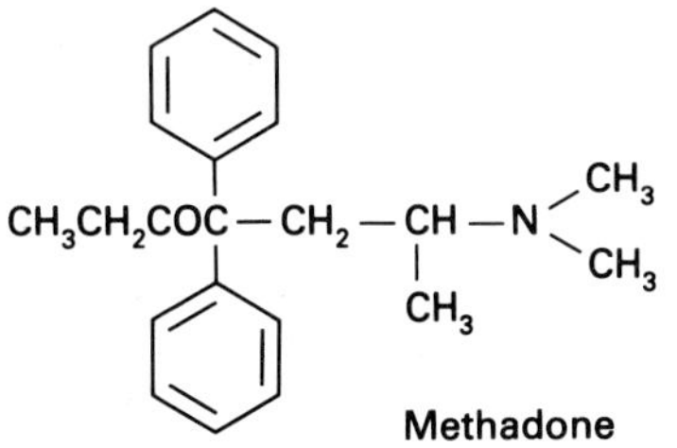

Methadone

Drugs of this type are powerful painkillers (analgesics) and morphine in particular, derived from the opium poppy, has been in use for this purpose for a very long time. The sort of pain that these agents are used to treat is the chronic nerve pain that often accompanies terminal cancer and recovery from operations. Less severe pain is countered by the use of non-steroidal anti-inflammatory drugs such as aspirin, which inhibit the biosynthesis of prostaglandins (see Chapter 7).

A large number of analogues have been synthesized in order to improve on the properties of morphine by reducing undesirable side-effects such as drowsiness, nausea, vomiting, constipation and, most serious of all, respiratory depression that can be lethal if a sufficiently high dose is given. In addition, there is also the grave problem of physical dependence on the opiate family of drugs.

The new structures fall into two families in the sense that they bind to two different receptors: morphine and congeners are in one group binding to receptors described as μ, while ketazocine and analogues are designated as binding to κ receptors. Agonists interacting with the κ receptor do not substitute for morphine in the morphine-dependent monkey, i.e. do not cause physical dependence as morphine does (see Hutchinson *et al.*, 1975). A further receptor, classified as δ, has been proposed on the basis of differences in response to various agonists between the guinea pig ileum and the mouse vas deferens. δ Receptors appear to exist in the latter but not the former organ (see Kosterlitz, 1985, for a review). To complete the present view a fourth type of receptor, known as σ, has been identified.

The finding that there were saturable, stereospecific binding sites for opioid drugs in the central nervous system prompted a search for endogenous ligands to bind at these receptors. Very soon two naturally occurring peptides (the enkephalins) were discovered in pig brain. Structurally they are very alike and are known as Met-enkephalin (Tyr-Gly-Gly-Phe-Met) and Leu-enkephalin (Tyr-Gly-Gly-Phe-Leu) differing only at the carboxyl terminus (Hughes *et al.*, 1975). Subsequently other, larger opiate peptides were isolated from the pituitary gland and called endorphins to distinguish them from the enkephalins (Li *et al.*, 1976; Chretien *et al.*, 1976). Figure 9.8 shows opioid structural relationships. It is interesting that Met-enkephalin forms the amino-terminus of the endorphins which in turn represent residues 61–91 at the carboxyl terminus of the pituitary hormone, β-lipotrophin, although β-lipotrophin does not possess any opiate properties.

A further 17 amino acid peptide was isolated by Goldstein *et al.* (1977) and named dynorphin. Dynorphin is Leu-enkephalin with an extension of 12 amino acid residues at the carboxyl terminus. There are at least four peptides with opioid activity coded for by the cDNA for the dynorphin precursor (Kakidani *et al.*, 1982). That these peptides are endogenous ligands for the opiate receptor was shown early on since they closely mimic the action of morphine and could be antagonized by the morphine antagonist naloxone (Frederickson, 1977, see Figure 9.10).

In order to establish the role of these opioid peptides as neurotransmitters and their relationship to the opiate receptors, a very large number of studies have been carried out. The close link of opioid peptides and opiate receptors is demonstrated by the correlation between immunofluorescent staining (enkephalin-like) with binding sites for diphenorphine (a morphine analogue) shown by autoradiography in rat brain (Simantov *et al.*, 1977). The main concentration of the receptors appears to be in the brain, spinal cord and the gastro-intestinal tract, and parallels the distribution of opiate-like material (reviewed in Kosterlitz and McKnight, 1980).

The endogenous ligands are not absolutely specific for the different receptors, for example β-endorphin appears to bind equally well to μ and κ receptors but has no affinity for δ. Leu-enkephalin, on the other hand, binds preferentially to the δ-site but binds less well to the μ-site. Dynorphin A

Aminoacid sequence of the opioid peptides.

Precursor	Peptides	Sequence
Proopiomelanocortin	∝-endorphin	Tyr-Gly-Gly-Phe-Met-Thr-Ser-Glu-Lys-Ser-Gln-Thr-Pro-Leu-Val-Thr-Leu: 17 residues
	४-endorphin	Tyr-Gly-Gly-Phe-Met-Thr-Ser-Glu-Lys-Ser-Gln-Thr-Pro-Leu-Val-Thr-Leu: 17 residues
	ß-endorphin	Tyr-Gly-Gly-Phe-Met-Thr-Ser-Glu-Lys-Ser-Gln-thr-Pro-Leu-Val-Thr-Leu-Phe-Lys-Asp-Ala-Ile-Ile-Lys-Asn-Ala-His-Lys-Lys-Gly-Gln: 31 residues
Proenkephalin	[Met]enkephalin	Tyr-Gly-Gly-Phe-Met: 5 residues
	[Leu]enkephalin	Tyr-Gly-Gly-Phe-Leu: 5 residues
	[Met]enkephalin-Arg-Phe: 7 residues	
	[Met]enkephalin-Arg-Gly-Leu: 8 residues	
Prodynorphin	Dynorphin A	Tyr-Gly-Gly-Phe-Leu-Arg-Arg-Ile-Arg-Pro-Lys-Leu-Lys-Trp-Asp-Asn-Gln: 17 residues
	Dynorphin B	Tyr-Gly-Phe-Leu-Arg-Arg-Ile: 8 residues
	∝-neoendorphin	Tyr-Gly-Gly-Phe-Leu-Arg-Lys-Tyr-Pro-Lys: 10 residues
	ß-neoendorphin	Tyr-Gly-Gly-Phe-Leu-Arg-Lys-Tyr-Pro: 9 residues

Tyrosine is always at the amino terminal end of the sequence. The classic sequence of the first five amino acids (Tyr-Gly-Gly-Phe-Leu-) can be seen running through the entire group of peptides. Further congruence can be seen within each of the three groups as we move towards the carboxyl terminus.

Melanocyte stimulating hormones (MSH), both α, β and γ, are also derived from proopiomelanocortin, and so is adrenocorticotrophic hormone (ACTH). α-MSH, indeed, is residues 1 to 13 of ACTH.

Figure 9.10 Amino acid sequence of the opioid peptides.

favours the κ site, while still giving measurable binding at μ. Morphine is also not entirely selective in that it favours the μ-site but still has measurable affinity for both κ and δ. Naloxone, likewise, is more potent at the μ site but can also bind at the δ- and κ-sites. This lack of selectivity allows the opioid peptides to appear similar to morphine in pharmacological terms although they do not necessarily favour the same receptor. Other agonists are being synthesized that are more specific (Kosterlitz, 1985).

Because we lack specific agonists and antagonists at each of the four major opiate receptors, the role of each receptor is still not entirely clear. Nevertheless, the μ-receptor seems to be the main receptor involved in pain suppression, since all the compounds that are the most effective painkillers do have a preference for this receptor (Kosterlitz and McKnight, 1980; Kosterlitz, 1985). Euphoria, respiratory depression and physical dependence may also be mediated by this receptor. The κ receptor may be responsible for painkilling and also for sedation, while the σ may control hallucinations and dysphoria. The actions of the δ receptor are not yet understood.

Opiate receptors come within the group of receptors that interact with GTP in two ways: either to couple receptors to adenylate cyclase in an inhibitory fashion (Rodbell, 1980), or to regulate the binding of agonists to receptors probably of the μ type (see Figure 9.1 in Section 9.1). The GTP binding protein that is inhibitory (known as N_I) normally couples with the catalytic unit of adenylate cyclase to lower cAMP production as a feature of opiate agonist binding. Consistent with this effect upon the adenylate cyclase system, opiate agonists stimulate GTP-ase activity (low K_m) in membranes from rat brain with a K_d of 9×10^{-8} M for morphine which is close to the binding affinity of morphine for the opioid receptor. The effect was antagonized by naloxone. Sodium ion is required both for the coupling of receptors to adenylate cyclase (Hsia *et al.*, 1984) and also for the inhibition of binding of agonists, but not antagonists to the opioid receptor (see Childers, 1984).

By this means the opiates can suppress the release of follicle-stimulating hormone and luteinizing hormone from the pituitary since the release of these hormones is caused by the activation of adenylate cyclase (see Grossman *et al.*, 1981). The importance of this effect relates to the inhibition of ovulation characteristic of morphine and related alkaloids.

9.9.1 Opioid dependence

Narcotic dependence is thought to relate in some way to the chronic inhibition of adenylate cyclase activity since levels of this enzyme are lowered in membranes from morphine-dependent rats and, as noted above, GTPase levels from these animals are lowered. Naloxone increases these levels towards the normal values (Barchfeld and Medzihradsky, 1984). It is interesting to note that methadone, although addictive itself, is often used to treat withdrawal symptoms of heroin users (Jaffe and Martin, 1985).

A possible mechanism for the derivation of withdrawal symptoms is suggested by the interaction of opioids with gonadotrophins. A long-acting analogue of Met-enkephalin was shown to inhibit the release of follicle-stimulating hormone and luteinizing hormone whilst naloxone produced a significant rise in levels of gonadotrophins in both male and female subjects (Grossman *et al.*, 1981). The symptoms of withdrawal and naloxone treatment are very similar to the symptoms of premenstrual syndrome (Reid and Yen, 1983) and opioid agonists and antagonists have their greatest effect on serum luteinizing hormone levels in the premenstruum. These symptoms may be a consequence of high gonadotrophin levels, resulting partly from inadequate feedback inhibition by ovarian steroids and also from continuous stimulation from the hypothalamus by release of luteinizing hormone releasing hormone. The sudden withdrawal of opiate narcotics, therefore, may sharply raise gonadotrophin levels that can give rise to the withdrawal symptoms (Coulson, 1986).

9.9.2 Pentazocine as analgesic

In order to develop pain-killing drugs that do not cause dependency and painful withdrawal symptoms when their use is discontinued, considerable synthetic effort has been expended. This led eventually, among other drugs, to the development of pentazocine — structurally a derivative of benzomorphan. When given orally this drug does not cause dependence, but when extracted from the tablets and injected intravenously with the antihistamine tripelennamine, it may do so. Nevertheless, pentazocine is still one of the least addictive morphine-like drugs (reviewed in Brogden *et al.*, 1973).

Pentazocine shows a mixed type of action at opiate receptors, being a moderately weak antagonist at μ and a relatively powerful agonist at κ and σ. The antagonism of μ receptors helps to explain why the compound is less likely to cause dependence while the agonism at κ receptors is probably responsible for its analgesic effects. This type of analgesia differs from that induced by morphine since it only affects the spinal cord whereas morphine is able to act at supra-spinal loci, presumably as a consequence of agonism at μ receptors. At high doses (60 to 90 mg in man), actions characteristic of stimulation of the σ receptors occur, namely hallucinations and dysphoria.

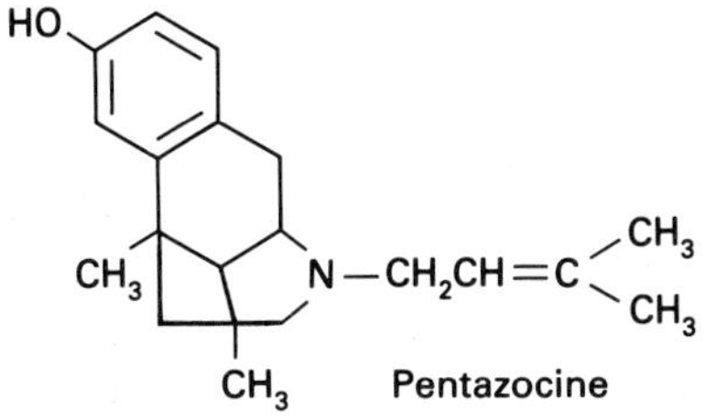

Pentazocine

9.9.3 Loperamide as anti-diarrhoeal

One of the most marked side-effects of opiate agonist usage, constipation, probably arises through action at the opiate receptor in the gastro-intestinal tract. This has been put to good use in the development of drugs for the treatment of diarrhoea. Morphine has the ability to increase muscle tone, diminish the amplitude of contractions and markedly reduce the propulsive activity in the intestine. Consequently, over-the-counter preparations of morphine adsorbed to kaolin are often effective in treating diarrhoea. Efforts have been made, however, to find an opiate that does not suffer from the central nervous system side-effects of morphine such as drowsiness etc.

Loperamide was developed as an answer to the problem (Awouters *et al.*, 1983).

Loperamide does not cross the blood–brain barrier and thus does not give rise to sedation, although it will bind to opioid receptors in brain homogenates. The drug slows gastro-intestinal motility (peristalsis) and is of particular value in cases where the bowel action is excessive — over-active bowel syndrome for example (see Hughes *et al.*, 1982).

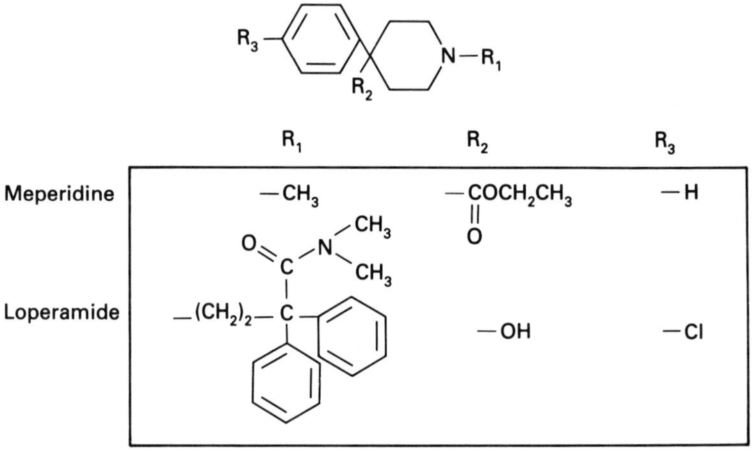

	R_1	R_2	R_3
Meperidine	$-CH_3$	$-COCH_2CH_3$ (C=O)	$-H$
Loperamide	$-(CH_2)_2-C$ (with O=C-N(CH_3)_2 and two phenyl groups)	$-OH$	$-Cl$

Diarrhoea, however, may arise not only through excessive peristalsis but also through reduced absorption or increased secretion of water and ions at the mucosal surface. Loperamide antagonizes the secretion induced by prostaglandin E_1, theophylline and cholera toxin (Awouters *et al.*, 1983) but not apparently that induced by vasoactive intestinal polypeptide (Schiller *et al.*, 1984). Both cholera toxin and prostaglandin E_1 induce the activation of adenylate cyclase, thus raising cAMP levels, and the evidence suggests that cAMP is the mediator in the release of chloride ion followed by water to maintain osmotic balance. Cholera toxin acts as an enzyme, albeit with a low turnover number, transferring adenosine-diphosphoribose from NAD to arginine residues on the stimulatory N_S component of the adenylate cyclase system (see Section 9.1 and Figure 9.1), thereby blocking the hydrolysis of GTP to GDP and switching on the enzyme irreversibly (reviewed in Coulson *et al.*, 1984).

The mechanism of action of loperamide is probably via the opioid receptor since naloxone antagonizes its action. On the other hand, a direct inhibitory effect of the opioid receptor on adenylate cyclase is unlikely because loperamide does not antagonize the rise in cAMP induced by both prostaglandin E1 and cholera toxin. The suggestion has been made that calmodulin may be involved (Awouters *et al.*, 1983) because:

(a) The secretory process is calcium-dependent and calmodulin mediates many of the intracellular actions of calcium in secretion.

(b) Loperamide binds to other moderately high affinity sites in the gut wall, where calmodulin is highly concentrated.
(c) Loperamide binds to calmodulin in the presence of calcium *in vitro* with a binding constant of $1 \cdot 2 \times 10^{-5}$ M (Zavecz *et al.*, 1982).

This situation remains unresolved at the present time. What is clear, however, is that loperamide suppresses the secretion of chloride ion from the mucosal cell into the gut lumen (Hughes *et al.*, 1982).

9.10 Histamine receptors

The response of mammals to histamine has been studied extensively ever since the work of Sir Henry Dale, one of the great pioneers of pharmacology (reviewed in Douglas, 1985). About 20 years ago it became clear that like the other neurotransmitters discussed in this chapter, histamine interacts with receptors which may be divided into at least two major groups, H_1 and H_2 (Ganellin, 1978). Histamine can cause many smooth muscles to contract in lung bronchi and gut and increases capillary permeability, actions which are governed by H_1 receptors. H_2 receptors control the action of histamine in inducing the secretion of gastric acid.

9.10.1 Histamine₁ receptor antagonists

H_1 antagonists, such as mepyramine and chlorpheniramine, have been available for several decades and are used to block the action of histamine released endogenously in allergic reactions, particularly in the upper respiratory tract as in hay fever. Skin allergies also respond favourably in some cases. Histamine is one of the mediators (autacoids) of the allergic response generating itching and some aspects of the inflammation that accompany this reaction. The lower part of the respiratory tract is involved in the genesis of asthma which does not respond to H_1 antagonists in man since leukotrienes are the major cause of allergic bronchoconstriction (see Section 10.6 for a discussion on asthma). In guinea pigs, however, histamine is the major effector and H_1 antagonists are protective (Douglas, 1985).

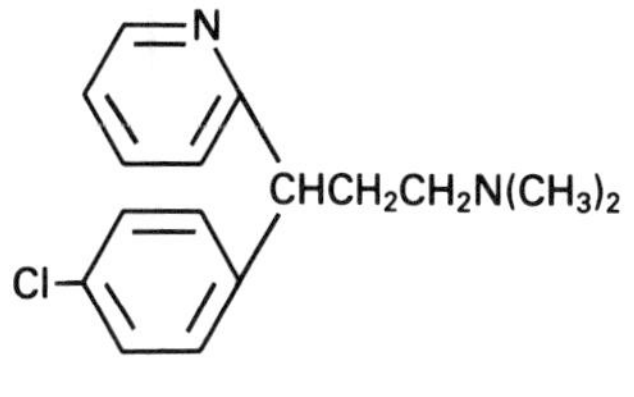

Chlorpheniramine 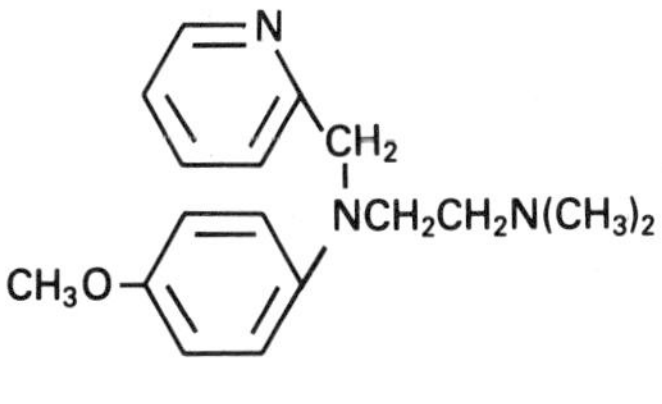   Mepyramine

Mepyramine has a low binding constant with H_1 receptors as measured by binding to homogenates of, for example, guinea pig cortex ($1 \cdot 5 \times 10^{-9}$ M;

Daum *et al.*, 1982) and rat cerebral cortex ($2 \cdot 1 \times 10^{-9}$ M; Hall and Ogren, 1984). Chlorpheniramine binding is in the same range as measured by competition with mepyramine (K_d is $8 \cdot 3 \times 10^{-10}$ M for guinea pig brain and $9 \cdot 1 \times 10^{-9}$ M for rat brain (Hill and Young, 1980).

Histamine$_1$ receptors are not directly coupled to adenylate cyclase, unlike H$_2$ receptors, but appear to be indirectly linked in that H$_1$ antagonists can inhibit and agonists potentiate the stimulation by adenosine of cAMP production in guinea pig cerebral cortex (Daum *et al.*, 1982). The H$_1$ receptor seems to activate turnover of phosphatidylinositol as measured in rabbit aorta and the effect is inhibited most by H$_1$ receptor antagonists such as mepyramine. Histamine stimulates smooth muscle contraction in this organ. The involvement of the phosphatidylinositol process (see Section 9.1) to activate calcium uptake into the cell has been proposed (Villalobos-Molina and Garcia-Sainz, 1983) so that muscle contraction may be stimulated by calcium.

The antagonism of histamine is competitive and reversible. The H$_1$ antagonists as a whole are lipophilic and demonstrate variations on the following structure:

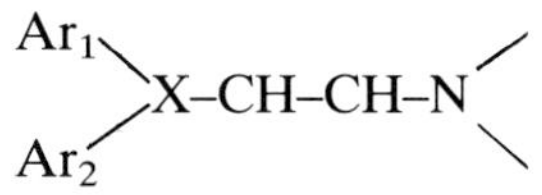

$$\begin{array}{c} Ar_1 \\ \diagdown \\ \diagup \\ Ar_2 \end{array} X\text{--}CH\text{--}CH\text{--}N \diagup \diagdown$$

where Ar is usually an aryl group and X is a nitrogen or carbon atom or a C–O–ether group (Douglas, 1985).

9.10.2 Histamine$_2$ receptor antagonists

The rational development of the drug cimetidine, including the importance of physico-chemical factors such as partition coefficient and ionization constant, makes an interesting story (Ganellin, 1978). The structure of histamine was taken as the starting point, and it was shown that 2-methylhistamine was a more selective agonist for the H$_1$ receptor, while 4-methylhistamine favoured the H$_2$ receptor. After separation of agonist and antagonist activities, burimamide was derived as the first true H$_2$ receptor blocker. Optimization of the H$_2$ receptor blocking activity resulted in the synthesis of metiamide. Various structural changes to eliminate unwanted side-effects finally led to the development of cimetidine.

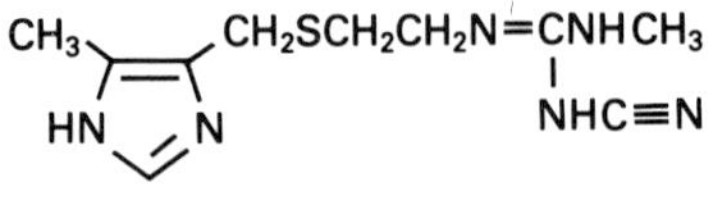

$$\begin{array}{c} CH_3 \\ \diagup \diagdown \\ HN \quad N \end{array} \quad CH_2SCH_2CH_2N{=}CNHCH_3 \\ \qquad\qquad\qquad\quad | \\ \qquad\qquad\qquad NHC{\equiv}N$$

Cimetidine

Histamine activates adenylate cyclase in brain and this has been shown to be mediated by H$_2$ receptors. Cimetidine antagonizes this activation in homogenates of the dorsal hippocampus of guinea pig brain with a K_i of

8.9×10^{-7} M (Kanof and Greengard, 1979). This compares well with a figure of 7.9×10^{-7} M against histamine stimulation of the rate of contraction of the right atrium of the guinea pig *in vitro* i.e. the positive chronotropic effect (Ganellin, 1978). The values for other H_2 antagonists, burimamide and metiamide, also agree, thus confirming the connection between adenylate cyclase activation and binding to the H_2 receptor. It should be noted that cimetidine is a compound that is much more hydrophilic than an H_1 receptor antagonist, and thus does not cross the blood–brain barrier to react with brain receptors, and so the correlation obtained above is all the more convincing.

CH$_2$CH$_2$CH$_2$CH$_2$NHCNHCH$_3$ ‖ S

Burimamide

CH$_3$ CH$_2$SCH$_2$CH$_2$NHCNHCH$_3$ ‖ S

Metiamide

Tiotidine binds to H_2 receptors in homogenates of guinea pig cerebral cortex with a binding constant of 1.7×10^{-8} M. That the H_2 receptor is involved in the binding is suggested by the correlation of binding with the antagonism of the chronotropic effect in guinea pig heart. In addition, antagonism of histamine-induced adenylate cyclase activity in guinea pig gastric mucosa also correlates very well with receptor binding for a number of antagonists including tiotidine (Gajtkowski *et al.*, 1983).

Cimetidine was put on the market in the mid 1970s for the treatment of gastric acid secretion in peptic ulcer patients, and effected a revolutionary improvement in the chemotherapy of ulcers. Most patients suffering from ulcers hypersecrete gastric acid, possibly as a result of stress or for some other reason. Gastric acid secretion is stimulated by food intake and by a variety of agents, known as secretagogues, such as histamine or caffeine, and by stimulation of the vagus nerve. Cimetidine will also block secretion induced by muscarinic agonists and by gastrin non-competitively, although these agents have their own separate receptors and do not act through histamine directly. The mechanism for this action is not fully understood but the inhibition of the action of these other secretagogues does indicate how effectively the drug inhibits acid secretion. Cimetidine thereby inhibits both basal and stimulated secretion (see Douglas, 1985). The drug both relieves the pain and allows the ulcer to heal.

Relapses are, however, quite common after treatment has stopped. In some cases the ulcer may not have healed completely although the patient may have no symptoms. It has been remarked that the problem is not how to heal an ulcer but how to keep it healed (Thomas and Misiewicz, 1984). If the ulcer is healed initially but the predisposing factor is not addressed, then it is not surprising that the ulcer may recur. Cimetidine may be used prophylactically as a maintenance therapy to prevent relapse. Pepsin secretion is also reduced by the drug.

$$CHNO_2$$

$$CH_2SCH_2CH_2NHCNHCH_3$$

$$CH_2N(CH_3)_2 \quad \text{Ranitidine}$$

$$CH_2SCH_2CH_2NHCNHCH_3$$
$$NC{\equiv}N$$
$$\text{Tiotidine}$$

After further synthetic optimization of H_2 receptor blocking activity, another more potent receptor blocker, ranitidine, has joined cimetidine on the market. Structurally the need for a small heterocyclic ring has been met by a furan, thus indicating that an imidazole, reminiscent of histamine, is not essential (moreover, tiotidine has a thiazole ring). The present evidence indicates that ranitidine is no more effective than cimetidine in healing ulcers although less material is required, but there are suggestions that ranitidine can cure some ulcers that have not previously responded to cimetidine (reviewed in Thomas and Misiewicz, 1984). Furthermore, ranitidine is more effective than cimetidine as maintenance therapy over a period of one year, as shown by a large trial in the United States (Silvis *et al.*, 1985).

References

Ahlquist, R.P. (1948). *Am. J. Physiol.*, **153**, 586–600.
Allan, A.M. and Harris, R.A. (1986). *Molec. Pharmacol.*, **29**, 497–506.
AMA Drug Evaluations (1983). 5th Edition. American Medical Association (Philadelphia: W.B. Saunders Co) p. 648.
Anden, N.E. and Stock, G. (1973). *J. Pharm. Pharmacol.*, **25**, 346–348.
Andren, L., Svensson, A., Dahlof, B., Eggertsen, R. and Hansson, L. (1983). *Acta Med Scand.*, **214**, 125–130.
Apperley, G.H., Daly, M.J. and Levy, G.P. (1976). *Brit. J. Pharmacol.*, **57**, 235–246.
Awouters, F., Niemegeers, C.J.E. and Janssen, P.A.J. (1983). *Annu. Rev. Pharmacol. Toxicol.*, **23**, 279–301.
Baldessarini, R.J. (1979) *Postgrad. Med.*, **65**, 123–128.
Barchfeld, C.C. and Medzihradsky, F. (1984). *Biochem. Biophys. Res. Commun.*, **121**, 641–648.
Barlow, R.B., Franks, F.M. and Pearson, F.D.M. (1972). *J. Pharm. Pharmacol.*, **24**, 753–761.
Baum, T. and Symbertz, E.J. (1983). *Fed. Proc.*, **42**, 176–181.
Bendetti, M.S. and Dostert, P. (1985). *Trends in Pharmacol. Sci.*, **6**, 246–251.
Bianchine, J. (1985). In: *The Pharmacological Basis of Therapeutics*, 7th Ed., Gilman, A.G., Goodman, L.S., Rall, T.W. and Murad, F., Eds (New York: Macmillan) Ch. 21.
Birdsall, N.J.M. and Hulme, E.C. (1983). *Trends in Pharmacol. Sci.*, **4**, 459–463.
Birdsall, N.J.M., Burgen, A.S.V. and Hulme, E.C. (1978). *Molec. Pharmacol.*, **14**, 723–736.
Birdsall, N.J.M., Burgen, A.S.V. and Hulme, E.C. (1979). *Brit. J. Pharmacol.*, **66**, 337–342.
Borison, R.L., Hitri, A., Blowers, A.J. and Diamond, B.I. (1983). *Clin. Neuropharmacol.*, **6**, 137–150.

Bourne, H.R., Lichtenstein, L.M. and Melmon, K.L. (1972). *J. Immunol.*, **108**, 695–705.
Bowery, N.G. (1984). *Trends in Pharmacol. Sci.*, **5**, 413–414.
Bradley, P.B., Engel, G., Feniuk, W., Fozard, J.R., Humphrey, P.P.A., Middlemiss, D.N., Mylecharane, E.J., Richardson, B.P. and Sakena, P.R. (1986). *Neuropharmacology*, **25**, 563–576.
Braestrup, C., Albrechtsen, R. and Squires, R.F. (1977). *Nature*, **269**, 702–704.
Braestrup, C., Nielsen, M., Krogsgaard-Larsen, P. and Falch, E. (1980). *Nature*, **280**, 331–333.
Brodde, O.-E. (1986). *J. Cardiovasc. Pharmacol.*, **8**, Suppl. 4, S16–S20.
Brogden, R.N., Speight, T.M. and Avery, G.S. (1973). *Drugs*, **5**, 6–91.
Brown, J.R. and Arbuthnott, G.W. (1983). *Neuroscience*, **10**, 349–355.
Buckner, C.B. and Abel, P. (1974). *J. Pharmacol. Exp. Ther.*, **189**, 616–625.
Bunney, B.S., Walters, J.R., Roth, R.H. and Aghajanian, G.K. (1973). *J. Pharmacol. Exp. Ther.*, **185**, 560–571.
Burges, R.A. and Blackburn, K.J. (1972). *Nature New Biol.*, **235**, 249–250.
Calne, D.B. (1978). *Postgrad. Med.*, **64**, 82–88.
Calne, D.B. (1984). *New Engl. J. Med.*, **310**, 523–524.
Campbell, W.C., Fisher, M.H., Stepley, E.D., Albers-Schonberg, G. and Jacob, T.A. (1983). *Science*, **221**, 823–828.
Carlsson, A. (1984). *Adv. Biochem. Psychopharmacol.*, **39**, 213–221.
Caulfield, M.P. and Stubley, J.K. (1982). *Brit. J. Pharmacol.*, **76**, 216P.
Caulfield, M.P. and Straughan, D.W. (1983). *Trends in Neurosci.*, **6**, 73–75.
Cawthon, R. M. and Breakfield, X. O. (1979). *Nature*, **281**, 692–694.
Charney, D. S., Menkes, D. B. and Heninger, G. R. (1981). *Arch. Gen. Psychiat.*, **38**, 1160–1180.
Childers, S. R. (1984). *J. Pharmacol. Exp. Ther.*, **230**, 684–690.
Chretien, M., Benjannet, S., Dragon, N., Seidah, N. G. and Lis, M. (1976). *Biochem. Biophys. Res. Commun.*, **72**, 472–478.
Christmas, A. J., Coulson, C. J., Maxwell, D. R. and Riddell, D. (1972). *Brit. J. Pharmacol.*, **45**, 490–503.
Church, M. K. and Warner, J. O. (1985). *Clin Allerg.*, **15**, 311–320.
Cooper, D. M. F., Bier-Laning, C., Halford, M. K., Ahlijanian, M. K. and Zahniser, N. R. (1986). *Molec. Pharmacol.*, **29**, 113–119.
Coppen, A. (1967). *Brit. J. Psychiat.*, **113**, 1237–1264.
Costa, E., Chuang, D. M., Barbaccia, M. L. and Gandolfi, O. (1983). *Experientia*, **39**, 855–858.
Coulson, C. J. (1986). *Med. Hypothes.*, **19**, 243–256.
Coulson, C. J., Nassau, P. M. and Tait, R. M. (1984). *Biochem. Soc. Trans.*, **12**, 184–187.
Creese, I., Schneider, R. and Snyder, S. H. (1977). *Eur. J. Pharmacol.*, **46**, 377–381.
Cullum, V. A., Farmer, J. B., Jack, D. and Levy, G. P. (1969). *Brit. J. Pharmacol.*, **35**, 141–151.
Daum, P.R., Hill, S.J. and Young, J.M. (1982). *Brit. J. Pharmacol*, **77**, 347–357.
Douglas, W. (1985). In: *The Pharmacological Basis of Therapeutics*, 7th Ed., Gilman, A.G., Goodman, L.S., Rall, T.W. and Murad, F., Eds. (New York: Macmillan) Ch. 26.
Duce, I.R. and Scott, R.H. (1985). *Brit. J. Pharmacol.*, **85**, 395–401.
Egerton, J.R., Ostlind, D.A., Blair, L.S., Eary, C.H., Suhayda, D., Cifelli, S., Riek, R.F. and Campbell, W.F. (1979). *Antimicrob. Ag. Chemother* **15**, 372–378.
Ehlert, F.J., Roeske, W.R. and Yamamura, H.I. (1981). *Fed. Proc.*, **40**, 153–159.
Fisher, S.K. and Bartus, R.T. (1985). *J. Neurochem.*, **45**, 1085–1093.
Fowler, C.J., Mantle, T.J. and Tipton, K.F. (1982). *Biochem. Pharmacol.*, **31**, 3555–3561.

Frederickson, R.C.A. (1977). *Life Sci.*, **21**, 23–42.

Fukuda, K., Kubo, T., Akiba, I., Maeda, A., Mishina, M. and Numa, S. (1987). *Nature*, **327**, 623–625.

Gajtkowski, G.A., Norris, D.B., Rising, T.J. and Wood, T.P. (1983). *Nature*, **304**, 65–67.

Gandolfi, O., Barbaccia, M.L. and Costa, E. (1985). *Life Sci*, **36**, 713–721.

Ganellin, G.R. (1978). *J. Appl. Chem. Biotechnol.*, **28**, 183–200.

Gavish, M. and Snyder, S.H. (1981). *Proc. Nat. Acad. Sci. (U.S.A.)*, **78**, 1939–1942.

Gilman, A.G. (1984). *Cell*, **36**, 577–579.

Giorgio-Conciato, M., Banister, S., Ferrari, P.A., Gaetani, M., Petrin, G., Sala, P. and Valentini, P. (1982). *Scand. J. Gastroenterol.*, **17**, Suppl. 81, 1–42.

Goldstein, A., Fischli, W., Lowney, L.I., Hunkapilla, M. and Hood, L. (1977). *Proc. Nat. Acad. Sci. (U.S.A.)*, **78**, 7219–7223.

Goodwin, F.K. and Post, R.M. (1983). *Brit. J. Clin. Pharmacol.*, **15**, 393S–405S.

Grossman, A., Moult, P.J.A., Gaillard, R.C., Delitalia, G., Toff, W.D., Rees, L.H. and Besser, G.M. (1981). *Clin. Endocrinol.*, **14**, 41–47.

Hall, H. and Ogren, S.-O. (1984). *Life Sci.*, **34**, 597–605.

Hammer, R., Berrie, C.P., Birdsall, N.J.M., Burgen, A.S.V. and Hulme, E.C. (1980). *Nature*, **283**, 90–92.

Harvey, J.N., Worth, D.P., Brown, J. and Lee, M.R. (1986). *Brit. J. Clin. Pharmacol.*, **21**, 53–61.

Harvey, S.C. (1985). In: *The Pharmacological Basis of Therapeutics*, 7th Ed., Gilman, A.G., Goodman, L.S., Rall, T.W. and Murad, F., Eds. (New York: Macmillan) Ch. 17.

Hill, S.J. and Young, J.M. (1980). *Brit. J. Pharmacol.*, **68**, 687–696.

Hsia, J.A., Moss, J., Hewlett, E.L. and Vaughan, M. (1984). *J. Biol. Chem.*, **259**, 1086–1090.

Hulme, E.C., Birdsall, N.J.M., Burgen, A.S.V. and Mehta, P. (1978). *Molec. Pharmacol.*, **14**, 737–750.

Hughes, J., Smith, T.W., Kosterlitz, H.W., Fothergill, L.A., Morgan, B.A. and Morris, H.R. (1975). *Nature*, **258**, 577–579.

Hughes, S., Higgs, N.B. and Turnberg, L.A. (1982). *Gut*, **23**, 974–979.

Hutchinson, M., Kosterlitz, H.W., Leslie, F.M., Waterfield, A.A. and Terenius, L. (1975). *Brit. J. Pharmacol.*, **55**, 541–546.

Iorio, L.C., Barnett, A., Leitz, F.H., Houser, V.P. and Korduba, C.A. (1983). *J. Pharmacol. Exp. Ther.*, **226**, 462–468.

Jaffe, J.H. and Martin, W.R. (1985). In: *The Pharmacological Basis of Therapeutics*, 7th. Ed., Gilman, A.G., Goodman, L.S., Rall, T.W. and Murad, F., Eds. (New York, Macmillan) Ch. 22.

Johnston, J.P. (1968). *Biochem. Pharmacol.* **17**, 1285–1297.

Kahn, R.A. and Gilman, A.G. (1984). *J. Biol. Chem.*, **259**, 6235–6240.

Kakidani, H., Fumitani, Y., Takahashi, H., Noda, M., Morimoto, Y., Hirose, T., Asai, M., Inayama, S., Nakanishi, S. and Numa, S. (1982). *Nature*, **298**, 245–249.

Kan, J.P., Malone, A. and Benedetti, M.S. (1978). *J. Pharm. Pharmacol.*, **30**, 190–192.

Kan, J.P. and Benedetti, M.S. (1981). *J. Neurochem.*, **36**, 1561–1571.

Kanof, P.D. and Greengard, P. (1979). *J. Pharmacol. Exp. Therapeut.*, **209**, 87–96.

Kass, I.S., Stretton, A.O.W. and Wang, C.C. (1984). *Molec. Biochem. Parasitol.*, **13**, 213–225.

Keane, P.E., Kan, J.P., Sontag, N. and Benedetti, M.S. (1979). *J. Pharm. Pharmacol.*, **31**, 752–754.

Kebabian, J.W., Agui, T., van Oene, J.C., Shigematsu, K. and Saavedra, J.M. (1986). *Trends in Pharmacol. Sci.*, **7**, 96–99.

Kendall, D.A. and Nahorski, S.R. (1985). *J. Pharmacol. Exp. Ther.*, **233**, 473–479.

Knoll, J. and Magyar, K. (1972). *Adv. Biochem. Psychopharmacol.*, **5**, 393–408.

Kosterlitz, H.W. (1985). *Proc. Roy. Soc. Lond.*, Series B, **225**, 27–40.
Kosterlitz, H.W. and McKnight, A.T. (1980). *Adv. Intern. Med.*, **26**, 1–36.
Langston, J.W. (1985). *Trends in Neurosci.*, **8**, 79–83.
Lariviere, M., Aziz, M., Weimann, D., Ginoux, J., Gaxotte, P., Vingtain, P., Beauvais, B., Derouin, F., Schulz-Key, H., Basset, D. and Sarfati, C. (1985). *Lancet*, **ii**, 213–225.
Lefkovitz, R.J. (1975). *Biochem. Pharmacol.*, **24**, 583–590.
Lefkowitz, R.J. and Caron, M.G. (1986). *J. Mol. Cell. Cardiol.*, **18**, 885–895.
Lew, M.J. and Angus, J.A. (1985). *J. Cardiovasc. Pharmacol.*, **7**, 401–408.
Leysen, J.E., de Chaffoy de Courcelles, D., De Clerck, F., Niemegeers, C.J.E. and Van Neuten, J.M. (1984). *Neuropharmacology*, **23**, 1493–1501.
Li, C.H., Chung, D. and Doneen, B.A. (1976). *Biochem. Biophys. Res. Commun.*, **72**, 1542–1547.
Lieberman, A.N. and Goldstein, M. (1985). *Pharmacol. Rev.*, **37**, 217–227.
Limbird, L.E., Speck, J.L. and Smith, S.K. (1982). *Molec. Pharmacol.*, **21**, 609–617.
Low, M.G., Ferguson, M.A.J., Futerman, A.H. and Silman, I. (1986). *Trends in Biochem. Sci.*, **11**, 212–215.
McPherson, G.A. and Summers, R.J. (1982). *Brit. J. Pharmacol.*, **77**, 177–184.
Manschrek, T.C. (1981). *New Engl. J. Med.*, **305**, 1628–1631.
Martin, R.J. (1987). *Biochem. Soc. Trans.*, **15**, 61–65.
Martini, C., Rigacci, T. and Lucacchini, A. (1983). *J. Neurochem.*, **41**, 1183–1185.
Martres, M.P., Sokoloff, P., Delandre, M., Schwartz, J.-C., Protois, P. and Costentin, J. (1984). *Naunyn-Schmiedeberg's Arch. Pharmacol.*, **325**, 102–115.
Matsui, H., Imafuki, J., Asakura, M., Tsukamoto, T., Ino, M., Saitoh, N., Miyamura, S. and Hasegawa, K. (1984). *Biochem. Pharmacol.*, **33**, 3311–3314.
Maycock, A.L., Abeles, R.H., Salach, J.I. and Singer, T.P. (1976). *Biochemistry*, **15**, 114–125.
Melamed, E., Rosenthal, J., Cohen, O., Globus, M. and Uzzan, A. (1985). *Eur. J. Pharmacol.*, **116**, 179–181.
Minneman, K.P. and Molinoff, P.B. (1980). *Biochem. Pharmacol.*, **29**, 1317–1323.
Minneman, K.P., Hegstrand, L.R. and Molinoff, P.B. (1979). *Molec. Pharmacol.*, **15**, 21–33.
Mitchell, H.W., Hau, H. and Denborough, M.A. (1979). *Eur. J. Pharmacol.*, **57**, 399–406.
Newberry, N.R. and Nicoll, R.A. (1984). *Nature*, **308**, 450–452.
Newsome, P.M., Burgess, M.N., Burgess, M.R., Holman, G.D. and Nahorski, S.R. (1984). *Biochem. Soc. Trans.*, **12**, 208–211.
Ogren, S.O., Ross, S.B., Hall, H., Hohn, A.C. and Renyi, A.L. (1981). *Acta Psychiat. Scand.*, **63**, Suppl. 290, 127–151.
Parkes, D. (1977). *Adv. Drug Res.*, **12**, 247–344.
Pearce, J.M.S. (1984). *Brit. Med. J.*, **288**, 1777–1778.
Pearce, L.B. and Roth, J.A. (1984). *Biochem. Pharmacol.*, **33**, 1809–1811.
Peroutka, S.J. (1984). *Neuropharmacology*, **23**, 1486–1492.
Pinder, R.M., Brogden, R.N., Speight, T.M. *et al.* (1977). *Drugs*, **1**, 321–352.
Pong, S.S. and Wang, C.C. (1982). *J. Neurochem.*, **38**, 375–379.
Rasmussen, H. (1986). *New Engl. J. Med.*, **314**, 1094–1101.
Reid, R.L. and Yen, S.S.C. (1983). *Clin. Obstet. Gynaecol.*, **26**, 710–718.
Richardson, B.P., Engel, G., Donatsch, P. and Stadler, P.A. (1985). *Nature*, **316**, 126–131.
Richelson, E. and El-Fakahary, E. (1981). *Biochem. Pharmacol.*, **30**, 2887–2891.
Rodbell, P. (1980). *Nature*, **284**, 17–22.
Rosenberg, T.L. (1975). *Adv. Enzymol.*, **43**, 103–218.
Roth, B.L., Nakaki, T., Chuang, D.-M. and Costa, E. (1986). *J. Pharmacol. Exp. Ther.*, **238**, 480–485.
Rudd, P. and Blaschke, T.F. (1985). In: *The Pharmacological Basis of Therapeutics*,

7th Ed., Gilman, A.G., Goodman, L.S., Rall, T.W. and Murad, F., Eds. (New York: Macmillan) Ch. 32.

Sanger, D.J. (1985). *Life Sci.*, **36**, 1503–1513.

Scadding, G.K. and Havard, C.W.H. (1981). *Brit. Med. J.*, **283**, 1008–1012.

Schiller, L.R., Santa Ana, C.A., Morawski, S.G. and Fordtran, J.S. (1984). *Gastroenterol.*, **86**, 1475–1480.

Seeman, P. (1982). *Biochem. Pharmacol.*, **31**, 2563–2568.

Seeman, P. and Grigoriadis, D. (1985). *Biochem. Pharmacol.*, **34**, 4065–4066.

Seeman, P., Ulpian, C., Grigoriadis, D., Pri-Bar, I. and Buchman, O. (1985). *Biochem. Pharmacol.*, **34**, 151–154.

Sengupta, S. and Lokhandwala, M.F. (1985). *J. Auton. Pharmacol.*, **5**, 289–294.

Shorr, R.G.L., Heald, S.L., Jeffs, P.W., Lavin, T.N., Strohsacker, M.W., Lefkovitz R.J. and Caron M.C. (1982). *Proc. Nat. Acad. Sci. (U.S.A.)*, **79**, 2778–2782.

Sidhu, A. and Fishman, P.H. (1986). *Biochem. Biophys. Res. Commun.*, **137**, 943–949.

Silverman, R.B. (1983). *J. Biol. Chem.*, **258**, 14766–14769.

Silvis, S.E. *et al.* (1985). *J. Clin. Gastroenterol.*, **7**, 482–487.

Simantov, R., Kuhar, M.J., Uhl, G.R. and Snyder, S.H. (1977). *Proc. Nat. Acad. Sci. (U.S.A.)*, **74**, 2167–2171.

Singer, T.P., Salach, J.I. and Crabtree, D. (1985). *Biochem. Biophys. Res. Commun.*, **127**, 707–712.

Smith, T.L. and Yamamura, H.I. (1985). *Biochem. Biophys. Res. Commun.*, **130**, 282–285.

Snyder, S.H. (1981). *Am. J. Psychiat.*, **138**, 460–463.

Snyder, S.H. (1986). *Nature*, **323**, 292–293.

Sokolovsky, M., Gurwitz, D. and Kloog, J. (1983). *Adv. Enzymol.*, **55**, 137–196.

Speight, T.M. and Avery, G.S. (1972). *Drugs*, **3**, 159–203.

Speth, R.I., Wastek, G.J., Johnson, P.C. and Yamamura, H.I. (1978). *Life Sci.*, **22**, 859–866.

Stahl, K.D., Van Bever, W., Janssen, P.A.J. and Simon, F.J. (1977). *Eur. J. Pharmacol.*, **46**, 199–205.

Sugrue, M.F. (1981). *Pharmacol. Ther.*, **13**, 219–247.

Sybertz, E.J., Sabin, C.S., Pula, K.K., Vander Vliet, G., Glennon, J., Gold, E.H. and Baum, T. (1981). *J. Pharmacol. Exp. Ther.*, **218**, 435–443.

Taylor, C.W. and Merritt, J.E. (1986). *Trends in Pharmacol. Sci.*, **6**, 238–242.

Thomas, J.M. and Misiewicz, G. (1984). *Clin. Gastroenterol.*, **13**, 501–541.

Van Zwieten, P.A. (1986). *J. Cardiovasc. Pharmacol.* **8**, Suppl. 4, S21–S28.

Vauholden, R., Lameire, N. and Ringoir, S. (1985). *Eur. J. Clin. Pharmacol.* **28**, 125–130.

Villalobos-Molina, R. and Garcia-Sainz, J.A. (1983). *Eur. J. Pharmacol.*, **90**, 457–459.

Wakade, A.R., Malhotra, R.K. and Wakade, T.D. (1986). *Nature*, **321**, 698–700.

Weinstein, D., Schenker, J.G., Gloger, I., Slonim, J.H., DeGroot, N., Hochberg, A.A. and Folman, R. (1981). *Fed. Eur. Biochem. Soc. Lett.*, **126**, 29–32.

Williams, L.T., Snyderman, P. and Lefkovitz, R.J. (1976). *J. Clin. Invest.*, **57**, 149–155.

Williams, M. and Yarbrough, G.G. (1979). *Eur. J. Pharmacol.*, **56**, 273–276.

Wilson, I.B., Harrison, M.A. and Ginsburg, S. (1961). *J. Biol. Chem.*, **236**, 1498–1500.

Wojcik, W.J. and Neff, N.H. (1984). *Molec. Pharmacol.*, **25**, 24–28.

Wright, D.J. (1987). *Biochem. Soc. Trans.*, **15**, 65–67.

Yu, P.H. (1981). *Can. J. Biochem.*, **59**, 30–37.

Zavecz, J.H., Jackson, T.E., Limp, G.L. and Yellin, T.O. (1982). *Eur. J. Pharmacol.*, **78**, 375–377.

Chapter 10

Membrane-active agents

10.1 Introduction

We consider in this chapter those examples of drugs which act directly on the cell membrane, whether mammalian or fungal. The processes concerned govern the transport of ions across the cell membrane in either direction, and are either channels through which ions pass, or require the hydrolysis of ATP to drive various ions across the membrane. Alternatively, the drug may disrupt the membrane of a pathogenic organism in such a way that it becomes highly permeable to small molecules, and consequently loses its effectiveness as a semi-permeable membrane. These effects may all be regarded as direct; indirect effects whereby a drug binds to a receptor which subsequently modulates membrane permeability are discussed in Chapter 9.

In order to understand more about how drugs can interact with membranes, it is important to know how present theories of membrane structure relate to transport of ligands and ions across membranes, whether this be carried out by enzyme or by ion channel. Furthermore, some understanding of the action potential (potential difference) that exists across the membrane is also important for an understanding of the mechanism of action of local anaesthetics and antiarrhythmic agents that act via blockade of sodium channels, and those antihypertensive drugs that intefere with calcium ion channels.

10.1.1 Membrane structure

Our present view of membrane structure in eukaryotic cells has been outlined by Singer and Nicolson (1972) and is depicted in Figure 10.1. The major matrix of the membrane comprises a double leaflet of phospholipid molecules. The polar headgroups of the two layers face out towards the aqueous environment of the cytoplasm and the extracellular medium respectively, while the non-polar fatty acyl hydrocarbon chains (two per molecule) are orientated towards the middle of the membrane. Cholesterol in mammalian cell membranes and ergosterol in fungal cell membranes are inserted into the lipid bilayer between phospholipid molecules. The 3-hydroxyl group common to both of the sterols is orientated towards the aqueous environments and

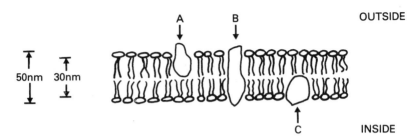

In the lipid bilayer, phospholipid molecules are shown with their polar head groups as small circles, and the non-polar tails as wavy lines. The leaflet is held together by the non-polar interaction between the fatty acyl sidechains. A is an integral protein bound in the extracellular surface of the membrane, while B is a transmembrane protein. C is an integral protein facing the cytoplasm. Both A and C penetrate only one half of the bilayer.

Figure 10.1 Membrane structure

interacts with the polar headgroups of the phospholipids, while the non-polar sterol skeleton and hydrocarbon tail are positioned so that they interact with the hydrocarbon chains of the phospholipid fatty acyl groups.

Proteins are also present in the membrane; the receptors for some hormones span the membrane, unlike adenylate cyclase which is positioned on the inside. Some transmembrane proteins Na^+/K^+ ATPase are the transport carriers for ions, while there are other protein carriers for small molecules such as glucose. These proteins have been shown to operate on both sides of the membrane. Thus for Na^+/K^+ ATPase, the sodium and ATP binding sites are on the cytoplasmic side of the membrane while the potassium and ouabain binding sites are situated on the extracellular side. As noted in Chapter 9 the GABA receptor contains a channel for chloride ions.

Membrane fluidity has a marked effect on the transport of small molecules across the membranes — the more fluid the membrane the more permeable it is. Fluidity is controlled to some extent by the type of fatty acid present in the phospholipids: saturated fatty acids decrease and unsaturated fatty acids increase the fluidity. Cholesterol also has the effect of reducing fluidity in areas that are rich in unsaturated fatty acids and *vice versa* in areas high in saturated fatty acids. The net effect is to produce regions of markedly differing permeability within the same membrane.

With regard to the mechanism of transport, Scarborough (1985) has suggested that when membrane-bound proteins bind ligands, they undergo a conformational change rather like the operation of a hinge so that the ligand is embedded within the protein. Scarborough views the open or ligand-free state as having a cleft (state 1 in Figure 10.2) within which the ligand binds (state 2). This binding induces a conformational change which opens an aqueous diffusion pathway from the binding site towards the far side of the membrane (state 3). The ligand is then free to escape to the opposite side (state 4). The conformation then changes back to the original to allow the process to begin again.

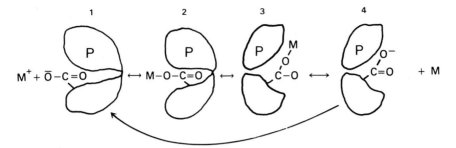

P represents the transmembrane protein that acts as a carrier.
M is the ligand to be transported; in this case a cation, and the carboxylate ion the binding site — probably the aspartate sidechain.
1-4 are the different states of the transport system.

Figure 10.2 Membrane transport

10.1.2 Dynamics of the heart beat

From a biochemical point of view the above description might be considered sufficient for an understanding of membrane events. The cell membrane has, however, been shown to support a considerable potential difference; 70–90 mV in cardiac cells and 200 mV in cells of the filamentous fungus *Neurospora crassa* (the inside of the membrane is negative with respect to the outside). The consequences that flow from changes in the electrophysiological state of the membrane are crucial for an understanding of the mechanism of action of three of the types of drug that are discussed in this chapter which modulate the transport of cations: local anaesthetics and type 1 antiarrhythmic agents both of which block the sodium channel, and calcium channel blockers that are also used for arrhythmias and for high blood pressure. In order to understand more about how these drugs exert their effects, an outline is presented here of the generation of the action potential in a pacemaker cell that controls the beating of the heart, specifically in the Purkinje fibres (Kumana and Hamer, 1979, described in Muhiddin and Turner, 1985, see Figure 10.3).

The sodium channel initiates the process in Phase 0 by allowing a rapid transport of sodium ion into the cell to occur, thus causing the membrane potential to depolarize from a potential of −70 mV to approximately +25 mV. These channels are then shut off or inactivated and cannot open again until the membrane has been repolarized. Next in phase 1 there is a sharp fall of about 20 mV probably due to chloride ion entry. In phase 2, calcium begins to enter through the slow calcium channels (called slow because the time constant for inactivation is 50 ms compared with 0·5 ms for the 'fast' sodium channel). Eventually chloride and calcium channels close and the major repolarization of the membrane is effected by a large outflow of potassium in phase 3. The potassium channels are inactivated when the membrane potential has reached about −80 to −90 mV, the resting potential.

A second outward potassium channel, known as the pacemaker current,

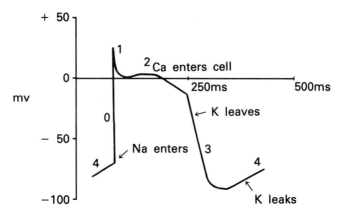

The phases of the normal action potential are shown; 0 — depolarization, 1,2,3 — repolarization and the diastolic phase (4).

Figure 10.3 Diagram of myocardial action potential

continues to operate for a short time. Eventually, the point is reached at which the inward sodium and calcium background current, which do not change with time, produce a net inflow of current. The overall effect in phase 4, therefore, is that of gradual depolarization as the background repolarizing current gradually falls away. When the membrane potential has dropped to −70 mV the reactivated sodium channel is triggered and the process repeats itself. It usually takes about 400 ms from start to finish in a pacemaker cell, which acts, as its name suggests, to initiate the process by spontaneously generating an action potential — a property known as automaticity. A group of pacemaker cells act together to initiate an impulse which spreads throughout the rest of the heart. Although the details of ionic involvement may differ in other cells in the heart, the requirement for fast sodium and slow calcium channels is constant.

Arrhythmias, or abnormal beating of the heart, can be caused either by faults in the mechanism for initiating the electrical impulse or by disturbances in the conduction of an impulse. Abnormal automaticity occurs when the membrane depolarizes too soon before phases 3 and 4 have fully run their course, and an extra impulse is generated that is propagated throughout the heart. The result can be almost random beating of the heart. This is often referred to as after-depolarization and if it occurs during phase 3 (early after-depolarization) is usually due to the opening of the slow calcium channels. The membrane potential has not fallen low enough to activate the fast sodium channels (Wit, 1985).

Alternatively, other cells that are not normally pacemaker cells may take it upon themselves to initiate an impulse — a form of automaticity that arises in the wrong place. This may be a consequence of disease affecting the responsiveness in that the sodium channel is oversensitive to stimulation. In arrhythmia caused by either situation, a drug that can partially block the initial sodium current is of value in steadying the heart beat and returning the process to normal. A partial blockade of the calcium channel is also found to be of value in some cases, particularly after depolarization.

The conduction of an impulse can also lead to arrhythmias if it re-enters a part of the heart that has already been excited. This normally requires that the rate of conduction be slowed, usually as a consequence of disease, and that a block in conduction occurs which can divert the impulse back into a region of the heart that it has already traversed (Wit, 1985).

10.2 *The sodium channel*

The mechanism of transmission of nerve impulses, and the propagation of electrical impulses in the heart that accompany beating, both require a sharp and rapid change in the trans-membrane potential difference. This is achieved by means of a swift transport of sodium cations into the cell through what is known as the sodium channel. Estimates of 10^7 ions per second or an electrical conductance of 2 to 8 pS (the unit of conductance is a Siemens abbreviated to S) per site have been made (reviewed in Catterall, 1980). This rate is too fast, apparently, to be associated with a mobile carrier, and it is likely that the sodium channel is a selective pore or hole in the membrane. It is not, however, open at all times and requires repolarization of the membrane to take place before it is reactivated. After a rapid transient influx of cations has taken place, the channel is closed by a mechanism that is not understood.

At least two types of drug interact with the electrically excitable sodium channel to block it; local anaesthetics, where the transmission of nerve impulses must be blocked, and anti-arrhythmic drugs that are needed to restore the beating of the heart to normal. An outline of the different sites in the sodium channel is essential for an understanding of the detailed mechanism of action of these drugs.

Much of our understanding of the mechanics of operation of the sodium channel derives from studies of three different families of toxins that bind at separate sites in or near the channel. Tetrodotoxin is found in the tissues of a number of aquatic creatures including puffer fish, newt and octopus, and is an agent that binds to a site designated as site 1. The toxin is a heterocyclic compound containing both a guanidine group and an ionizable hydroxyl (pK_a about 8·0) that are essential for its action, probably because they represent the binding site for the toxin receptor.

The toxin blocks the increase in sodium ion permeability caused by a slight depolarization of the membrane, but has no effect either on the ion permeability of the unstimulated axon of a nerve cell or on another current generated by depolarization, namely the potassium outward current (Catterall, 1980). Tetrodotoxin binds equally well to resting, activated or inactivated channels since the binding is unrelated to voltage. This effect takes place at concentrations below 10^{-7} M and is saturable, as shown by classical hyperbolic binding curves. It is suggested that one molecule of toxin is sufficient to block one sodium channel (Catterall, 1980).

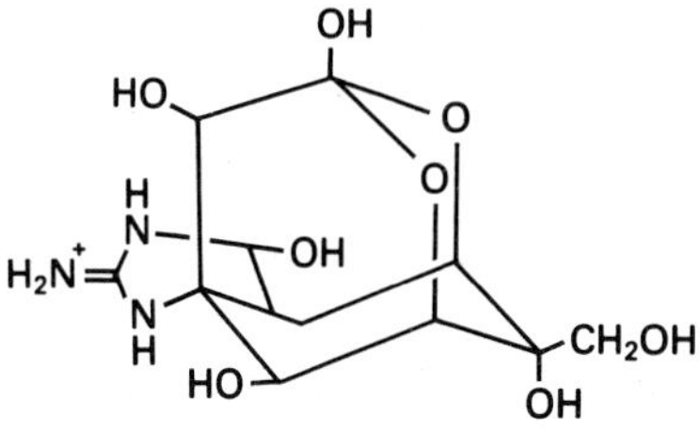

Tetrodotoxin

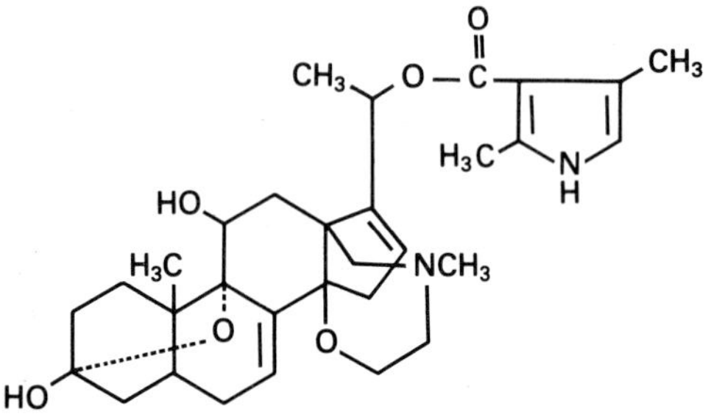

Batrachotoxin

A second type of toxin, exemplified by batrachotoxin, is a steroid alkaloid derivative found, amongst other sources, in the Colombian poison dart frog; it produces a completely opposite effect to site 1 ligands. The oxygen bridge between the A and B rings and the dimethylpyrrole carboxylate ester moiety are essential for activity. This toxin binds to a second site in the channel and causes it to become greatly sensitized by shifting the voltage dependence of activation some 30 mV to more negative potentials, so that the channels become sensitive to activation at a potential much closer to the resting potential. Indeed, at the resting potential, which is not altered, some of the channels become permanently activated. Batrachotoxin is, therefore, acting as a full agonist of the channel. Tetrodotoxin, on the other hand, is a non-competitive inhibitor of batrachotoxin action, which is consistent with the view that the two toxins may act at different sites.

Another totally different structural type, the polypeptide toxin obtained from scorpion venom, binds to a third site in the channel. Scorpion toxin interacts in an allosteric fashion with site 2 ligands, by shifting the ligand binding curves as much as 20-fold to lower concentrations i.e. to increase the binding. The electrophysiological effect is to enhance persistent activation of sodium channels. The binding of scorpion toxin to its receptor is highly voltage-dependent and as the membrane potential falls, the binding of the toxin increases. Tetrodotoxin also antagonizes the action of scorpion toxin in a non-competitive fashion.

10.2.1 Sodium channel blockers — local anaesthetics

Local anaesthetics block both the generation and the transmission of sensory nerve impulses by binding to sodium channels in the nerve cell membrane. Normally, a slight depolarization of the membrane opens the sodium channel to allow a rapid but transient inward current. Local anaesthetics raise the threshold for electrical excitation of the membrane to such an extent that the permeability of the membrane becomes so low as to result in a block of nerve conductance. Small nerve fibres are blocked more rapidly than thick ones, possibly because the former are demyelinated, unlike the latter.

The detailed mechanism of action of these drugs has long been a subject for controversy. The potency of compounds increases with increasing lipophilicity, and it was thought initially that the drug simply dissolved in the membrane thereby blocking the channel in a non-specific way. Indeed, local anaesthetics will inhibit other membrane enzymes such as, Na^+/K^+ ATPase, but at rather high concentrations of 10^{-4} M and above (Agarwhal and Kalra, 1984). Recently, however, a number of lines of circumstantial evidence have accumulated which indicate a more specific binding to the sodium channel:

(a) The affinity of the local anaesthetics for the channel is dependent on the voltage which implies that a particular receptor becomes available as the channel is opened and the membrane potential falls.

(b) Quaternary amine local anaesthetics act more rapidly when applied from the intracellular side of the axonal membranes than from the extracellular side, suggesting that the site of action is located nearer the cytoplasm.

(c) Sodium currents can be blocked more rapidly and completely when the channels are repeatedly activated (use-dependent inactivation or block). This has been explained by assuming that the ligand binds more rapidly to open states of the channel whilst more tightly to inactivated states. This is difficult to explain with non-specific binding. This use-dependent block also occurs with anti-arrhythmic agents (see Section 10.2.2).

(d) If the non-specific binding hypothesis were correct, we would expect the potency of these drugs to decrease with cooling. This is because the hypothesis states that the effect of the anaesthetic is proportional to the ligand concentration in the membrane, and as the temperature is lowered the partition coefficient of ligand between lipid and aqueous medium falls, thus reducing the ligand concentration in the membrane (see Chapter 1 for discussion of partition coefficients). In fact, no change in activity was found by Bradley and Richards (1984) on lowering the temperature of isolated sciatic nerve from 20°C to 4°C. With some drugs the activity was actually enhanced.

Perhaps the most detailed evidence for a specific receptor site has been provided by Creveling *et al.* (1983) and Postma and Catterall (1984) who found that site 2 was almost certainly the binding site for local anaesthetics. Batrachotoxin induces a time- and concentration-dependent depolarization of a guinea pig cerebral cortex neurone, with $K_d = 1 \cdot 1 \times 10^{-8}$ M. When seven local anaesthetics were examined they were found to antagonize this effect with K_i from 9×10^{-7} M for dibucaine to $7 \cdot 8 \times 10^{-4}$ M for lidocaine. These seven inhibition constants were identical, with the exception of lidocaine, to those obtained from inhibition of binding of ^{3}H-batrachotoxin to voltage-sensitive channels in vesicles obtained from guinea pig cortex (Creveling *et al.*, 1983). A series of 11 local anaesthetics was tested for correlation between inhibition of toxin binding and blockade of channel transmission and a correlation coefficient of 0·84 was obtained (Postma and Catterall, 1984). Interestingly, local anaesthetics had no effect on the binding of scorpion toxin thus confirming the distinct nature of the binding site.

With respect to the correlation of increasing potency of local anaesthetics with lipophilicity, it is probably a reflection of how efficiently the drug can be transported across membranes into the cell in the neutral state, since it is unlikely that the cationic form would be able to be transported in such a fashion. Subsequent activity inside the (aqueous) sodium channel would require the cationic form to enter the channel from the cytoplasm. Consistent with this view, all the commonly used local anaesthetics contain a secondary or tertiary amino nitrogen which has a pK_a between 8 and 9, so that between 5 and 20% of the drug will be in the form of a neutral amine at pH 7·4 (pK_a is the pH at which 50% ionization occurs. As pH is a logarithmic scale, at one unit removed from the pK_a, only 5% of one component and 95% of the other are present). Indeed, the fact that the quaternary nitrogen forms of these compounds are active tends to underline the importance of the cationic forms of these compounds. This is not to say, however, that the neutral form of the drug is not active since it could migrate through the lipid portion of the surrounding cell membrane (see discussion on anti-arrhythmic agents and Hille, 1977).

The drugs have a strictly limited life in the circulation since they are all esters and are hydrolysed by plasma esterases. This ensures that one of the major requirements of a local anaesthetic is met, namely a transient and reversible anaesthesia. In principle, however, it is possible to see that such drugs could be of use as anti-arrhythmics if their duration of action were sufficient; in procainamide, an amide link replaces the ester group in procaine and the reduced hydrolysis by esterases allowed the drug to be used as an anti-arrhythmic. Lidocaine is an exception in being used both as a local anaesthetic and anti-arrhythmic, but in the latter case it is usually during acute myocardial infarction or cardiac surgery where a long duration of action is not required.

$$H_2N-\!\!\left\langle\bigcirc\right\rangle\!-CO-NH(CH_2)_2N\!\!\begin{smallmatrix}{}^{C_2H_5}\\{}_{C_2H_5}\end{smallmatrix}$$

Procainamide

$$\begin{smallmatrix}CH_3\\ \end{smallmatrix}\!\!\left\langle\bigcirc\right\rangle\!-NH-CO-CH_2-N\!\!\begin{smallmatrix}{}^{C_2H_5}\\{}_{C_2H_5}\end{smallmatrix}\;\begin{smallmatrix}\\CH_3\end{smallmatrix}$$

Lidocaine

Local anaesthetics find a use in dentistry and a variety of surgical procedures to reduce the pain suffered by the patients, without causing unconsciousness. Examples of this procedure include the injection of a local anaesthetic to block a particular nerve such as the median or ulnar at the elbow to allow surgery at the wrist (nerve block anaesthesia); infiltration anaesthesia where a particular tissue is to be removed, and surface anaesthesia where an operation is to be carried out on the mucous membranes of, for example, nose, mouth and throat (Ritchie and Green, 1985).

10.2.2 Anti-arrhythmic agents

Anti-arrhythmic drugs may be classified into five groups (Muhiddin and Turner, 1985); sodium channel blockers such as lidocaine and procainamide belong to Class 1; sympatholytic agents such as β-blockers noted in Chapter 9 form Class 2; Class 3 is composed of agents such as amiodarone that prolong the repolarization phase; calcium channel antagonists form Class 4 (see Section 10.3.1 this chapter) and chloride channel antagonists Class 5, discussed in Section 10.4 this chapter. It is the agents of Class 1 with which we are concerned in this section.

Class 1 anti-arrhythmic drugs block sodium channels in both nerve and cardiac cells but their efficacy is greater on the latter by a factor of between 10 and 200 times, and so local anaesthetic action is likely to make only a very modest contribution while the drugs are being used for heart conditions. Three states of the cardiac sodium channel are postulated: resting, activated (open) or inactivated (closed) after use. Furthermore, the idea that if the drug binds to a channel conductance is shut off is confirmed. These different states of the channel have different affinities for the drug, and the receptor site is on or very close to the ionic channel. The concept of a use-dependent block, noted above for local anaesthetics, also holds good here in that if the rate of beating of the heart is sufficiently fast, the effect of the drug in inhibiting the transport of sodium ion across the membrane increases with every beat. This concept has been termed the "modulated receptor hypothesis" (Hondeghem and Katzung, 1984).

Repolarization of the membrane closely parallels the activation of the channel in that when the membrane potential has returned to the resting level, the channels are reactivated. Inactivated channels (channels in a depolarized membrane) have a much greater affinity for blocking drugs, and recover much more slowly from blockade, than do resting channels. The Class 1 anti-arrhythmics have, as their main effect, the reduction of the maximum rate of depolarization. In addition, they may raise the threshold of excitability, making the cardiac cell less sensitive to being triggered. These drugs may also prolong the period during which no action potential can be initiated, known as the refractory period. They produce no change in the resting potential (Muhiddin and Turner, 1985).

A similar argument can be applied to pH behaviour of sodium channel blocking anti-arrhythmic agents (Class 1 — Muhiddin and Turner, 1985) as to local anaesthetics (see Section 10.2.1) as the drugs are almost all weak bases with pKa values in the range from 7·5 to 9·5. In the case of anti-arrhythmic drugs it has been proposed that both neutral and cationic forms of the drugs, which will both be present at neutral pH, have sodium channel blocking activity; the cations can enter the channel only from the cytoplasm while the neutral form can pass through the lipophilic portions of the membrane into the channel, and then ionize. This is particularly possible if the channel is closed (Hille, 1977).

Recent studies on the effect of charged and uncharged analogues of lidocaine on the use-dependent block of mammalian cardiac fibres has indicated that the charged form of the drug is likely to be the active one, since benzocaine (which is uncharged) has no effect (Gintant *et al.*, 1983). The effect of lidocaine on sodium channels has also been discussed by Bean *et al.* (1983) who showed that the results were consistent with the "modulated receptor hypothesis" noted above, in that lidocaine block increases with increasing frequency of channel use. This reflects block of inactivated channels rather than open ones. The concentration at which the channel was half-blocked was 1×10^{-5} M where inactivation was nearly complete.

10.2.3 Diuretics

The type of sodium channel discussed in the previous two sections is that present in tissue which is electrically excitable. In epithelial tissue that is not electrically excitable, such as that lining the surface of the kidney or cornea, there are three other types of sodium channel which are characterized by insensitivity to toxins such as tetrodotoxin even at high concentrations, and by a slower rate of ion transport than the fast sodium channels of the heart. One group of these channels is characterized by a high trans-epithelial potential difference and resistance; these are known as "tight" epithelia and occur in the toad urinary bladder. Another group, which includes the proximal tubule of the kidney (see Figure 10.4), have a very low or negligible trans-epithelial potential and are able to transport large volumes of isotonic fluid (leaky epithelia). Intermediate between these extremes are the channels with moderate trans-epithelial potential difference such as those in the cells of the distal portion of the kidney tubule (reviewed in Cuthbert, 1977) and occur in the hatched area for amiloride and triamterene action (see below) in Figure 10.4.

The role of these sodium channels in the kidney is to reabsorb sodium ion which, together with chloride ion and other solutes, has been filtered from the blood in the glomerular portion of the kidney (see Figure 10.4). At the same time, in order to maintain equal osmotic pressures either side of the kidney cell membrane, fluid is absorbed.

The transport of sodium across the tight epithelia is reasonably well understood in that sodium passes across the membrane from the bladder into the cell, passively down an electrochemical gradient. Sodium pumps located on the other side of the cell (known as the baso-lateral) remove the sodium by an

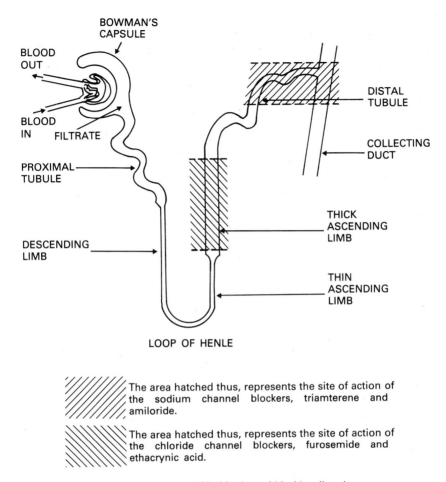

The area hatched thus, represents the site of action of the sodium channel blockers, triamterene and amiloride.

The area hatched thus, represents the site of action of the chloride channel blockers, furosemide and ethacrynic acid.

Figure 10.4 Site of action of sodium and chloride channel blocking diuretics.

active transport process across the membrane coupled to an enzyme hydrolysing ATP. The sodium ion flow may help to carry chloride ions in addition, usually against the electrochemical gradient of the latter. Transport systems in leaky and intermediate epithelia are less well understood at present although it is probable that an electroneutral sodium exchange for protons may take place as in the renal proximal tubule.

Amiloride and triamterene are both commonly used diaminopyrimidine diuretics which block the sodium channel in the distal portion of the kidney tubule (see Figure 10.4). Amiloride has been shown to bind to the exterior surface of the cell membrane instantaneously and reversibly. The drug thereby prevents sodium transport, and increases the electrical resistance of the toad bladder that is used as a model of a tight epithelium. The reduction in the negative charge of the membrane (depolarization) is blocked, and consequently the secretion of potassium ion into the lumen of the tubule, which normally serves to maintain the membrane potential, does not occur (reviewed in Gussin, 1977; Kokko, 1984; Lant, 1985a).

Amiloride blocks sodium transport in tight epithelia with an I_{50} less than 10^{-6} M. It also inhibits with a higher IC_{50} electroneutral exchange processes (Benos, 1982). Kinetic studies of the inhibition of sodium uptake indicate that the process is saturable and is amenable to study using Michaelis-Menten analysis. Amiloride inhibition is non-competitive with sodium ion. Hill plot analysis yields a Hill coefficient significantly less than 1 (0·82), which suggests negative cooperativity between amiloride binding sites (see Appendix for discussion). The proposal has been made that amiloride binds to a receptor that modifies the mouth of the channel in some way that prevents the ingress of sodium ions (Benos, 1982). It is also possible that chloride ion is required for amiloride action, but not for its binding (see Benos, 1982).

The precise nature of the amiloride receptor is not known but structure–activity relationships indicate that any group that prevents the guanidine side-chain taking up a position co-planar with the pyrimidine ring reduces activity. Moreover, benzyl groups can be attached to the guanidine nitrogen and activity is increased. The active form of the drug is the cation in which the positive charge resides on the guanidine cation. The pK_a is 8·7, and at pH below 5 and above 11, amiloride ceases to have any activity. Above pH 11 the drug is deprotonated, while below 5 it is proposed that amiloride competes with protons for a binding site with a pK_a of around 5 (Benos, 1982).

Amiloride exerts its effects on the sodium channel and not on the outward potassium channel, which indicates that the two alkali metal cations are transported by different channels — not surprising in view of their different sizes. The drug is also active, unlike triamterene, on sodium channels in the proximal tubule and the cortical collecting tubule (Gussin, 1977) in addition to those on the distal tubule (see Figure 10.4). The proximal tubule channels, however, appear to be of the sodium/hydrogen electroneutral exchange type where the drug is less effective, and it is not clear how the drug is working in this case.

Amiloride and triamterene do not give rise to excessive potassium secretion (spironolactone acts similarly, see Chapter 6) — thus preventing hypokal-emia. The drug's major action is to prevent sodium (and concomitantly chloride ion) reabsorption together with fluid. The urine is thereby increased in volume, sodium and bicarbonate ion excretion is increased (the pH of the urine rises) and there is a fall in potassium excretion.

There are some conditions where there is an excess of water (oedema) in certain peripheral tissues of the body and also the lung as, for example, after heart failure, and when it is therapeutically advantageous to block the reab-sorption of fluid. This can be carried out by blocking the re-uptake in the kidney of sodium ion through specific channels, as with amiloride and triam-terene, or chloride ion in a similar fashion by furosemide and ethacrynic acid (see Section 10.4 on chloride channel blockers), although the specific agents operate at different sites in the kidney tubule. These drugs all help to reduce excess fluid in the tissues, and thereby reduce the workload of the heart — clearly an important contribution to the therapy of heart disease.

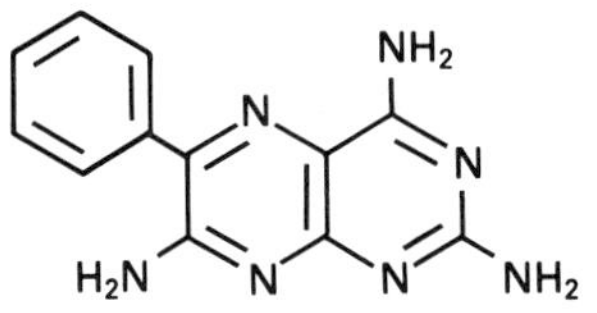

Triamterene 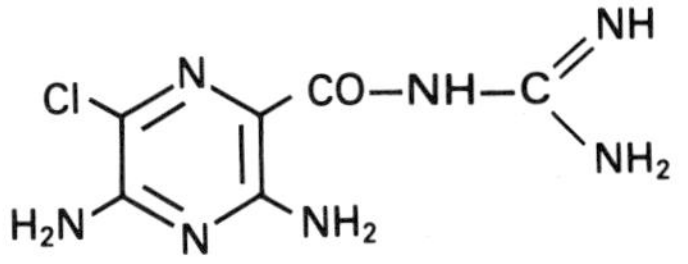   Amiloride

10.3 The calcium channel

Calcium is indispensable for the initiation of a number of essential cellular processes, notably muscle contraction as for example in the heart, secretion of neurotransmitters and hormones from neurone and endocrine cells, and rhythmic firing of heart and nerve cells. Some of the calcium required for this activation may be found intracellularly, bound to endoplasmic reticulum, mitochondria and to the calcium binding proteins such as calmodulin. The quantity of calcium available in these sites is sufficient to support only a short burst of activity, however, and for longer sustained activity calcium has to be drawn into the cell from outside through specific ion channels (reviewed in Droogmans *et al.*, 1985), except for the platelet which can sustain the release of ADP from intracellular stores of calcium.

Calcium channels are in a continuous state of flux, opening and closing repeatedly, typically remaining open for 1 ms and then closing for between 1 and 100 ms. The channels may be opened by the binding of a neurotransmitter or hormone to the cell surface, in which case they are termed receptor-dependent. Neurotransmitters activate the channel not by increasing the number of ions flowing through per unit time (conductance), but rather by increasing the length of time that the channel remains open (and reducing the time that it is shut) (reviewed in Schramm and Towart, 1985). Catecholamines, such as noradrenaline, usually act through cAMP and its attendant protein kinases to phosphorylate a membrane protein to open the channel while, in contrast, phosphatases dephosphorylate the protein during inactivation (see Glossman *et al.*, 1982).

Alternatively, other calcium channels may be influenced by changes in the membrane potential, and are known as voltage-dependent channels. Using the patch-clamp technique, whereby a single cell can be maintained at a given potential difference across its membrane, the kinetics of opening and closing (known as "gating"), conductance and sensitivity to various drugs can be determined.

Recent studies indicate that there are three types of voltage operated calcium channels to be found in sensory neurones and two types in cardiac cells (Nowycky *et al.*, 1985; Nilius *et al.*, 1985). They differ primarily in that different reductions in the membrane potential are required to affect their gating. As the membrane is depolarized a channel is opened (called T for transient) that shows a moderate current and rapid decay. Stronger depolarization opens a channel termed L (for longer lasting) as it decays only slowly, while another channel termed N (neither T nor L) is also opened by strong depolarization and grows more strongly with increasing depolarization. L is apparently the channel which is sensitive to 1,4-dihydropyridine drugs such as nifedipine (Nowycky *et al.*, 1985) — see below.

As with all biologically active molecules, mechanisms must exist for terminating their action. In the case of calcium it clearly cannot be metabolized and so it has to be removed from the cell cytoplasm. One mechanism is via the calcium pump — a calcium/calmodulin activated membrane enzyme which uses the energy from the hydrolysis of ATP to drive calcium ions out of the cell. This enzyme is activated as soon as the cellular level of free calcium ion rises much above the resting level of approximately 2×10^{-7} M. The maximum activation of the contractile proteins in smooth muscle cells, for example, is achieved at 10^{-5} M, and this level also induces secretion in hormone-producing cells that are activated by calcium ions.

Another cellular mechanism for removing calcium from the cytoplasm is by the sodium/calcium exchange where calcium is pumped out of the cell in return for sodium entry. Cardiac glycosides exert their pharmacological effects on this exchange (see Section 10.5.1). The extracellular level of calcium ion is approximately 10^{-3} M, and the gradient is maintained to a large extent by the very low permeability of the cell membrane to the doubly charged alkaline earth metal.

10.3.1 Calcium channel antagonists

In addition to the patch-clamp technique noted above, pharmacological intervention has led to great advances in our understanding of the calcium channel, although so far no toxins have been discovered to delineate binding sites in the channel. A particular class of ligands known as the 1,4-dihydropyridines have been found to bind selectively and specifically to the inward (cytoplasmic) side of the calcium channel and to prevent the transport of calcium ions. These ligands include a number of drugs used for the control of angina pectoris and of high blood pressure. Nimodipine, for example, shows a very tight binding to the channels in guinea pig brain and has a K_i of 7×10^{-10} M (Glossman *et al.*, 1982).

	R_1	R_2	R_3	R_4	R_5
Nifedipine	H	NO_2	CH_3	CO_2CH_3	CH_3
Nimodipine	NO_2	H	$CH(CH_3)_2$	$CO_2(CH_2)_2OCH_3$	H
Bay k 8644	H	CF_3	CH_3	NO_2	CH_3

Verapamil is an example of a second chemical class of calcium antagonist which interferes with nimodipine binding in a biphasic fashion. A third type is exemplified by diltiazem which potentiates nifedipine binding allosterically (Glossman *et al.*, 1982).

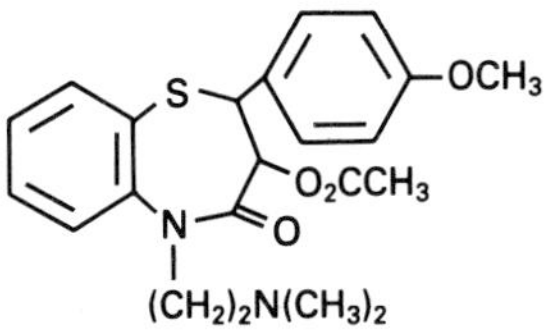

Diltiazem

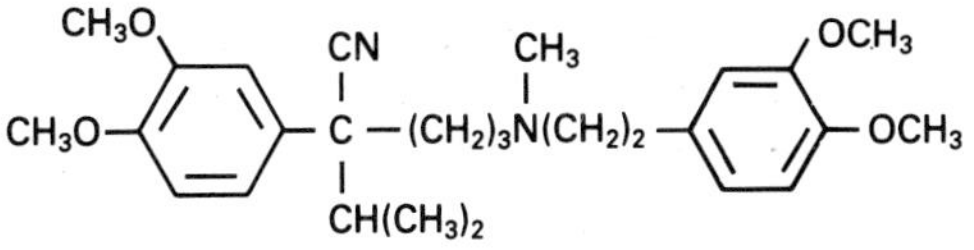

Verapamil

One of the problems to be taken into account is that calcium antagonists usually interact with the open form of the channel but not with the closed or resting form (Spedding, 1985). Thus in the reverse fashion to neurotransmitters and the receptor-operated channels, calcium antagonists prolong the closed period, and reduce the length of time that the channel is open. The equation describing the interaction is:

$$C_r \rightleftharpoons C_o + B \rightleftharpoons C_oB$$

where C_r is the resting state, C_o the open state, C_oB the blocked state and B the antagonist drug. If the channel is opened by potassium-induced membrane depolarization, the equation simplifies to that governing receptor/ligand binding. Hence competitive binding may be frequently observed in smooth muscle, which is continuously depolarized, but not so frequently in myocardial tissue.

Various other attempts have been made to classify calcium antagonists on the basis of their pharmacological action rather than chemical structure and binding characteristics. Fleckenstein-Grun *et al.* (1984) propose that the more active agents such as verapamil, nifedipine and diltiazem should be classified in one group since they reduce the calcium-dependent activation of myocardium by at least 90% without affecting the fast sodium channels. Their second group, which includes the diphenylalkylamines such as prenylamine and cinnarizine, are less specific and start to block the sodium channel when the block of the calcium channel has reached 50–70%. This may be a consequence of differential action of the optical isomers; the + isomer interfering with the sodium channels while the isomer preferentially blocks the calcium channels (Bayer *et al.*, 1975).

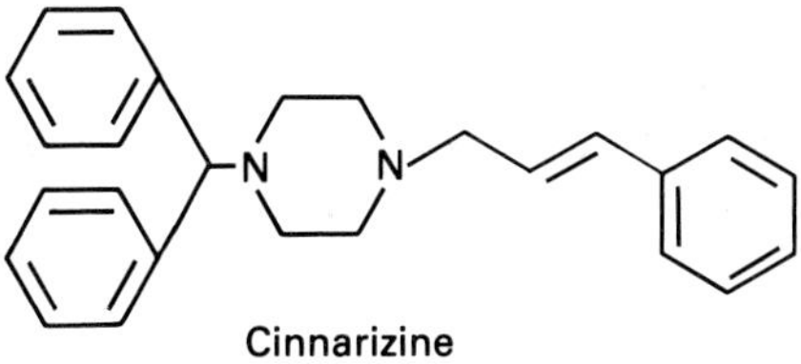

Cinnarizine

Spedding (1985), on the other hand, considers that these drugs may be classified into Group 1 – nifedipine and other 1,4-dihydropyridines; Group 2 – verapamil and diltiazem and Group 3 – diphenylalkylamines partly on the grounds that the presence of salicylate has no effect on nifedipine antagonism of calcium-induced contractions of potassium-depolarized preparations from the guinea pig caecum. In contrast, salicylate antagonizes Group 2 drug action and potentiates Group 3. The mechanism of these interactions is not entirely clear however, and it may be most useful in the long run to adopt the classification of Glossman *et al.* (1982), based on both structure and binding effects.

Binding studies on the molecular architecture of the calcium channel are at present in the early stages. It appears that a 1,4-dihydropyridine receptor has been isolated from various tissues and its molecular weight estimated at from 100 to 300 kDa (see Schramm and Towart, 1985). The reason for this discrepancy is not clear. Further biochemical studies on calcium uptake are not easy to carry out since the proportion of calcium entering the myocardium, for example, as a consequence of excitation/contraction coupling is only 2% of the total. Nevertheless, it has been shown using ^{45}Ca that nisoldipine, an analogue of nifedipine, can inhibit this influx into rabbit aorta in a fashion that correlates with inhibition of smooth muscle contraction (see Rahwan, 1983).

The effects of calcium entry blockade are more readily seen on the whole heart. Verapamil was first shown by Fleckenstein's group to mimic the effects of calcium withdrawal, as the drug reduced (a) calcium-dependent contraction without affecting sodium-dependent aspects of the action potential and (b) calcium-dependent myofibrillar ATPase activation and oxygen consumption of the beating heart (see Fleckenstein-Grun *et al.*, 1984).

The situation in which these effects may be particularly useful is when the oxygen available to part of the heart is greatly reduced as, for example, after a myocardial infarction. In this condition a blockage occurs in a branch of an artery supplying the heart, cutting off part of its blood supply. The area involved rapidly uses up the available oxygen and becomes hypoxic. It is believed that catecholamines released as a result will greatly increase calcium uptake into the myocardial cell, sending the muscular apparatus into a contractile spasm and using up ATP in the process. Mitochondria try to take up calcium from the cytoplasm and use up more ATP. Eventually the calcium load damages the structure of the organelle and uncouples oxidative phosphorylation. The loss of ATP ultimately shuts off the plasma membrane calcium pump, thus allowing calcium to accumulate further and increase the damage. Calcium channel blockers, although less effective against neurotransmitter-operated channels may still be of use in a cardioprotective fashion (Kendall and Horton, 1985), and indeed are already used before cardiac surgery (Fleckenstein-Grun *et al.*, 1984).

At present, however, the main uses of calcium channel antagonists are for the treatment of angina pectoris, hypertension and arrhythmias (see

Schramm and Towart, 1985). Angina pectoris is characterized by dull chest pain brought on by conditions of stress such as over-exercise, over-indulgence, smoking etc. and is the result of insufficient oxygen for the oxidation of substrates in the heart muscle. Calcium antagonists are able to reduce the oxygen demand of the heart muscle by reducing the amount of muscular activity and are particularly useful for the treatment of angina pectoris.

Another type of angina known as variant angina is characterized by a reduction in blood flow in the heart rather than lack of oxygen. Calcium channel blockers are of use in this condition because they are particularly effective at relaxing cardiac smooth muscle in the coronary arteries, thereby increasing the diameter of the artery. As a consequence, the effort the heart has to make to drive the blood through the vessels is reduced and the blood pressure falls. In addition, calcium channel antagonists relax peripheral blood vessels, thereby reducing blood pressure, and can be used to treat both acute hypertensive crises and chronic hypertension.

The type of arrhythmia in which there is a change in the normal beating of the heart by a group of cardiac cells beating in a disorganized fashion (when pacemaker groups of cells arise in the 'wrong' part of the heart) responds to verapamil, but not nifedipine (Spedding, 1985). In some cases these arrhythmias can arise from a partial depolarization of the membrane causing the excitation process to depend on the slow calcium channel rather than a fast sodium channel (Fleckenstein-Grun *et al.*, 1984).

Another potential therapeutic use is in atherosclerosis. This disease has long been attributed to the development of fatty streaks, and eventually larger deposits called plaques, in coronary arteries as a consequence of the intracellular deposition of cholesteryl esters. Subsequently, a fibrin clot forms at the place where the artery wall is weakened, the artery becomes clogged and myocardial infarction is the result. Fleckenstein-Grun *et al.* (1984) suggest that the presence of calcium in the arterial wall (hardening of the arteries) may be a factor contributing to the damage, particularly in those over 60 years of age. In animal experiments, calcium antagonists are able to prevent the uptake of calcium into the arterial wall and subsequent damage resulting.

10.3.2 Calcium channel agonists

Recent work has identified some compounds that activate the slow calcium channel and are known, therefore, as calcium channel agonists. They have contributed considerably to our understanding of the nature of those channels. Although they are not on the market as drugs, they have potential value as positive inotropic agents (i.e. to increase the force of contraction of the heart) for the treatment of heart failure. This requires, however, their effects on heart and blood vessels to be separated; most of the known agonists such as Bay k 8644, a 1,4-dihydropyridine analogue, act to constrict blood vessels, which is distherapeutic (reviewed in Schramm and Towart, 1985).

10.4 Coupled sodium/chloride ion channels

As noted in the section on sodium ion channels (10.2.3) chloride is reabsorbed in the kidney by active transport systems of which that in the ascending limb of Henle's loop is particularly active (see Figure 10.4). The transport system has been proposed as a sodium/chloride co-transport with sodium entering the cell down its electrochemical potential and chloride moving against its electrochemical gradient. The chloride ion exits across the opposite border of the cell down a favourable potential difference (Frizzell *et al.*, 1979). In earlier years it was thought that chloride uptake alone was the target of the drugs discussed in this section and so much of the work has been directed to the transport of chloride but not sodium (hence the emphasis on chloride in the studies reported below). Nevertheless, it is now clear that the co-transport of the ions is the major target.

Two structurally unrelated diuretics, furosemide and ethacrynic acid, among others, act to inhibit this coupled transport and are known as "loop" diuretics. Up to 25% inhibition of chloride and sodium reabsorption is possible with these agents which is a far higher proportion than that obtainable with other diuretics — hence the term "high ceiling" is used to describe these diuretics. Perhaps not surprisingly, such a powerful effect is characterized by swift onset in about 15 minutes and rapid cessation in 4 to 6 hours (Kokko, 1984; Lant, 1985a).

Furosemide acts from within the tubular lumen, i.e. the fluid side, not the intracellular side, to inhibit active chloride transport (see Odlind, 1984). The drug reaches the lumen by being transported from the proximal tubule cell through a non-specific channel that secretes organic acids, thus using the normal kidney transport pathways to reach its site of action (Kokko, 1984; Lant, 1985a). Furosemide demonstrates competitive kinetics with chloride ion uptake and non-competitive with sodium ion which suggests, but does not prove, that the drug may bind at or near the chloride uptake site (Ludens, 1982). Coupled transport exhibits saturation kinetics as would be expected of an active transport system. Furthermore, ouabain blocks active transport of chloride which suggests that the process is linked to a Na^+/K^+ ATPase as this drug is a specific inhibitor (see Section 10.5). It is impossible, however, to be more precise at present because no definitive receptor has yet been identified for furosemide.

Frog cornea, as another epithelial system, is believed to contain similar coupled transport systems to the kidney, although the cornea secretes chloride whereas absorption occurs in the ascending limb of Henle. Electrochemical studies have confirmed the selective inhibitory effect of furosemide on chloride uptake, and have shown that the drug increases the electrochemical gradient of chloride across the apical membrane of frog cornea. It prevents equalization of chloride concentrations either side of the membrane, leading to hyperpolarization of the apical membrane, and depolarization of the basolateral membrane (Patarca *et al.*, 1983). The net effect is to reduce the permeability of the membrane to chloride transport.

Ethacrynic acid   Furosemide

The receptor for ethacrynic acid too has not yet been identified. It is likely, however, bearing in mind the differences in structure between furosemide and ethacrynic acid, that the receptors for the two drugs differ. Ethacrynic acid is effective at blocking chloride transport across the epithelial membrane of the ascending loop of Henle with a concomitant decrease in potential difference. A decrease is found in this instance because the potential difference across the luminal membrane is maintained by the chloride transport (Burg and Green, 1973). Interestingly, the cysteine adduct of ethacrynic acid, which appears to be a normal excretion product, is far more effective than the drug itself; this adduct may therefore be the active form of the drug *in vivo*.

The contention that the loop of Henle is the sole site of action of these drugs has been challenged by Caffruny and Itskovitz (1982) who demonstrated that both drugs can act to block chloride transport in the proximal tubule, but the importance of this observation is not entirely clear. Obviously, there is much that we do not yet understand in the precise molecular mode of action of the sodium/chloride channel blocking diuretics.

10.5 Membrane-bound ATPases

There are a variety of enzymes in eukaryotic and prokaryotic systems that use the energy of hydrolysis of ATP to drive the transport of protons and other cations across membranes that are otherwise impermeable to these ions. These enzymes are all dependent upon magnesium, presumably because the true substrate is an ATP/magnesium complex. One group that is part of the process of oxidative phosphorylation is characterized by the reverse reaction of ATP synthesis coupled to oxidation and is found in the membranes of mitochondria, bacteria, chloroplasts etc. The mitochondrial enzyme is crucial to the chemi-osmotic principle of Mitchell which states that phosphorylation of ADP is driven by the energy of a proton gradient across the mitochondrial membrane. These enzymes are often referred to as F_0/F_1 ATPases because they consist of two portions: one is responsible both for anchoring the enzyme in the membrane (F_1) and for the translocation of hydrogen ion, while F_0 is the binding site for the interconversion of adenine nucleotides. General inhibitors of these enzymes are oligomycin which acts on F_0 and blocks proton transport and dicyclohexylcarbodiimide (DCCD), which reacts with a carboxyl group crucial for the enzyme reaction (reviewed in Pedersen, 1982).

Another group of ATPases, bound to the plasma membrane of the cell, is concerned with the pumping of cations against a concentration gradient out of the cell; notable examples are: (a) the enzyme that catalyses the exchange of sodium for potassium in the cardiac cell membrane particularly, although it is also found in other tissues, (b) the calcium pump which drives calcium out of secretory cells, among others and (c) the proton pump which exchanges protons for potassium ion in the parietal cells of the stomach and thereby produces the low gastric pH. In general, the cation pumps consist of at least one polypeptide with molecular weight in the region of 100 kDa which usually acts as a channel for the cation transport.

Oligomycin does not inhibit these enzymes but DCCD does, again because there is a carboxyl group in a glutamate side chain which is essential for activity. A phosphate-enzyme intermediate is an essential step in the mechanism, the phosphate being attached to the active site aspartate. Vanadate is a characteristic inhibitor of these enzymes probably because it can replace the phosphate in the intermediate, forming a vanadyl-enzyme intermediate. A considerable degree of homology between cation pumps has been discovered, which suggests that they may have evolved from a common ancestor (see Serrano *et al.*, 1986).

Mitchell has suggested a charge relay type mechanism in which the transport of a proton in one direction can be balanced by the transport of an ion such as potassium in the opposite direction. The aspartyl-phosphate conveys a proton outwards with the phosphate group, while potassium is translocated inwards together with another phosphate on the same aspartyl group. The aspartyl phosphate group is assumed to rotate and go from a high energy to a low energy conformation (reviewed in Pedersen, 1982). Not all the cation pumps are electroneutral, however, since the Na^+/K^+ ATPase pumps three sodium ions out of the cell in exchange for the entry of two potassium ions for every ATP molecule hydrolysed, resulting in a change of potential difference across the membrane.

10.5.1 Sodium/potassium ATPase

Heart cells, in common with most eukaryotic cells, need to maintain gradients of the alkali metal cations — potassium levels must be kept high inside the cell and sodium low, in order to maintain the cell's integrity. To do this, the cell requires a pump to eject sodium ions from the cell in exchange for potassium ions. This will counterbalance the effect of sodium influx and potassium efflux through the respective ion channels during the early part of the action potential which accompanies a heart beat. The transmembrane ATPase that effects this exchange has been purified from a number of sources, and, when inserted into lipid vesicles, can catalyse the exchange transport of alkali metal cations at a rate approaching that seen *in vivo* (see Jorgensen, 1982).

The enzyme is made up of two subunits: α of molecular weight about 100 kDa and β about 40 kDa, although minor variations in the α-subunit may be found between tissues to a sufficient extent to warrant the designation of isozymes (see Lytton, 1985). The enzyme spans the membrane with the catalytic and sodium activator sites facing the cytoplasm and the potassium binding site on the extracellular surface. All these sites are located on the α-subunit: no functional role has yet been described for the β-subunit. Both the amino and carboxyl termini of the α-subunit face the cytoplasm and the polypeptide chain traverses the lipid bilayer several times in between. The mode of transport of the cations is not known, although it has been suggested that the ionic channel may pass either through the α-subunit or between adjacent α-subunits (Jorgensen, 1982).

In the presence of sodium ions and Mg/ATP on the intracellular side of the enzyme, the carboxyl side-chain of an essential aspartate group is phosphorylated to yield a high-energy phosphate. A conformational change takes place as ADP is released ($E_1P \rightarrow E_2P$; where E_1 and E_2 represent different conformational forms of the enzyme). Potassium ions enter the α-subunit from the other side of the membrane and bind to E_2P causing it to lose the phosphate group. K^+E_2 breaks down to free potassium and enzyme to complete the catalytic cycle.

The conformational change induced by ATP phosphorylation is potentiated by sodium, and dephosphorylation is activated by potassium. In separated and purified enzyme preparations sodium ion stabilizes E_1 and potassium E_2. ATP binds with high affinity to E_1Na in a hydrophobic pocket and only weakly to E_2K in an open structure. Magnesium ion is required for activity as, with most enzymes that use ATP, Mg/ATP is the true substrate. This rather complicated reaction scheme may be envisaged as follows (after Jorgensen, 1982):

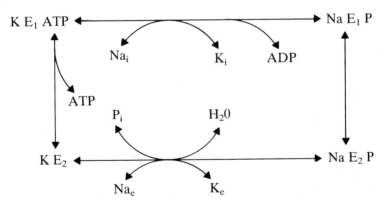

Na^+_i, K^+_i represent cytoplasmic cations; Na^+_e, K^+_e extracellular.

Cardiac glycosides are used in the treatment of congestive heart failure, and appear to act by inhibiting the sodium pump (Hansen, 1984). They are composed of a large steroid-like structure with an unsaturated lactone ring attached to the 17-position and a sugar, or chain of sugars, linked to the 3-position. All these parts of the molecule are essential for activity. These drugs were discovered as part of the extract of foxgloves, digitalis, which was recommended for the treatment of heart failure as long ago as 1785. Digitoxin is the major constituent of digitalis while digoxin is also present to a lesser extent. The original preparations from foxgloves have now been replaced by chemical preparations because the dose is much easier to control and is not subject to variable composition. The cardiac glycosides are still very useful drugs for the treatment of cardiac failure despite occasionally serious incidents of kidney toxicity. Digoxin is probably the most widely used agent (Counihan, 1982).

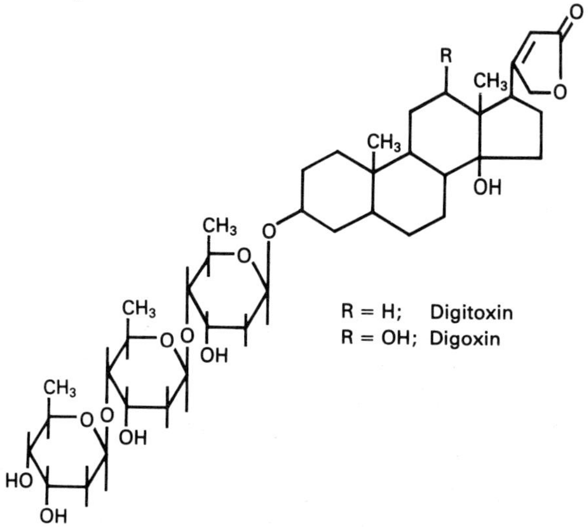

R = H;   Digitoxin
R = OH; Digoxin

Experimentally ouabain is probably the most studied of the cardiac glycosides, and has been shown to bind tightly but reversibly to the α-subunit of the enzyme on the extracellular face with an I_{50} of 10^{-7} M. The affinity of the drug for the isozyme α^+ of brain and kidney is much less with an I_{50} of 10^{-4} M (Sweadner, 1979). The enzyme undergoes a large conformational change on ouabain binding which can be partially antagonized in a non-competitive fashion by potassium, suggesting that the latter binds at a different site. Ouabain binds preferentially to the E_2P conformation of the enzyme but the affinity of that form is reduced for the drug if it is complexed with potassium. Ouabain/enzyme complexes will also bind ATP but with greatly reduced affinity (Hansen, 1984).

Ouabain is thus able to block the translocation of alkali metal cations in and out of the cell at concentrations about 10^{-6} M, and rate of ouabain binding parallels inhibition of rate of pumping. If the pump is inhibited, intracellular sodium levels rise. Consequently, calcium levels also rise as a result of a transmembrane sodium/calcium exchange. The raised calcium level activates the heart to contract more forcefully (the positive inotropic effect). There have been suggestions, however, that this positive inotropic effect can be dissociated from pump inhibition, since after the cardiac glycoside has been washed out of the heart the enzyme was still significantly inhibited whereas the positive inotropic effect was no longer present. One explanation may be that another unknown mechanism operates in addition to pump blockade (reviewed in Godfraind, 1984). On the other hand, the degree of enzyme inhibition required to show an observable effect on the pumping of the heart may be very high.

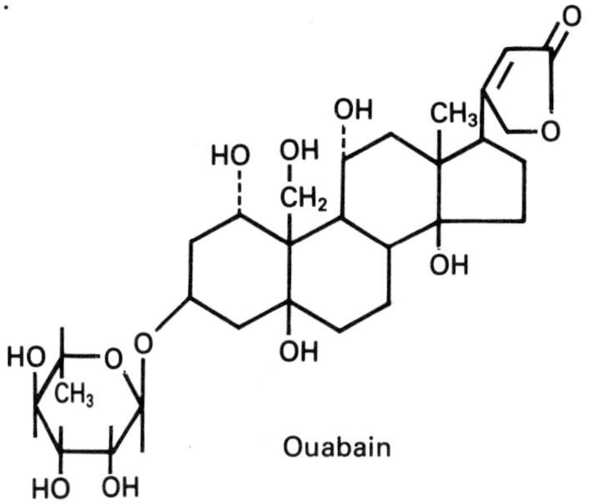

Ouabain

Other studies by Brown and Erdmann (1984), however, have cast doubt on this hypothesis. They obtained biphasic Scatchard plots and suggested that there were two sets of binding sites for ouabain on rat and guinea pig cardiac membranes — a high affinity, low capacity site the occupancy of which was associated with the positive inotropic effect, and a low affinity, high capacity site which was apparently the sodium pump. The high affinity site was not identified. With cat and human cardiac membranes the position was much less clear-cut with only a very slight curvature of the Scatchard plots. Bearing in mind that biphasic Scatchard plots can arise for other reasons than multiple sets of binding sites, the detailed mode of action of the cardiac glycosides is still a subject for debate.

The action of the cardiac glycosides is particularly valuable in congestive heart failure which is characterized by, among other things an enlarged heart and a high venous pressure as a consequence of blood flow congestion. The increased contraction forces the blood out of the heart and it reduces in size. As a consequence the sympathetic activity is reduced and the venous pressure falls back towards normal. In addition, the drugs slow the ventricular rate of beating by rendering this part of the heart less sensitive to electrical stimuli known as a refractory state.

Atrial fibrillation, which is a condition characterized by rapid and very irregular beating, and flutter that is a circular impulse in the atrium, will both respond to digoxin therapy by virtue of the drug's blockade of the atrio-ventricular node (see Hoffman and Bigger, 1985, for a detailed discussion of the effect of cardiac glycosides on the heart).

10.5.2 Potassium/hydrogen ATPase

Recently, an enzyme has been discovered which pumps protons out of the parietal cells that release acid into the stomach and, indeed, is believed to be the terminal link in the chain of acid secretion. This proton pump is the target of a new type of anti-ulcer drug that is exemplified by omeprazole, a substituted benzimidazole.

The enzyme appears to be unique to the parietal cell of the gastric mucosa, and is the last link in the chain of the transport of hydrogen ions into the small channels (canaliculi) that eventually lead to the surface of the cell leading into the gut. These canaliculi form a highly acidic space in the cell, and exhibit a convoluted membraneous structure with short microvilli projections. The pump is found in a particular type of vesicle which is isolated from the microsomal fraction of gastric parietal cells. This fraction is probably derived from the plasma membrane and vesicles which have a smooth surface. Normally, there are a large number of these vesicles (about 0·2 μm in diameter) in the cytoplasm of the resting cell. When acid secretion is stimulated by histamine, for example, these vesicles coalesce with the canaliculi and allow protons to be transported to the cell exterior (Olbe *et al.*, 1979; see Sachs, 1986 for review).

Potassium ion and ATP are required for the action of the proton pump, in fact the enzyme is a proton-stimulated ATPase which uses the energy of ATP hydrolysis to drive the exchange of protons for potassium across the vesicle membrane. ATP can thus generate a pH gradient across the canalicular membrane in parietal cells. Studies on the purified enzyme have revealed that it is similar to the Na^+/K^+ ATPase of the cardiac cell membrane discussed in Section 10.5.1, in that a phosphorylated form of the enzyme appears to be an obligatory intermediate in the catalytic reaction. Indeed, the similarity extends to the fact that the phosphorylated form is active in the transport of potassium ion while the dephosphorylated form is not.

Magnesium ion is also required for activity, as is common for enzymes where ATP is a co-substrate, and it is likely that a Mg/ATP complex is again the true substrate. In the absence of potassium there is a basal level of magnesium-stimulated ATPase activity which is composed of magnesium-dependent phosphorylation of a carboxyl group of an aspartate residue in the enzyme by ATP, followed by a slow hydrolysis to inorganic phosphate and free enzyme. The following scheme has been proposed for the overall reaction (see Wallmark *et al.*, 1980; Sachs, 1986):

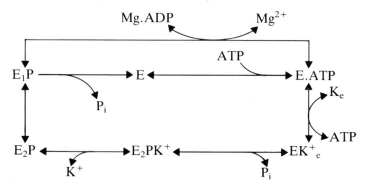

K^+_i is internal potassium K^+_e is external potassium

With regard to ligand binding sites, there appear to be two for potassium, one of low affinity situated on the cytosolic face of the enzyme and one of higher affinity on the luminal face. Occupancy of the latter site enhances the dephosphorylation of the phospho-enzyme intermediate, while the former site appears to reduce the binding of ATP (a similar effect to the cardiac proton pump). ATP also appears to have two binding sites on the enzyme varying in affinity by an order of magnitude. This enzyme is a single polypeptide of molecular weight 94 kDa (see Sachs, 1986), but otherwise rather less is known about it in molecular detail than about the cardiac proton pump.

The mechanism of proton pumping is outlined in the following way by Sachs (1986). ATP binds to the cytosolic surface of the enzyme and allows a proton to bind to a site in the middle of the channel. The aspartyl group in the channel is phosphorylated and an opening to the luminal side is created by this process. The affinity of the proton site is sharply reduced and the proton moves out into the lumen. The binding of K^+ to the lumen results in dephosphorylation and protons can no longer move outwards.

Omeprazole, a benzimidazole, is stable at neutral pH, but is rapidly transformed into another compound of unknown structure at pH levels below 6. Omeprazole is in fact a pro-drug as it requires this activation by acid to show any inhibitory effect on acid secretion. This is shown by the lack of effect of the drug if the gastric mucosa cells are treated with a weak base, such as 4-aminoantipyrine, which neutralizes the acid compartment of the gland. Furthermore, the drug requires prior incubation with acid *in vitro* (pH 5·2 for example) in order to show any inhibitory effect on ATPase activity (Sachs, 1986; Wallmark, 1986). No inhibition is observed on either ATPase activity or aminoantipyrine uptake if the drug is incubated under neutral conditions. Interestingly, omeprazole appears to act *in vivo* in rats by reacting covalently with the enzyme since synthesis of fresh enzyme is required to overcome the inhibition (Im *et al.*, 1985).

A thiol group on the luminal surface of the membrane is probably the target for the activated drug, both *in vitro* and *in vivo* (see Sachs, 1986). It is not possible to measure the secretion of acid from gastric glands directly because of the very small amount of acid released and indirect methods have to be used. Thus the uptake of a radiolabelled weak base such as ^{14}C-aminopyrine has shown the correlation between benzimidazole concentration and inhibition of both aminopyrine uptake and proton pump (Fellenius *et al.*, 1981). Moreover, Wallmark (1986) has shown that *in vivo* there is close correlation between the secretion of acid from rat stomach, the activity of ATPase and the formation of phospho-enzyme intermediate.

Gastric mucosal ATPase is the final step in the induction of acid secretion by various secretagogues; histamine is the most potent and acts through adenylate cyclase and the formation of cAMP; acetylcholine stimulates the uptake of calcium ion into the cytoplasm and has no effect on cyclic nucleotide levels while the peptide gastrin, which is the physiological mediator of gastric acid secretion in response to feeding, appears to release calcium from intracellular stores. It is not yet known whether the latter two agonists stimulate the formation of diacylglycerol and the activation of the inositol phosphate pathway (see Rasmussen, 1986, and Section 9.1 for a discussion on this pathway).

Clearly, it is possible to inhibit the secretion of acid by (a) blocking the histamine receptor, as has been demonstrated by the anti-ulcer drugs cimetidine and ranitidine and (b) the acetylcholine (muscarinic) receptor as the drug pirenzepine does. Furthermore, the histamine$_2$ receptor blockers inhibit gastrin and acetylcholine induced secretion which are not mediated through histamine receptors (as noted in Section 9.8.2). The proton pump, however, is the last link in the chain of proton secretion and blockers of this may be expected to be at least as effective as blockers of either of the two receptors described above. In addition, the expected blockade of both histamine, dibutyryl cAMP and pentagastrin induced secretion in human gastric glands *in vitro* and *in vivo* was found by Olbe *et al.* (1982, Olbe *et al.*, 1986). Pentagastrin is a synthetic pentapeptide containing the C-terminal tetrapeptide amide moiety of gastrin, and is often used in tests to measure inhibition of acid secretion as it has appreciable secretory activity.

In clinical trials studies, omeprazole was shown to heal duodenal ulcers more rapidly than ranitidine; in some cases that were resistant to prolonged treatment with H$_2$ receptor blockers, omeprazole was effective. As we might expect from its mode of action, less drug and less frequent administration is required in the case of omeprazole. Ranitidine and omeprazole were equally effective at healing gastric ulcers. There is concern, however, that the prolonged reduction in acid content of the stomach that long-term treatment with omeprazole produces may lead to cancer, since rats treated for two years with the drug showed some evidence of stomach tumours (Clissold and Campoli-Richards (1986) review the therapeutic potential of omeprazole).

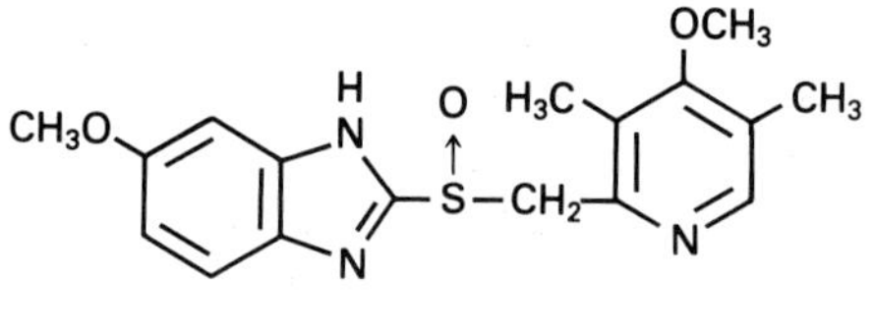

Omeprazole

10.6 *Cromoglycate — calcium antagonist or membrane stabilizer?*

Another drug whose mechanism of action is not fully understood, but which does initially act at a membrane site is cromoglycate — a drug introduced over 20 years ago for the treatment of asthma (see Church and Warner, 1985, for a clinical review). The mode of action of the drug was originally thought to be simply via inhibition of the release from mast cells of inflammatory mediators called autacoids — principally histamine — which were held to be responsible for generating the symptoms of asthma. Cromoglycate was believed to do this by stabilizing the mast cell membrane in some unknown fashion (Foreman and Garland, 1976). Although this action has been demonstrated on many occasions, there is still controversy on a number of points including:

(a) the molecular details of the release of histamine from mast cells
(b) the importance of mast cell release mechanisms to the aetiology of asthma
(c) the importance of histamine relative to other mediators of inflammation such as the leukotrienes.

10.6.1 Mast cell degranulation

Histamine is stored in mast cells bound by ionic forces to heparin and protein. These complexes are visible in the form of granules, and the release process is often known as degranulation. In allergic individuals, IgE antibodies bind selectively to proteins on the mast cell surface, and when the antigen, whether pollen protein or other agent, binds to and bridges the gap between adjacent IgE molecules, the granules fuse together. These larger granules then fuse with the plasma membrane and discharge their contents into the extracellular space. The details of the process are not worked out but it is known that calcium is required (whether from external or internal sources or both) and metabolic energy. This type of process is described as stimulus-secretion coupling.

Histamine is not the only mediator to be released: prostaglandins, leukotrienes, serotonin and various other agents such as enzymes (proteases and β-glucuronidase) and eosinophil chemotactic factor may also be involved to various extents in different species (see Foreman and Garland, 1976). The enzymes and other agents released from mast cells can also cause other materials to be liberated from neighbouring tissue. Indeed, leukotrienes, which are responsible for the characteristic narrowing of the bronchioles (bronchoconstriction) and spasm of acute asthma, may be the most important agents in allergic bronchoconstriction in man (Douglas, 1985). The inflammation in chronic cases may also be a result of leukotriene action (Church and Warner, 1985).

This overall initial response to antigen is referred to as the immediate hypersensitivity reaction. There is also a reaction of a similar type which occurs later, and is known as the delayed hypersensitivity reaction — perhaps not surprisingly — which is not a function of the mast cell.

The mechanism of the secretory process is at present far from clear (see reviews by Pearce 1984, 1985). The involvement of calcium (as in all secretory processes) suggests that channels or gates in the membrane are opened as a consequence of antigen/antibody complexation to allow calcium to enter. This is not a voltage-operated calcium channel as calcium entry blockers such as verapamil and nifedipine at 5×10^{-5} M concentration have no effect on histamine release induced by antigen in rat mast cells. In contrast, quercetin (a flavonoid) inhibits by 51% at 1×10^{-5} M (Middleton *et al.*, 1981), but the exact site of its action is not known. Subsequently, calmodulin or some other calcium binding protein may be recruited to activate a protein kinase to phosphorylate a membrane protein as part of the release process. There may also be a mobilization of calcium from the intracellular stores in mitochondria and endoplasmic reticulum.

Cyclic nucleotides modulate the reaction, and β_2-adrenergic agents (isoprenaline and salbutamol, for example), that raise cAMP levels inhibit, whilst cholinergic agonists, such as carbamylcholine, that act at muscarinic receptors and raise cyclic GMP (cGMP) levels, augment the reaction (Bourne *et al.*, 1972). Theophylline, which has been used to treat asthma for many years, also raises cyclic nucleotide levels at high concentrations by virtue of its inhibition of cyclic nucleotide hydrolysis to the 5'-monophosphates. The drug does, however, inhibit cGMP hydrolysis almost as effectively as cAMP by partially purified enzymes from human lung (Bergstrand *et al.*, 1977). Theophylline may also act as an antagonist at adenosine receptors to interfere with the allergic process. Indeed, those compounds that were the most effective antiallergics were significantly better inhibitors of cGMP hydrolysis than of cAMP by a membrane-bound preparation from human lung microsomes (Coulson *et al.*, 1977). The relative role of the two cyclic nucleotides is not entirely clear at present.

10.6.2 Cromoglycate action

Cromoglycate has at least two interesting properties which may ultimately shed light on its mode of action and, by the same token, on the detailed mechanism of mediator release. The drug does not need to enter the mast cell in order to exert its anti-allergic effect since it is active when covalently linked to polyacrylamide beads (Mazurek *et al.*, 1980), and further work on basophil leukaemia cells has implicated a cromoglycate binding protein which may be a calcium gate (but see Pearce, 1985, for a critical review of this work). Since cromoglycate can also prevent the mobilization of intracellular calcium it is not clear what contribution the former inhibition makes. It is probable though that external calcium is required for a sustained release of inflammatory mediators to occur.

The drug is a dicarboxylic acid with pKa's both in the region of 2. It is therefore an anion at neutral pH. Consistent with the extracellular action noted above, this means that the drug is very unlikely to be able to cross membranes effectively in the absence of an active transport system, and no such systems have been identified for a bis-chromone dicarboxylic acid.

The other interesting finding with respect to cromoglycate action regards phosphorylation of a membrane protein. The secretory process itself induces the immediate phosphorylation (in less than 10 seconds) of three proteins of molecular weights 68, 59 and 42 kDa. Phosphorylation of a fourth protein of molecular weight 78 kDa occurs after 30 to 60 seconds when the secretion process is ending. This later phosphorylation may be associated with the shutting down of release and restabilization of the cell. In contrast, cromoglycate induces the 78 kDa protein phosphorylation within 10 seconds, and it is possible that the drug is utilizing a natural cell mechanism for terminating the secretion process. This phosphorylation also shows tachyphylaxis (i.e. repeated administration of the drug results in reduced effectiveness) as does inhibition of histamine release (Wells and Mann, 1983).

Cromoglycate

Another curious feature of phosphorylation of the 78 kDa protein is that agents that elevated cAMP levels do not induce phosphorylation whereas dibutyryl cGMP does at concentrations at which it inhibits the release process. If cGMP is acting as a negative feedback inhibitor, which these results imply, then this is consistent with the greater effectivenss of cGMP hydrolysis inhibitors rather than cAMP observed by Coulson *et al.* (1977).

The relevance of mast cell degranulation to all forms of asthma, however, is debatable, particularly for those attacks in which antigen is not believed to play a part, notably those induced by exercise or cold air; in both of these situations cromoglycate is effective (Church and Warner, 1985). In addition, the delayed hypersensitivity response to bronchial provocation by inhalation of antigen is also blocked by cromoglycate, but not by salbutamol, which suggests that some other mechanism may be more important *in vivo*.

In sharp contrast, salbutamol is rather more effective at inhibiting histamine release from human lung tissue than is cromoglycate. Indeed, early studies indicated that the latter requires higher concentrations to be effective *in vitro* than are used clinically. More recently, this test has employed mast cells that have been washed out from the bronchioles and small airways of the human lung, and cromoglycate has been shown to be more effective at inhibiting degranulation of these cells than of human lung tissue *in vitro* (reviewed in Pearce, 1984; Church and Warner, 1985). These discrepancies may, of course, also reflect the greater importance of leukotrienes than histamine in the human condition.

The very great effort expended in recent years trying to find new antiallergic agents based on mast cell degranulation tests *in vitro* has so far yielded many new compounds that are more active than cromoglycate but show little advantage *in vivo*; indeed some are inactive. Although in some cases this has been attributed to serum binding and different pharmacokinetic profiles amongst other things, the lack of correlation in other cases is likely to be a consequence of the use of an inappropriate *in vitro* test.

Cromoglycate is most useful in cases of mild to moderate asthma — when given in advance as a prophylactic treatment. The acute attack is treated by salbutamol or occasionally by isoprenaline (isopropylnoradrenaline). Severe cases respond only to steroids (Douglas, 1985; Church and Warner, 1985). Cromoglycate may also be used to treat hay fever.

10.7 Sterol ligands — polyene antibiotics

Fungal infections have become progressively a more serious problem in recent years. This has been attributed partly to our mastery of bacterial infections leaving a vacuum that pathogenic fungi take the opportunity to fill, and partly to the fact that a large number of hospital patients — particularly those with cancer and transplant patients — have a partially suppressed immune system as a consequence of drug treatment. Thus there has been a great increase in the number of systemic infections by fungi, particularly opportunist infections such as thrush caused by *Candida albicans*, and they now present a serious medical problem. It has been known, however, for a long time that *Candida* infections can be lethal. Gooday (1977) notes that a paper of 1665 records at the height of the plague: 'The diseases and casualties this week, London, Sept. 12 to the 19; Consumption, 129 . . . Fever, 332 . . . Plague, 6544 . . . Thrush, 6'.

For nearly thirty years systemic fungal infections have been treated with amphotericin B as the drug of choice (indeed for much of that period the only drug) because it is very active against a wide range of pathogenic organisms, is fungicidal not fungistatic and does not induce resistance (reviewed in Medoff *et al.*, 1983). The drug's major drawback, and it is a serious one, is kidney toxicity; in about 80% of the patients treated the glomerular filtration rate falls to about 40% of normal. The effect is usually reversible when treatment is withdrawn, but in some cases there is some residual impairment of kidney function. Amphotericin B, however, was for many years the only polyene antibiotic sufficiently safe to be used systemically in man. Nystatin, a close analogue, may be given orally for *Candida* infections of the mouth or intestinal tract but since it is almost completely non-absorbed it is not used for systemic infections. Mepartricin may now be an alternative (see Petrou and Rogers, 1985).

Polyene antibiotics are natural products produced by the actinomycetes (notably the streptomycetes) and are characterized by a macrolide ring closed by an internal ester or lactone. As their name implies, polyenes have a system of conjugated double bonds (normally between 4 and 7) in the *trans* configuration, which gives rise to a characteristic ultraviolet absorption spectrum with a number of peaks in the visible and near ultraviolet. Amphotericin B is a heptene (seven double bonds) and, additionally, contains an amino sugar (mycosamine) attached to the ring. The ring size can range from 12 to 37 carbon atoms and is substituted by a number of hydroxyl groups. The combination of several polar hydroxyl groups and a carboxyl on the one hand, and, on the other, several hydrophobic double bonds gives a dual character to the molecule, particularly as these groups reside on opposite sides of the ring (Norman *et al.*, 1976).

10.7.1 Amphotericin action

The specific target of the polyenes is the sterol in the plasma membrane of the eukaryotic cell; ergosterol in fungi and cholesterol in protozoal and mammalian cells. The evidence for this may be summarized as follows:

(a) Bacteria do not contain sterol and are resistant to amphotericin. Protozoa, fungi and mammalian cells are all sensitive. The mycoplasma, *Acholeplasma laidlawii*, can grow whether sterol is present in the culture medium or not and incorporates it into the membrane if it is present. The microorganism is sensitive to the polyene only when sterol is present in the membrane.

(b) Free sterols in the culture medium antagonize the effect of the drug on microorganisms, such as *Saccharomyces cerevisiae* and *Candida albicans*.

(c) A complex between amphotericin and sterols which can be demonstrated by various techniques, such as ultraviolet difference spectroscopy, e.s.r, and circular dichroism, is sufficiently tightly bound to explain the antimicrobial action.

(d) Fungal cells die as a result of damage to the membrane where the drug/sterol complexes are found. The membrane begins to leak small ions, notably ammonium and potassium.

The binding of polyenes to sterol can be demonstrated in model membrane systems such as vesicles composed of sterol and phosphatidylcholine. The binding constant for the amphotericin B/vesicle complex is 9.7×10^{-7} M when cholesterol is used and 1.52×10^{-6} M with ergosterol. Although this suggests that the drug might interfere more with animal cell membranes, the number of binding sites is important — ergosterol containing vesicles have a larger number of binding sites for the drug than those made with cholesterol, giving a ten-fold factor in favour of the former in terms of product of binding sites and binding constant (Witzke and Bittman, 1984). Furthermore, circular dichroism studies suggest that there is more than one conformer of the drug present in membrane vesicles at higher drug concentrations but one conformer predominates at lower concentrations whichever sterol is present (Vertut-Croquin *et al.*, 1983).

The fact that cholesterol binds polyenes tightly (indeed amphotericin is very unusual in binding more effectively to ergosterol) probably explains why these compounds are toxic to mammals. There is no direct correlation.

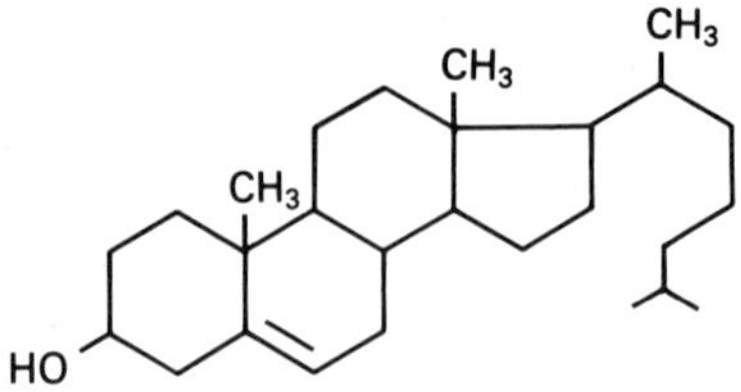

Cholesterol 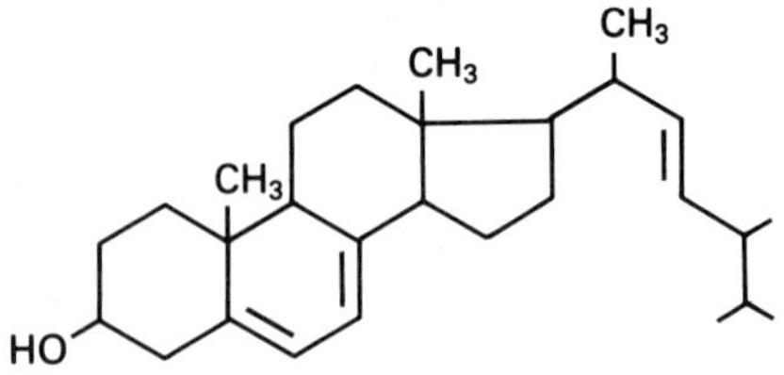   Ergosterol

between anti-*Candida* activity and binding affinity in model vesicles for a small group of four polyenes including amphotericin B, but the group may be too small to expect correlations. Alternatively, the nature of the phospholipid in the vesicles may be important as phosphatidylcholine faces the extracellular side of the plasma membrane in that its polar head groups are pointing in that direction while phosphatidyl-ethanolamine and -inositol face the cytoplasm and so a different composition in the vesicles might give a better correlation. It is also possible that the leakiness induced by binding may not be directly related to the extent of binding, but rather to the particular complexes made by the polyenes and how they disrupt the membrane architecture (Witzke and Bittman, 1984).

A number of features on the sterol nucleus are essential for the interaction to take place, namely a 3-β-hydroxyl group and a Δ-22 double bond together with a cholestane skeleton. An intact polyene is also required — if the ring is broken no interaction will take place (Norman *et al.*, 1976).

Membrane fluidity plays a key role in the action of the polyenes. They can induce permeability in the "solid" or crystalline form of the membrane in the absence of cholesterol and above the transition temperature when the membrane is in the liquid form, amphotericin can still have a small effect, but the addition of cholesterol markedly increases it (reviewed in Medoff *et al.*, 1983).

The precise molecular nature of the interaction of polyenes with membrane components, and how this induces permeability, is still obscure even on the model membrane front. Electron microscope pictures of erythrocyte membranes treated with amphotericin show a series of pits and protuberances but no obvious pores, and so it is difficult to accept the original view that the polyene/sterol complex forms aqueous channels or pores in the membrane through which the ions pass. An alternative hypothesis is that the complex may alter the local membrane fluidity by removing the sterol from its interaction with membrane phospholipids, and thereby induce leakage of cell constituents. Particularly sensitive points may occur where there is a change in the membrane from solid to liquid phase (reviewed in Medoff *et al.*, 1983). At present the situation is still unresolved.

Amphotericin at low concentrations *in vivo* can stimulate cell growth — indeed macrophages are stimulated to kill microorganisms, and the drug can markedly increase the number of antibody-forming cells in the mouse spleen and lymph node. Not surprisingly, therefore, the drug has a stimulatory effect on the immune system *in vivo*, but the clinical relevance of this finding is as yet unknown. In view of the immune suppression of many of the patients with fungal infections this property could be very useful.

In conclusion, therefore, amphotericin B is still the drug of choice for life-threatening fungal infections by primary pathogens such as *Histoplasma capsulatum* and *Paracoccidioides brasiliensis* as well as opportunist pathogens such as *Candida albicans* (Warnock and Speller, 1982). The problem of the toxicity of amphotericin may be countered by the concomitant use of flucytosine which can allow the dose of amphotericin to be reduced (Cohen, 1982). Therapy with amphotericin, however, is not always straightforward because, in addition to toxicity, there can be a lack of clinical response even when the organism is susceptible to the drug *in vitro*. In some cases this is because the drug is unable to penetrate in sufficient quantity the particular part of the body which harbours the pathogen. This is the case in infections of the central nervous system with *Coccidioides* species which cause meningitis (Medoff *et al.*, 1983).

There is clearly room for another drug which is widely disseminated on systemic administration, is fungicidal and does not produce resistance (but does not have the side-effects of amphotericin B). It is possible that the related polyene, mepartricin (the methyl ester of partricin, a polyene with an aromatic ring) may be such a drug. Possibly less nephrotoxic than amphotericin, it is more active and so has a better therapeutic index (Petrou and Rogers, 1985).

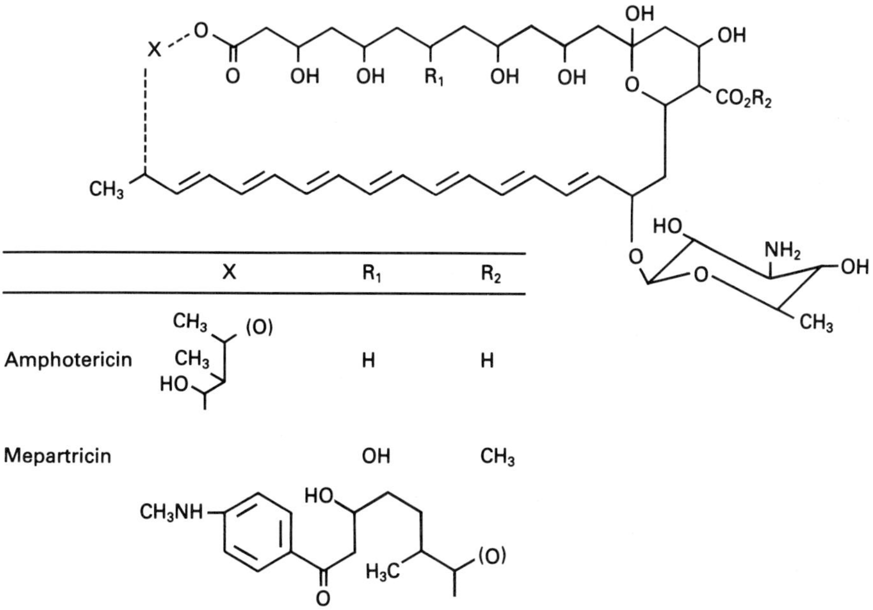

	X	R_1	R_2
Amphotericin		H	H
Mepartricin		OH	CH_3

References

Agarwhal, N. and Kalra, V. K. (1984). *Biochem. Biophys. Acta*, **764**, 316–323.

Bayer, R., Kalusche, D., Kaufmann, R. and Mannhold, R. (1975). *Naunyn-Schmiedeberg's Arch. Pharmakol.*, **290**, 81–97.

Bean, B.P., Cohen, C.J. and Tsien, R.W. (1983). *J. Gen. Physiol.*, **81**, 613–642.

Benos, D. J. (1982). *Am. J. Physiol.*, **242**, C131-145.

Bergstrand, H., Kristofferson, J., Lundquist, B. and Schurmann, A. (1977). *Molec. Pharmacol.*, **13**, 38–43.

Bourne, H. R., Lichtenstein, L. M. and Melmon, K. L. (1972). *J. Immunol.*, **108**, 695–700.

Bradley, D. J. and Richards, C. D. (1984). *Brit. J. Pharmacol.*, **81**, 161–167.

Brown, L. and Erdmann, E. (1984). *Basic Res. Cardiol.*, (Suppl.), 50–55.

Burg, M. and Green, N. (1973). *Kidney Int.*, **4**, 301–308.

Caffruny, E. J. and Itskovitz, H. D. (1982). *J. Pharmacol. Exp. Ther.*, **223**, 105–109.

Catterall, W. A. (1980). *Annu. Rev. Pharmacol. Toxicol.*, **20**, 14–43.

Church, M. K. and Warner, J. G. (1985). *Clin. Allerg.*, **15**, 311–320.

Clissold, S. P. and Campoli-Richards, D. M. (1986). *Drugs*, **32**, 15–47.

Cohen, J. (1982). *Lancet*, **ii**, 533–537.

Coulson, C. J., Ford, R. E., Marshall, S., Walker, J. L., Wooldridge, K. R. H., Bowden, K. and Coombs, T. J. (1977). *Nature*, **265**, 545–547.

Counihan, T. B. (1982). *Eur. Heart J.*, **3** (Suppl. D), 143–147.

Creveling, C. R., McNeal, E. T., Daly, J. W. and Brown, G. B. (1983). *Molec. Pharmacol.*, **23**, 350–358.

Cuthbert, A. W. (1977). *Progr. Med. Chem.*, **14**, 1–50.

Douglas, W. W. (1985). In: *The Pharmacological Basis of Therapeutics*, 7th Ed., Gilman, A. G., Goodman, L. S., Rall, T. W. and Murad F. M., Eds. (New York: Macmillan) Ch. 26.

Droogmans, G., Himpens, B. and Casteels, R. (1985). *Experientia*, **41**, 895–900.

Fellenius, E., Berglindh, T., Sachs, G., Olbe, L., Elander, B., Sjostrand, S.-E. and Wallmark, B. (1981). *Nature*, **290**, 159–161.

Fleckenstein-Grun, G., Frey, M. and Fleckenstein, A. (1984). *Trends Pharmacol Sci.*, **5**, 283–286.

Foreman, J. C. and Garland, L. G. (1976). *Brit. Med. J.*, **i**, 820–821.

Frizzell, R. A., Field, M. and Schultz, S. G. (1979). *Am. J. Physiol.*, **236**, F1-8.

Gintant, G. A., Hoffman, B. F. and Naylor, R. E. (1983). *Circ. Res.*, **52**, 735–746.

Glossman, H., Ferry, B. R., Lubbecke, F., Mewes, R. and Hoffman, F. (1982). *Trends Pharmacol. Sci.*, **3**, 431–437.

Gooday, G. W. (1977). *J. Gen. Microbiol.*, **99**, 1–11.

Godfraind, T. (1984). *Eur. Heart, J.*, **5**, Suppl. F, 303–308.

Gussin, R. Z. (1977). *J. Clin. Pharmacol.*, **17**, 651–661.

Hansen, O. (1984). *Pharmacol Rev.*, **36**, 143–163.

Hille, B. (1977). *J. Gen. Physiol.*, **69**, 497–515.

Hoffman, B. F. and Bigger, J. T. (1985). In: *The Pharmacological Basis of Therapeutics*, 7th Ed., Gilman, A. G., Goodman, L. S., Rall, T. W. and Murad, F. M., Eds. (New York: Macmillan) Ch. 30.

Hondeghem, L. M. and Katzung, B. G. (1984). *Annu. Rev. Pharmacol. Toxicol.*, **24**, 387–423.

Im, W. B., Blakeman, D. P. and Davis, J. P. (1985). *Biochem. Biophys. Res. Commun.*, **126**, 78–82.

Jorgensen, P. L. (1982). *Biochim. Biophys. Acta*, **694**, 27–68.

Kendall, M. J. and Horton, R. C. (1985). *J. Royal Coll. Phys. Lond.*, **19**, 85–89.

Kokko, J. P. (1984). *Am. J. Med.*, **77**, 5A, 11–17.

Kumana, C. and Hamer, J. (1979). In: *Drugs for Heart Disease*, J. Hamer, Ed. (London: Chapman & Hall) p. 61.

Lant, A. (1985a). *Drugs*, **29**, 57–87.
Lant, A. (1985b). *Drugs*, **29**, 162–188.
Ludens, J. W. (1982). *J. Pharmacol. Exp. Ther.*, **223**, 25–29.
Lytton, J. (1985). *Biochem. Biophys. Res. Commun.*, **132**, 764–769.
Mazurek, N., Berger, G. and Pecht, I. (1980). *Nature*, **286**, 722–723.
Medoff, G., Brajtburg, J., Kobayashi, G. S. and Bolard, J. (1983). *Annu. Rev. Pharmacol. Toxicol.*, **23**, 303–330.
Middleton, E., Orzewiecki, G. and Triggle, D. (1981). *Biochem. Pharmacol.*, **30**, 2867–2869.
Muhiddin, K. A. and Turner, P. (1985). *Postgrad. Med. J.*, **61**, 665–678.
Nilius, B., Hess, P., Lausman, J. B. and Tsien, R. W. (1985). *Nature*, **316**, 443–446.
Norman, A. W., Spielvogel, A. M. and Wong, R. G. (1976). *Adv. Lipid. Res.*, **14**, 127–170.
Nowycky, M. C., Fox, A. P. and Tsien, R. W. (1985). *Nature*, **316**, 440–443.
Odlind, B. (1984). *Acta Pharmacol. Toxicol.*, **54**, Suppl. 1, 5–15.
Olbe, L., Berdglindh, T., Elander, B., Helander, H., Fellenius, E., Sjostrand, S.-E., Sundall, G. and Wallmark, B. (1979). *Scand. J. Gastroenterol.*, **55**, Suppl., 131–135.
Olbe, L., Haglund, U., Leth, R., Lind, T., Cederberg, C., Ekenved, G., Elander, B., Fellenius, E., Lundborg, P. and Wallmark, B. (1982). *Gastroenterol.*, **83**, 193–198.
Olbe, L., Lind, T., Cederberg, C. and Ekenved, G. (1986). *Scand. J. Gastroenterol.*, **21**, S 118, 105–107.
Patarca, R., Candia, O. A. and Reinach, P. S. (1983). *Am. J. Physiol.*, **245**, F660-669.
Pearce, F. L. (1984). *Trends Pharmacol. Sci.*, **5**, 5.
Pearce, F. L. (1985). *Trends Pharmacol. Sci.*, **6**, 389–390.
Pedersen, P. (1982). *Ann. N.Y. Acad. Sci.*, **401**, 1–20.
Petrou, M. A. and Rogers, T. R. (1985). *J. Antimicrob. Chemother.*, **16**, 169–178.
Postma, S. W. and Catterall, W. A. (1984). *Molec. Pharmacol.*, **25**, 219–227.
Rahwan, R. G. (1983). *Med. Res. Rev.*, **3**, 43–72.
Rasmussen, H. 1986). *New Engl. J. Med.*, **314**, 1164–1170.
Reuter, H. (1985). *Nature*, **316**, 391.
Ritchie, J. M. and Green, N. M. (1985). In: *The Pharmacological Basis of Therapeutics*, 7th Ed., Gilman, A. G., Goodman, L. S., Rall, T. W. and Murad, F., Eds, (New York: Macmillan) Ch. 15.
Sachs, G. (1986). *Scand. J. Gastroenterol.*, **21**, 1–10.
Scarborough, G. A. (1985). *Microbiol. Rev.*, **49**, 214–231.
Schramm, M. and Towart, R. (1985). *Life Sci.*, **37**, 1843–1860.
Serrano, R., Kielland-Brandt, M. C. and Fink, G. R. (1986). *Nature*, **319**, 689–693.
Singer, S. J. and Nicolson, G. I. (1972). *Science*, **175**, 720–731.
Spedding, M. (1985). *Trends Pharmacol. Sci.*, **6**, 109–114.
Sweadner, K. J. (1979). *J. Biol. Chem.*, **254**, 6060–6067.
Vertut-Croquin, A., Bolard, J., Chabbert, M. and Gary-Bobo, C. (1983). *J. Biol. Chem.*, **22**, 2939–2944.
Wallmark, B. (1986). *Scand. J. Gastroenterol.*, **21**, S 118, 11–17.
Wallmark, B., Stewart, S. B., Rabon, E., Saccumari, G. and Sachs, G. (1980). *J. Biol. Chem.*, **255**, 5315–5319.
Warnock, D. W. and Speller, D. C. E. (1982). *J. Antimicrob. Chemother.*, **10**, 159–171.
Wells, E. and Mann, J. (1983). *Biochem. Pharmacol.*, **32**, 837–842.
Wit, A. L. (1985). *Clin. Physiol. Biochem.*, **3**, 127–134.
Witzke, N. M. and Bittman, R. (1984). *Biochemistry*, **23**, 1668–1674.

Chapter 11

Microtubule assembly

11.1 Introduction

Microtubules are cytoplasmic organelles which play a crucial role in the process of cell division (mitosis) in eukaryotic cells as part of the mitotic spindle. Furthermore, various other activities including intracellular transport and secretion are mediated by these organelles. Indeed, the secretion of cytoplasmic granules of hormonal mediators of the asthmatic process (see section 10.6) also requires the involvement of microtubules (Kaliner, 1977). Interestingly, the cilia of the protozoon, *Tetrahymena pyriformis*, that the organism uses for swimming are mainly composed of microtubules.

During the stage of mitosis known as prophase the chromosome doublets appear, linked by a body known as a centromere. The two centrioles, normally located near the nucleus, migrate to opposite ends of the cell. Around each centriole a system of radiating fibres, known as the aster, develops. In the next phase, metaphase, the centrioles organize microtubules into the mitotic spindle which links the centrioles. The centromeres become attached to a spindle fibre and migrate to the median position between the poles. The loose ends of the chromosomes are oriented in a random fashion but the centromeres of each chromosome lie exactly in a plane called the equatorial plate. In anaphase which follows, the centromeres duplicate themselves. The two resulting centromeres move apart towards the poles, still attached to the spindle fibre, with their attached chromosome "dragging behind". Subsequently, in telophase a nuclear membrane begins to encircle the uncoiling chromosomes, while a furrow appears at the equator, eventually deepening and splitting the daughter cells apart.

The spindle is composed of a number of microtubules which have the general structure of long hollow cylinders with inside and outside diameters of 15 and 25 nm respectively. Microtubules are composed of the protein tubulin which itself is constructed of two closely related subunits, α- and β-tubulin, with molecular weights in the region of 55 kDa. The functional form of tubulin is an α/β dimer with molecular weight 110 kDa and sedimentation coefficient ($S_{20,w}$) of 5·8 S (reviewed in Timasheff and Grisham, 1980).

Tubulin from brain, the most prolific source, can be assembled *in vitro* into microtubules by raising the temperature from 4° to 37°C. The polymerization may be measured by an increase in turbidity. Usual requirements are (a) the presence of a family of associated proteins, known not surprisingly as micro-tubule-associated proteins (or MAPs) which assist the process in a way that is not understood, (b) GTP which is hydrolysed to GDP and (c) magnesium ion. Tubulin will assemble in the absence of MAPs but usually requires a high magnesium concentration or the presence of a polycation. Calcium, on the other hand, causes the microtubules to depolymerize, probably by binding to the wall of the microtubule and inducing an increase in the rate of subunit dissociation from the ends of the microtubules (Weisenberg and Deery, 1981). Reduction of the temperature to 10°C also causes microtubules to disassemble and repeated cycles of assembly and disassembly are usually used to purify tubulin.

Tubulin contains two GTP binding sites per dimer. One of these sites is exchangeable with free GTP, the E-site, and the other is non-exchangeable (N-site). GTP bound at the E-site is the one that is hydrolysed when microtu-bules are formed, and is linked to the addition of a tubulin dimer to the growing microtubule (David-Pfeuty *et al.*, 1978). GTP and GDP stabilize different conformations of tubulin (Timasheff and Grisham, 1980).

In the absence of MAPs, the process of assembly requires a sufficiently large polymeric aggregate before assembly will begin. A threshold concentra-tion of tubulin is required for this stage. Once the reaction is initiated, 5·8 S dimers add sequentially until the polymer reaches its final length within 10–20 minutes at 37°C. The process terminates if the tubulin dimer level falls below the critical threshold value. Nevertheless, the process is an equilibrium process, both *in vitro* and *in vivo*, with tubulin dimers falling off at one end and rejoining the cylinder at the other, provided that there is also sufficient GTP available (Timasheff and Grisham, 1980).

Various drugs interfere with the process of assembly. In general, they all have similar actions *in vitro* in that they block mitosis. Their uses, however, differ widely in a medical sense from anti-fungal agents like griseofulvin, anthelminthic drugs such as the benzimidazole family to the vinca alkaloids, vincristine and vinblastine, which are used as anti-tumour agents. Colchicine, one of the best known of the microtubule ligands is used for the treatment of gout, but there is some doubt as to whether its therapeutic efficacy relies only, if at all, on inhibition of microtubule polymerization.

11.2 *Colchicine for gout*

Colchicine is probably the most studied of the microtubule ligands, binding as it does to tubulin dimers (5·8 S) in a 1:1 complex. The precise site at which the drug binds has yet to be determined, but both the α- and β-subunits have been suggested as the target. Indirect evidence implicating the β-subunit has been obtained by using N,N′-ethylenebisiodoacetamide, a bifunctional alkylating agent, which links together the side-chains of cysteine 239 and 354 of the β-chain, thereby preventing microtubule assembly (Luduena and Roach, 1981).

Colchicine protects the protein from alkylation, and this is put forward as an argument for the drug to be binding at or near these residues (Little and Luduena, 1985). On the other hand, colchicine produces a conformational change in the protein when binding, and this could have rendered the sulphydryl residues inaccessible to the alkylating agent even though the drug binding site was nowhere nearby.

The α-subunit has also been proposed as the target for colchicine since Serrano *et al.* (1984) have shown that pre-incubation of tubulin with [^{3}H]col-chicine, followed by proteolysis and separation of the fragments, yielded a 16 kDa segment with the drug attached at the C-terminal end of the α-subunit. It is more likely that this direct binding study reveals the true binding site for colchicine.

The binding study noted above only works because the drug binds to the protein in a time-dependent fashion which resembles an irreversible process in that a conformational change takes place in the protein. The colchicine bound to the tubulin is, consequently, not in equilibrium with free drug, i.e. it is non-exchangeable. The process is not irreversible in practice because colchicine gradually dissociates from the complex with a half-life at 37°C of approximately 12 hours (Wilson, 1970). Estimates of the binding constant vary between 0.1 to 3.0×10^{-6} M, depending on how much time has been allowed for equilibration. Other less tight binding sites have also been identi-fied with a binding constant of 6.5×10^{-4} M by nuclear magnetic resonance (Ringel and Sternlicht, 1984). The authors claim, however, that these sites may have a role in inhibition of microtubule assembly can only relate to very high concentrations.

The drug inhibits the assembly of microtubules by binding to soluble tubulin, forming a complex that adds to microtubule ends, and sharply reduces the affinity of the microtubule for fresh tubulin dimers. The tubulin addition rate to the growing microtubule is thereby greatly reduced (Margolis *et al.*, 1980). Colchicine acts, in fact, as a type of chain terminator. A conformational change takes place as the drug binds, as indicated by, among other things, the fact that the reversible dissociation of tubulin dimers into its respective α- and β-subunits is greatly inhibited by the binding of colchicine (Detrich *et al.* 1982).

Furthermore, colchicine stimulates the GTPase activity of tubulin perhaps as a result of drug binding to a site which is responsible for inducing polymeri-zation-dependent GTPase activity. Colchicine, therefore, induces a confor-mational change in the absence of microtubule assembly that mimics assembly by leading to increasd GTPase activity. Vinblastine, an indole alkaloid, inhibits this activation of GTPase activity by colchicine (David-Pfeuty *et al.*, 1978).

Colchicine is used in the treatment of acute gouty arthritis to reduce pain and inflammation. The swellings on joints and tendon sheaths composed mainly of sodium urate crystals (tophi), characteristic of gout, are reduced in size and eventually disappear. The mechanism of action of the drug in ameliorating the symptoms of gout is controversial. Malawista (1975) argues strongly that microtubules are involved in the mode of action of colchicine against gout.

One major difficulty with this view, however, is the activity of trimethylcolchicinic acid against gout; it is almost as active as colchicine itself but is no inhibitor of microtubule assembly (Wallace, 1975). In Malawista's view, trimethylcolchicinic acid could be converted *in vivo* to colchicine (or an analogue which can bind to microtubules). This could give rise to low levels of colchicine in the circulation, which are as high as is needed for colchicine action *in vivo* (namely 5×10^{-8} M) but still difficult to detect (Wallace, 1975). Although this concentration is below the apparent binding constant for tubulin/colchicine complex formation, the existence of a pseudo-irreversible complex suggests that time-dependent inactivation could occur. Low levels of the drug would then have the desired effect. This biotransformation has not been confirmed, however, and the question remains unresolved.

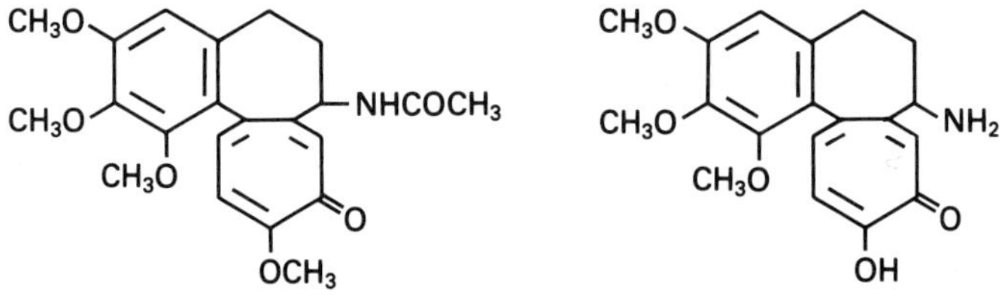

Colchicine   Trimethylcolchicinic acid

A second difficulty arises from the fact that colchicine is only a weak anti-inflammatory agent in conditions induced by substances other than urate crystals, and so it is possible that there is a particular step in crystal-induced inflammation which is more important for that type of inflammation than it is for any other. The release of a neutrophil-derived chemotactic factor has been proposed in this context (Spilberg *et al.*, 1979). Whether microtubules can be implicated in this secretory process is not entirely clear, although they are known to be involved in a number of other secretory processes such as the release of histamine from human lung tissue after antigen challenge (Kaliner, 1977). In conclusion, inhibition of microtubule assembly remains the only plausible explanation for the mode of action of colchicine in the absence of any other convincing hypothesis.

11.3 Vinca alkaloids as antitumour agents

Vincristine and vinblastine are two members of the family of indole alkaloids derived from the periwinkle (*Catharanthus* or *Vinca rosea*) — hence the name vinca alkaloids. These drugs bind to tubulin in a different fashion from colchicine. There are two binding sites for vinblastine per tubulin dimer, instead of one for colchicine. Moreover, vinblastine will bind instantaneously and reversibly, unlike colchicine, to binding sites on the microtubule, in the absence of soluble dimer, with a binding constant of 1.9×10^{-6} M. The number of high-affinity binding sites that is available, however, is far less than the total number possible if one site on each individual subunit were freely accessible (Wilson *et al.*, 1982).

Inhibition of *de novo* polymerization has yielded I_{50} of 4·3, and 3.2×10^{-7} M for vinblastine and vincristine respectively (Owellen *et al.*, 1976). If the experiment is carried out by measuring the inhibition of addition of tubulin dimers to microtubules at steady state, known as "freewheeling", in contrast to polymerization *de novo*, a figure of 1.38×10^{-7} M for half-maximal inhibition is obtained for vinblastine (Wilson *et al.*, 1982). Under these "freewheeling" conditions, 1·16 molecules of vinblastine are detected per microtubule, implying that the addition of one vinblastine molecule at the assembly end of a microtubule is enough to have a very marked effect on the polymerization process. This is referred to as "substoichiometric poisoning" since the concentration of drug is well below that of the tubulin (although greater than that of the microtubule ends). No effect on depolymerization is noted at these concentrations.

At higher concentrations (i.e. $>10^{-6}$ M) the drug binds to a greater number of sites on the microtubule and appears to loosen its structure. As a consequence, splaying and peeling of protofilaments at microtubule ends occurs together with active depolymerization (Wilson *et al.*, 1982).

Since freewheeling is more sensitive to drug inhibition than is *de novo* polymerization, the ability of four vinca alkaloids to inhibit freewheeling has been compared with their ability to inhibit cell proliferation in tumour cell cultures *in vitro*. Although all four drugs were very potent in both tests, no correlation was found between orders of potency on the two tests (Jordan *et al.*, 1985). It was suggested that this may be because the relative potency of the alkaloids may vary with respect to tumour type studied. Alternatively, it may be suggested that four compounds are hardly enough on which to base firm conclusions. The effect of the vinca alkaloids may be seen at a higher level of organization in HeLa cells where vincristine blocks mitosis by arresting the cell in metaphase (George *et al.*, 1965).

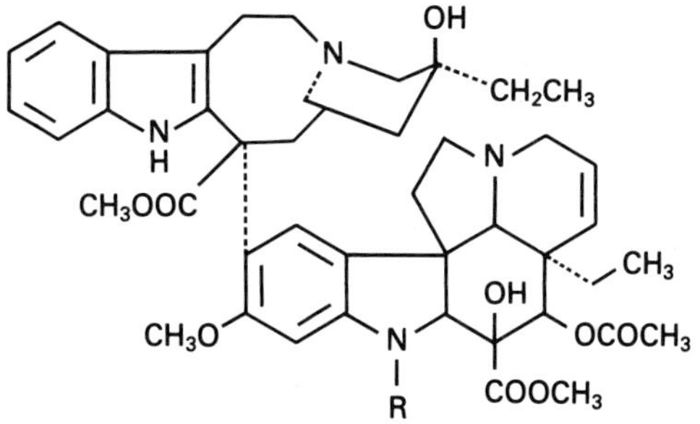

Vinblastine: R = CH₃
Vincristine: R = CHO

Vinblastine and vincristine are used normally in combination with other antitumour agents for the treatment of Hodgkin's and non-Hodgkin's lymphomas, breast carcinoma, cancers of the head and neck and others. One curious feature is that cross-resistance between the vinca alkaloids is not observed. This may reflect a difference in dosage rather than a true lack of cross-resistance (AMA Drug Evaluations, 1983), or conversely, that these drugs have more than one mode of action.

11.4 Griseofulvin as an antifungal agent

Griseofulvin is a secondary metabolite elaborated by the fungus, *Pencillium griseofulvium dierckx*. The action of griseofulvin is to arrest the target fungal cell in metaphase (Gull and Trinci, 1973). As a consequence, multinucleate hyphal cells are formed, frequently together with a Y-shaped metaphase equatorial plate. The microtubules maintain their normal morphological appearance but are disorientated within the spindle (Grisham *et al.*, 1973). The overall effect is thus antimitotic, although the site of action differs from that of colchicine. Griseofulvin does not prevent the binding of colchicine to tubulin nor does it affect the ability of the vinca alkaloids to stabilize the binding of colchicine to tubulin (Wilson, 1970).

Originally it was thought that the drug bound to the MAPs (Roobol *et al.*, 1977), thereby inhibiting microtubule assembly both in rate and extent (Roobol *et al.*, 1976). More recently, however, radioactive griseofulvin has been found to bind directly to tubulin dimer (0·83 mole/mole). Initiation of polymerization requires microtubule-associated proteins but subsequent elongation does not, and griseofulvin blocked this later stage. Moreover, griseofulvin also induces depolymerization of preformed microtubules at 37°C (reviewed in Kerridge, 1986).

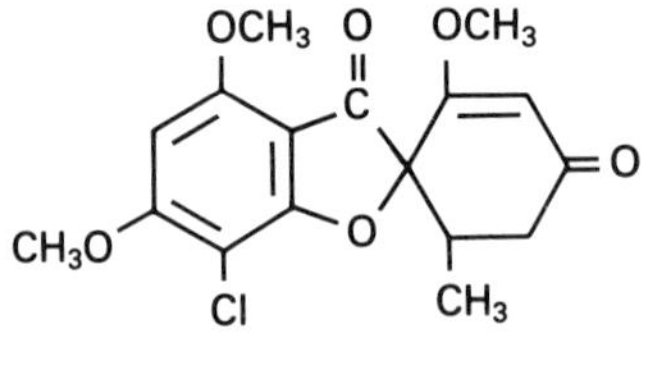

Griseofulvin

Keates (1981) has drawn attention to the need to carry out studies *in vitro* with griseofulvin at low concentrations since the solubility limit is 3×10^{-5} M in aqueous buffers. Attempts to introduce higher levels into solution may induce artifacts by precipitating proteins out of solution. Keates finds that griseofulvin inhibits both polymerization and depolymerization with an I_{50} of $1·25 \times 10^{-5}$ M, but the equilibrium position is unchanged.

Griseofulvin is fungistatic *in vitro* for a number of fungi that cause skin infections (dermatophytes, such as *Trichophyton, Micromonospora species,* and has been used for many years to treat infections such as athlete's foot (caused by *Trichophyton mentagrophytes*). The drug owes its efficacy partly to the fact that it is concentrated in the skin in keratin precursor cells, binding particularly to keratin so that the fungus is unable to gain a foothold on new hair and nails (Gull and Trinci, 1973). Secondly, many dermatophytes concentrate the drug particularly if they are sensitive to it (El-Nakeeb and Lampen, 1965). Griseofulvin is still being used as one of the standard drugs with which to compare the newer agents, such as ketoconazole, and appears to be just as effective (Stratigos *et al.*, 1983).

11.5 Benzimidazoles as anthelminthics

The benzimidazole family of drugs has long been used to treat helminthic infections of sheep, cattle, goats etc. More recently, however, the use of two members of the family, mebendazole and flubendazole, to treat river blindness in man caused by the helminth *Onchocerca volvulus* has been proposed (Dominguez-Vasquez *et al.*, 1983). Another benzimidazole, nocodazole, has been experimentally effective in the treatment of tumours.

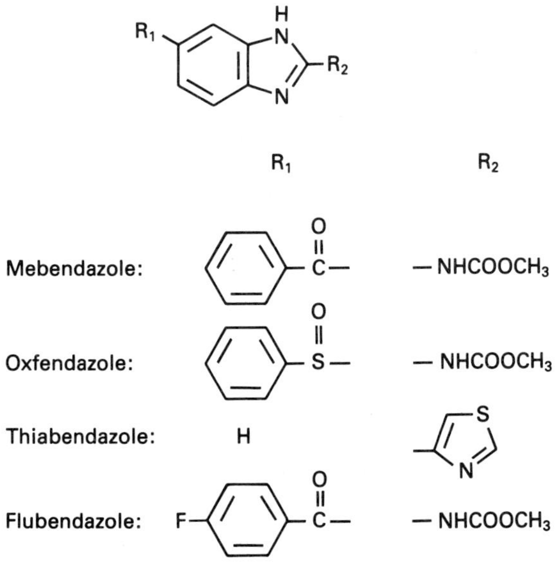

Several benzimidazoles inhibit the assembly of microtubules from a nematode, *Ascaridia galli*, which infects chickens, with I_{50} of 5×10^{-6} M and 6×10^{-6} M for mebendazole and oxfendazole respectively. Some but not all of the drugs bind to mammalian tubulin at similar concentrations, although thiabendazole and oxfendazole had I_{50} values in excess of 10^{-4} M. Electron microscopy of microtubules, polymerized under benzimidazole inhibition, showed a reduction in both the number and the length of microtubules formed (Dawson *et al.*, 1984). Correlation of *in vitro* with *in vivo* results is not possible for all the compounds, perhaps because of pharmacokinetic and/or metabolic considerations. Nevertheless, evidence from benzimidazole-resistant helminths such as *Trichostrongylus colubriformis* suggests that tubulin has been altered to a form which shows less drug affinity. Furthermore, resistant organisms show no change in the appearance of their cytoplasmic microtubules when treated with thiabendazole, unlike intestinal cells from the drug-sensitive organisms, that lose almost all their organelles (Sangster *et al.*, 1985).

Antagonism of colchicine binding can also be used as an index of microtubule ligand affinity for the benzimidazole drugs because they bind at the colchicine binding site, despite their obvious difference in structure. Mebendazole has been shown to interfere with colchicine for the binding site on tubulin from bovine brain with an inhibitor constant of 3.6×10^{-6} M. One binding site per tubulin dimer was found by Scatchard analysis (Laclette *et al.*, 1980) — see the Appendix for discussion on Scatchard analysis. Furthermore, an I_{50} of 1×10^{-6} M was found for albendazole by inhibition of [^{3}H]colchicine binding to tubulin from the intestinal cells of *Ascaris suum*, a nematode that infects pigs (Barrowman *et al.*, 1984). In confirmation of these findings, benzimidazole resistance results in lack of colchicine binding to microtubules (Sangster *et al.*, 1985).

The wide variety of structures that will bind to tubulin even at the same site argues for considerable heterogeneity on the protein surface, and would make a very interesting X-ray study. Furthermore, the variety of the uses of microtubule ligands is intriguing and suggests that other structural types may be found that bind to tubulin and show pharmacological activity.

References

AMA Drug Evaluations (1983). AMA Drug Division, 5th Ed. (Philadelphia: W. B. Saunders) pp. 1537–1538.

Barrowman, M. M., Marriner, S. E. and Bogan, J. A. (1984). *Biochem. Pharmacol.*, **33**, 3037–3040.

David-Pfeuty, T., Simon, C. and Pantaloni, D. (1978). *Nature*, **272**, 282–283.

Dawson, P. J., Gutteridge, W. E. and Gull, K. (1984). *Biochem. Pharmacol.*, **33**, 1069–1074.

Detrich, H. W., Williams, R. C. and Wilson, L. (1982). *Biochemistry*, **21**, 2392–2400.

Dominguez-Vasquez, A., Taylor, H. R., Ruvalcaba-Macias, A. M., Murphy, R. P., Greene, B. M., Rivas-Alcala, A. R. and Beltran-Hernandez, F. (1983). *Lancet*, **i**, 139–143.

El-Nakeeb, M. A. and Lampen, J. O. (1965). *J. Gen. Microbiol.*, **39**, 285–293.

George, P., Journey, L. J. and Goldstein, M. N. (1965). *J. Nat. Cancer Inst.*, **35**, 355–375.

Grisham, L. M., Wilson, L. and Bensch, K. (1973). *Nature*, **244**, 294–296.

Gull, K. and Trinci, A. P. J. (1973). *Nature*, **244**, 292–294.

Jordan, M. A., Himes, R. H. and Wilson, L. (1985). *Cancer Res.*, **45**, 2741–2747.

Kaliner, M. (1977). *J. Clin. Invest.*, **60**, 951–959.

Kerridge, D. (1986). *Adv. Microbial Physiol.*, **37**, 1–72.

Keates, R. A. B. (1981). *Biochem. Biophys. Res. Commun.*, **102**, 744–752.
Laclette, J. P., Guerra, G. and Zetina, C. (1980). *Biochem. Biophys. Res. Commun.*, **92**, 417–428.
Little, M. and Luduena, R. F. (1985). *EMBO J.*, **4**, 51–56.
Luduena, R. F. and Roach, M. C. (1981). *Biochemistry*, **20**, 4444–4450.
Malawista, S. E. (1975). *Arthr. Rheum.*, **18**, Suppl. 6, 835–846.
Margolis, R. L., Rauch, C. T. and Wilson, L. (1980). *Biochemistry*, **19**, 5550–5557.
Owellen, R. J., Hartke, C. A., Dickerson, R. M. and Hains, F. O. (1976). *Cancer Res.*, **36**, 1499–1502.
Ringel, I. and Sternlicht, H. (1984). *Biochemistry*, **23**, 5644–5653.
Roobol, A., Gull, K. and Pogson, C. I. (1976). *Fed. Eur. Biochem. Soc. Lett.*, **67**, 248–251.
Roobol, A., Gull, K. and Pogson, C. I. (1977). *Fed. Eur. Biochem. Soc. Lett.*, **75**, 149–153.
Sangster, N., C., Prichard, R. F. and Lacey, E. (1985). *J. Parasitol.*, **71**, 645–651.
Serrano, L., Avila, J. and Maccioni, R. B. (1984). *J. Biol. Chem.*, **259**, 6607–6611.
Spilberg, I., Mendell, B., Mehta, J., Simchowitz, L. and Rosenberg, D. (1979). *J. Clin. Invest.*, **64**, 775–780.
Stratigos, I., Zissis, N. P., Katsambes, A., Koumentakis, E., Michelopoulos, M. and Flemetakis, A. (1983). *Dermatologia*, **166**, 161–164.
Timasheff, S. N. and Grisham, L. M. (1980). *Annu. Rev. Biochem.*, **49**, 565–591.
Wallace, S. L. (1975). *Arthr. Rheum.*, **18**, Suppl. 6, 847–851.
Weisenberg, R. C. and Deery, W. J. (1981). *Biochem. Biophys. Res. Commun.*, **102**, 924–931.
Wilson, L. (1970). *Biochemistry*, **9**, 4999–5007.
Wilson, L., Jordan, M. A., Rouse, A. and Margolis, R. L. (1982). *J. Mol. Biol.*, **159**, 125–149.

Chapter 12

Hormonal modulators

12.1 Introduction

In this chapter we consider the drugs that modulate the action of two types of hormone: insulin and analogues of gonadotrophin hormone releasing hormone (GnRH). Although we do not know exactly how the drugs that lower blood glucose work or, indeed, the detailed mechanism of action of insulin itself (for example, the nature of the second messenger, if any) it is necessary to include the important area of anti-diabetic drugs. In fact it is interesting to compare our limited knowledge of the mode of action of insulin, which has been investigated for several decades, with the way in which GnRH analogues, that have only been known for just over a decade, have broadened our knowledge of the relationship between hypothalamus, pituitary and gonads.

12.2 Diabetes mellitus

The condition of diabetes is still a very serious medical problem which can only partially be alleviated by pharmacological intervention. There are two broad categories of diabetes mellitus; one in which the patient is totally dependent on insulin injections because the β-cells of the pancreas which normally secrete insulin under the influence of glucose have been destroyed, and the other in which the patient has circulating levels of insulin but is resistant to the action of the hormone. The former, insulin-dependent, type is less frequent than the latter but is more serious and usually occurs early in life (hence the term 'juvenile onset' is used to describe it). It may result from viral infection — Coxsackie B virus has been implicated in at least one case (Yoon *et al.*, 1979). Non-insulin dependent diabetes of which several sub-types exist, is more often a disease of later life ('maturity onset') and is frequently associated with obesity.

The basic metabolic defect lies in the fact that glucose cannot enter cells, either because there is no insulin in the plasma to induce transport, or because the insulin that is present (and its level may be higher than normal) is rendered ineffective possibly by inadequate or insufficient receptors or by a post-receptor lesion. Consequently, excess glucose is found in the plasma (hyperglycaemia) and is excreted into the urine together with copious amounts of water (polyuria). The patient becomes dehydrated, and drinks to assuage the thirst (polydipsia). Glucose cannot enter the appetite-regulating cells of the hypothalamus so the patient always feels hungry and frequently eats (polyphagia).

The overall consequence is an increase in weight leading to frank obesity in some cases; the production of vast quantities of sweet-tasting urine (mel is the Latin for honey and diabetes is related to the Greek work for syphon); and fatigue through the lack of available energy (Maurer, 1979).

The degree of severity of the condition may be studied by the administration of a glucose 'meal' (50 or 100 grams), otherwise known as a glucose tolerance test. In non-diabetics, the levels of blood glucose rise to a peak in 45 minutes and return to the starting level in about $1\frac{1}{2}$ hours. In those with mild disease, the glucose level rises much higher and takes up to 4 hours to return to the baseline, followed by an overshoot before returning to normal. Others with more serious disease show elevated glucose levels in the absence of food and very high levels of glucose for a long time after a glucose meal (Montgomery *et al.*, 1983).

If the condition is uncontrolled, the patient can relapse into a coma and die. This is partially a consequence of the dehydration and partly from metabolic acidosis. Because the cells cannot utilize glucose for the production of energy, they break down triglyceride fat instead. Acetyl CoA levels are raised, leading to the production of acetone and acetoacetate and β-hydroxybutyrate, which are secreted into the blood by the liver. The latter two can be used by the brain, heart and skeletal muscle as a source of energy to a limited extent but then they begin to accumulate in the plasma. Acetoacetate and β-hydroxybutyrate are relatively strong acids and eventually the buffering capacity of the blood is overcome, causing acidosis. The kidneys excrete these anions together with cations, such as sodium, and water, thus contributing to the dehydration, until eventually the patient goes into a coma as a result of the combination of acidosis and dehydration (Montgomery *et al.*, 1983).

12.2.1 The action of insulin

The action of insulin in cellular terms is responsible for the stimulation of glycogen and lipid synthesis and glucose oxidation concomitantly with the inhibition of glycogenolysis, lipolysis and gluconeogenesis. Any lesion in insulin action, whether it occurs at the receptor on the cell surface or subsequently, will cause increases in lipolysis and glycogenolysis, with elevation of glucose and acetyl CoA levels leading to the formation of acetone etc. Extracellular levels of glucose will rise since the major route of glucose removal from the blood is into the peripheral muscle tissue (see Pollet, 1983).

The intracellular effects of insulin are initiated possibly by means of an, as yet, unidentified second messenger (see Malchoff *et al.*, 1987). Molecules of glucose carrier protein are transferred from the Golgi apparatus to the cell surface ready to transport glucose into the cell. The major series of metabolic changes noted above requires the phosphorylation status of the key enzymes of glycogen breakdown, lipolysis and lipogenesis to be altered:

Dephosphorylation	*Result*	*Insulin*
Glycogen synthase	Glycogen synthesis activated	+
Glycogen phosphorylase	Glycogen breakdown inhibited	+
Pyruvate dehydrogenase	TCA cycle activity increased	+
HMG CoA reductase	Cholesterol synthesis activated	+
Phosphorylation	*Result*	
Glycogen synthase	Glycogen synthesis inhibited	−
Glycogen phosphorylase	Glycogen breakdown activated	−
ATP citrate lyase	Lipogenesis activated	+
Acetyl CoA carboxylase	Lipogenesis activated	+

These events follow from insulin binding to its receptor. The insulin receptor is a glycoprotein heterodimer ($\alpha_2 \beta_2$) of molecular weight 430 kDa, composed of two 125 kDa α-subunits and two β-subunits of 90 kDa. The receptor contains two functional binding sites and is stabilized by disulphide bonds between the subunits. It is one of the family of tyrosine kinases, and phosphorylates the hydroxyl of a tyrosine residue within the receptor complex (reviewed in Houslay *et al.*, 1986). This appears to lead to activation of a specific guanine nucleotide regulatory protein that may in turn activate a specific phospholipase C in a similar fashion to the other neurohormone receptors discussed in Chapter 9 (p. 130). A special pool of inositol phospholipid may be the target for this enzyme (Houslay *et al.*, 1986).

The binding of hormone to receptors in liver and muscle also causes the receptors to cluster in small groups followed by endocytosis of a receptor cluster; subsequent fusion with lysosomes leads to the degradation of insulin and the final event is recycling of the receptors. Auto-phosphorylation of the β-subunit of the receptor is also associated with binding but the significance of this event is not yet clear. A reduction in the number of available receptors on the cell surface (down-regulation) is a consequence of high levels of external hormone binding to, and causing the internalization of, a higher than normal proportion of receptors (Pollet, 1983).

12.2.2 Insulin therapy

From a therapeutic point of view, diabetes can be approached in a number of ways. Insulin-dependent diabetics, by definition, require insulin which in earlier years was extracted from animal pancreas (usually that of pigs or cows). In 1979, Goeddel *et al.* reported the expression of the human insulin gene in *E. coli*. They chemically synthesized the genes for the A chain (21 amino acids) and the B chain (30 amino acids) and inserted them into the plasmid pBR 322. The insulin genes were attached to the carboxy terminus of the gene for β-galactosidase so that the human insulin would be efficiently transcribed and translated as part of one long protein chain. The separate insulin chains were subsequently cleaved from the β-galactosidase using cyanogen bromide. The sulphydryl groups were converted to S-sulphonyl groups, the chains were mixed and were allowed to combine by first reducing the mixture followed by air oxidation. The presence of insulin was confirmed by radioimmunoassay. The production of sufficient quantities of insulin for human treatment is described by Johnson (1983) — an interesting story of the problems faced by the development of the first genetically engineered health care agent to be produced on a commercial scale.

Long-acting insulin preparations are produced by the reaction of insulin with the basic protein protamine in the presence of zinc. When the preparation is injected subcutaneously in aqueous suspension, it dissolves slowly at the injection site and is thus absorbed slowly.

Periodic injections of insulin, however, tend to cause the blood level of glucose to oscillate wildly (Bressler, 1978) because the action of insulin is normally balanced by glucagon, whose role is to raise blood glucose levels by increasing liver glucose production and opposing liver glucose storage (as glycogen). The net result in the normal subject is to maintain levels of glucose close to the norm. The peaks in glucose level in the diabetic treated with periodic insulin injections are believed to contribute to the complications associated with diabetes. Recently, studies have been carried out with insulin infusion pumps taped on to the wall of the abdomen, such that the point of needle lies in the subcutaneous tissue. Insulin is then pumped through the syringe from a reservoir at a steady basal level, with additional doses just before meals (see, for example, Mecklenburg *et al.*, 1982).

In patients with non-insulin-dependent diabetes that is relatively mild, the patients are likely to be insulin-resistant in the sense that they have a reduced number of receptors. In this case the maximum response to the hormone is merely attained at higher concentrations. Treatment with a drug that lowers serum glucose levels (a hypoglycaemic agent), causes the number of receptors usually to return to normal.

With patients more seriously affected by the disease, down-regulation of the receptors is less likely to be of importance since the circulating insulin levels range from slightly elevated to below normal. A marked post-receptor defect is almost certainly present in these patients since an increased insulin level does not return glucose levels to normal. The nature of this defect is not at present known (Pollet, 1983; Lockwood and Amatruda, 1983).

Glucagon is another peptide secreted by the pancreas, in this case by the α-cells. It has been proposed that too much glucagon as well as too little insulin may be necessary for insulin-dependent diabetes to manifest (see Unger and Orci, 1981b), since the actions of glucagon are antagonistic to those of insulin.

These actions are inhibition of cholesterol synthesis, and stimulation of lipolysis, glycogenolysis and gluconeogenesis, with eventual hyperglycaemia. Also produced in the pancreas is the 14-amino acid peptide somatostatin which, amongst other actions, inhibits the release of both glucagon and insulin. Indeed, since the α-cells of the pancreas produce glucagon, the β-cells insulin and the δ-cells somatostatin, an intrapancreal control mechanism has been proposed (Unger and Orci, 1981a) — see Figure 12.1. Consistent with this suggestion, there is at least one report of combined treatment with insulin and somatostatin which has given better results than insulin alone (Raskin and Unger, 1978). Somatostatin probably works by suppressing glucagon release (Figure 12.1).

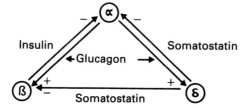

Arrows indicate an interaction between the hormone secreted from one cell type and the product of another. + indicates a stimulation, − indicates an inhibition.

Figure 12.1 The interactions between pancreatic hormones.

Although this is a book about the uses of pharmaceuticals to combat disease, it is not the intention of the author to imply that drugs are the ultimate panacea for all illness. In no disease is this more relevant than in non-insulin dependent diabetes. In many cases, dietary control with concomitant weight loss, coupled with exercise, will often control the condition. The rationale for this approach lies in the fact that many patients, particularly those who are obese, have elevated levels of insulin together with a lowered number of receptors, as has been seen in some diabetic animals (Roth *et al.*, 1975). A reduction in weight causes the number of receptors to rise and the level of circulating insulin to fall as is likely to be the consequence if the intake of mono- and di-saccharides is restricted. Intake of polysaccharides apparently need not be restricted presumably because these give rise to less available mono- and di-saccharides. Saturated fat intake should also be lowered in view of the greatly increased risk of heart disease with diabetes (American Diabetic Association, 1979). Weight reduction, even to a modest extent, greatly improves the responsiveness of obese subjects to insulin (see Lockwood and Amatruda, 1983). Exercise is also recommended, as it helps to activate the transport of glucose across the cell membranes even in the absence of insulin (see Larner, 1985).

The main goal of therapy is to lower the serum glucose level since not only does this reduce the symptoms of diabetes, but also because a number of serious complications associated with diabetes are related to excess serum glucose. One such complication is damage to the kidney by virtue of thickening of the basement membrane, which can lead to chronic renal failure. Retinal damage leading to oedema and haemorrhage makes blindness 25 times more common among diabetics. Cataracts also are more frequent in these patients. Atherosclerosis which predisposes to heart attacks, is another complication. Both types of diabetes lead to these sequelae, although non-insulin dependent diabetes may result in them less frequently.

The metabolic defect in insulin level or action is almost certainly the main cause of these complications. It is likely that hyperglycaemia is the cause of damage to kidney and retina, possibly because glucose glycosylates membrane proteins. A causal relationship between hyperglycaemia and vascular complications is not yet proven (see Brownlee and Cerami, 1981, for a review).

12.3 Sulphonylureas as hypoglycaemic agents

The circulating levels of glucose in non-insulin-dependent diabetes can be lowered by a family of drugs based on the sulphonylurea moiety. Tolbutamide and chlorpropamide were the earliest to be introduced over 30 years ago, while recently a second generation of drugs have been developed which include glibenclamide (glyburide) and glipizide. These drugs may, however, be regarded as only an adjunct to dietary control and exercise (Jackson and Bressler, 1981).

The effectiveness of these agents seems to depend on two actions. The first is an effect on the β-cells of the pancreas to improve the secretion of insulin. The drugs need a functioning pancreas to be fully effective and are thus of no value in insulin dependent disease. The precise mechanism of this action is not understood but it appears to involve drug binding to membranes of the β-cell leading to an alteration in ion transport. In resting β-cells the membrane potential is maintained by an efflux of potassium ions. The drugs reduce the permeability of the membrane to potassium, causing a depolarization of the membrane because of the retention of the positive charge of the cation. As the membrane potential difference falls, so voltage-dependent calcium channels are activated and calcium influx is greatly stimulated. The release of insulin which requires calcium ion is thereby facilitated (Garrino *et al.*, 1986). In addition, these drugs suppress glucagon release in diabetics but not in normal subjects (Lebovitz, 1984).

The link between the binding to the β-cell and hypoglycaemic activity has been demonstrated by Geisen *et al.* (1985), who used a rat β-cell tumour to which ^{3}H-glibenclamide bound tightly, reversibly and in a saturable fashion with a K_d of 5×10^{-11} M. A number of analogues of glibenclamide were

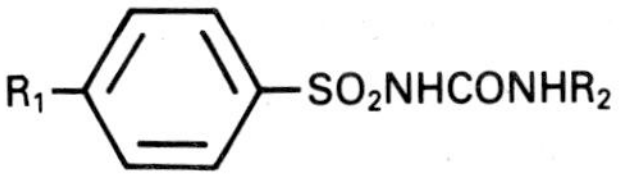

$$R_1 - \!\!\!\!\!\!\!\bigcirc\!\!\!\!\!\!\! - SO_2NHCONHR_2$$

	R_1	R_2
Tolbutamide	CH_3	C_4H_9
Chlorpropamide	Cl	C_3H_7

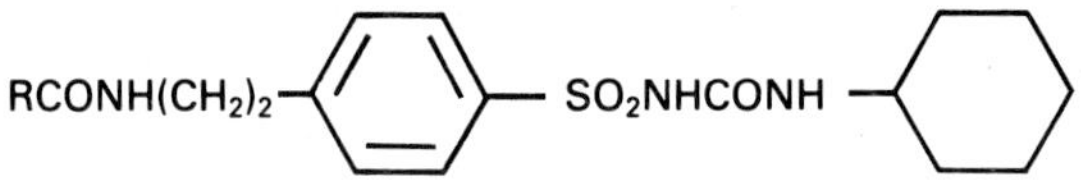

$$RCONH(CH_2)_2 - \!\!\!\!\!\!\!\bigcirc\!\!\!\!\!\!\! - SO_2NHCONH - \bigcirc$$

Glibenclamide

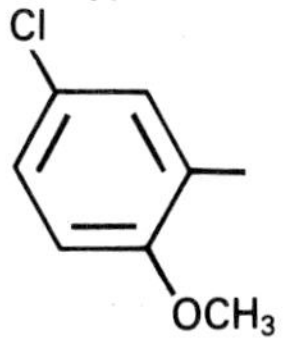

R, Cl, OCH$_3$ substituted ring structure

Glipizide CH_3 — pyrazine ring structure

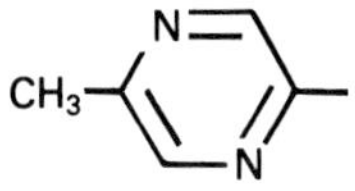

compared with respect to both inhibition of radioactive ligand binding and hypoglycaemic activity in rabbits. A good correlation was observed, although a number of compounds with binding constants between 10^{-7} and 10^{-8} M had to be omitted from the analysis because they failed to show sufficient hypoglycaemic activity — probably because they were not absorbed sufficiently or were too heavily serum-bound to be effective.

The second action of the hypoglycaemic agents is potentiation of the action of insulin. They increase the absorption of glucose engendered by a given dose of insulin; whether this is by increasing the number of insulin receptors through preventing down-regulation or by a post-receptor effect is not known (Lebovitz, 1984).

Chronic administration of these drugs eventually results in only normal (or even reduced) insulin secretion (with the possible exception of glipizide). In contrast, acute dosing raises insulin levels. It is likely, therefore, that the extrapancreatic effects are more important in the long term, although in some cases the drugs may gradually lose their efficacy.

There is no clear indication that the second generation drugs are any more effective therapeutically than the first, although the former are more active at low concentrations. The number of failures with each type is almost identical, and so there seems to be little to choose between them (Kreisberg, 1985).

12.4 Gonadotrophin hormone releasing hormone (GnRH) analogues

A growing area of pharmacological intervention has been opened up for a number of cases of hormonal aberrations, and cancers that depend on sex steroid hormones for growth, by a greater understanding of the axis that links the hypothalamus to the gonads via the pituitary gland (see Figure 12.2). The hormones that carry the information from pituitary to gonads are collectively known as the gonadotrophins; luteinizing hormone (LH) and follicle-stimulating hormone (FSH) in females — LH is referred to as interstitial cell stimulating hormone (ICSH) in males.

Gonadotrophin hormone releasing hormone (GnRH or gonadorelin) is released from the hypothalamus and acts on the anterior portion of the pituitary gland to release gonadotrophins, although there are suggestions that FSH may also be released by another as yet undiscovered hormone. GnRH secretion is under the control of higher centres in a way that we do not yet understand, although it is clear that in women some form of 'biological clock' must operate in order to maintain the functioning of the menstrual cycle.

GnRH is released in short pulses of millisecond duration. This can be recognized by following the pulsation in plasma levels of LH, since LH release is very sensitive to GnRH pulsation and the half-life of LH in plasma is shorter than the time between pulses. FSH release is much less sensitive to GnRH levels and does not pulse because the response is much slower and takes longer to die away (reviewed in Lincoln *et al.*, 1985).

GnRH acts via receptors in the pituitary which have not been purified but an estimate of molecular weight of 136 kDa has been made. Occupation of the receptors subsequently induces the opening of calcium gates in the membrane followed by the extrusion of granules which contain LH. The binding constant for GnRH at the receptor is $3{\cdot}0 \times 10^{-9}$ M (Conn *et al.*, 1985). The release process is linked to the phosphatidylinositol cycle, requiring the ingress of calcium and the activation of protein kinase C by diacylglycerol (reviewed in Catt *et al.*, 1985); see Section 9.1 for a detailed discussion.

In the female, FSH stimulates the growth of follicles in the ovary which are responsible for the production of the main oestrogen, the steroid hormone 17-β-oestradiol (oestrone and oestriol are derived from oestradiol in the liver and are less active as oestrogens). By the time that ovulation is about to occur, various follicles are almost mature. There is then a surge in the level of LH which initiates ovulation in one or more follicles while in some way causing the others to regress. The follicle bursts and discharges the egg. The follicle then undergoes a re-organization of its cells developing into a yellow body known as the corpus luteum (a process therefore called luteinization) that secretes progesterone still under the influence of LH.

Progesterone prepares the female for pregnancy by stimulating the development of the lining of the uterus, and inhibiting contraction of the smooth muscle of the uterine wall. It also prevents the development of a new follicle. Both progesterone and oestrogen act to close down FSH and LH release from the pituitary and possibly GnRH release from the hypothalamus in a negative feedback loop. Furthermore, oestrogen exerts a secondary, but positive, feedback effect on the pituitary in conjunction with GnRH which is believed to be responsible for the pre-ovulatory surge in LH production (see Figure 12.2). The use of GnRH analogues played a part in the discovery of these interactions (Schally, 1978; Asch *et al.*, 1983).

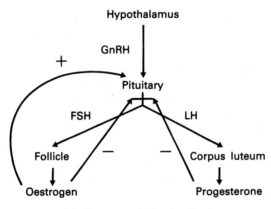

+ indicates a positive feedback
− indicates a negative feedback

Figure 12.2 Hypothalamus–Pituitary–Gonad axis

FSH and LH bind to membrane fractions from ovary tissue and exert their biochemical action through the stimulation of adenylate cyclase. Cleavage of the side-chain of cholesterol is induced, the rate-limiting step in steroid biosynthesis, in a fashion similar to the events in the adrenal cortex (Funkenstein *et al.*, 1983) — see Figure 6.2 in Chapter 6. Oestrogen and progesterone production are thereby greatly increased (Figure 12.3).

In the male LH stimulates the Leydig cells of the testis to secrete testosterone. As these cells are also known as interstitial cells, LH is sometimes referred to as interstitial cell stimulating hormone or ICSH. FSH, however, is required only for the maturation of the spermatids into spermatozoa in some way that is not fully understood. Nevertheless, there is a similarity with the female in that testosterone exerts a negative feedback inhibition on LH secretion from the pituitary.

LH and FSH are glycoproteins of molecular weight about 16 kDa and are

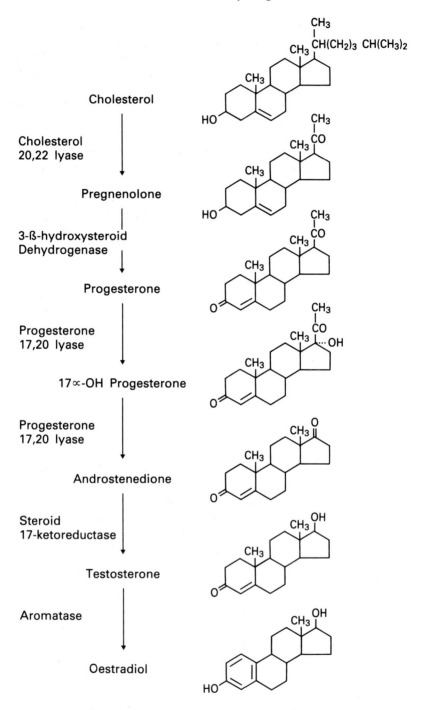

Figure 12.3 Sex hormone production from cholesterol.

dimers of non-identical subunits α and β. The α subunits are nearly identical in the two hormones (and in thyroid-stimulating hormone), whereas the β-subunits show some differences — though homology is still 80% (Pierce and Parsons, 1981). It is therefore the β-subunit that defines the biological activity.

There are intriguing similarities between the structures of the hormone and that of cholera toxin, which have led to the speculation that cholera toxin may exert its stimulatory effect on adenylate cyclase and induction of water loss from the cell by mimicking the action of a naturally occurring hormone. At present the evidence is, however, inconclusive (Pierce and Parsons, 1981).

GnRH is a decapeptide (see structures). The part of the molecule that is responsible for binding to the receptor (the R site) is made up of residues from the C and N terminal ends, namely PyroGlu[1] and Gly[10], and modification of these residues destroys binding. On the other hand, His[2] and Trp[3] appear to be essential to stimulate secretion of LH (M site) which occurs as a consequence of receptor occupancy. Agonists of GnRH such as the peptide leuprolide contain both M and R sites (Gly[10] is substituted by an ethylamide group) while antagonists contain only R. In addition, calcium is required for full agonist activity (Conn *et al.*, 1985), as one would expect for a secretory process.

GONADORELIN (GnRH):

PyroGlu - His - Trp - Ser - Tyr - Gly - Leu - Arg - Pro - GlyNH$_2$

LEUPROLIDE:

PyroGlu - His - Trp - Ser - Tyr - D Leu - Leu - Arg - Pro CONHC$_2$H$_5$

Another interesting feature of the agonist/receptor interaction is that the occupied receptors dimerize as a necessary stage in the LH release process, as shown by the agonist activity of an agent prepared by the dimerization of an antagonist (for details see Conn *et al.*, 1985). It is proposed that two occupied receptors come together to form a ring about a previously closed calcium channel which opens as a consequence.

Furthermore, agonist action induces another effect of importance to the therapeutic use of GnRH and its analogues, namely the reduction in the number of available receptors on the cell surface (down-regulation). This effect occurs in cultured pituitary cells (Catt *et al.*, 1985), and also *in vivo*. This is a calcium-independent action, unlike LH release. Antagonists, however, merely block GnRH binding to the receptor and do not cause down-regulation.

The conditions for which GnRH or its analogues are finding, or are likely to find, therapeutic usage are numerous (Cutler *et al.*, 1985). The major use of the native hormone is to induce ovulation in women who suffer from infertility because of insufficient gonadotrophin. This has given rise, on occasion, to multiple pregnancies because the high level of hormones released from the pituitary causes a large number of follicles to mature at the same time. Other hormone abnormalities in men, such as delayed puberty (hypogonadotrophic hypogonadism) and delayed descent of the testicles (cryptorchidism) have also been shown to respond to treatment. These latter treatments will probably need to be given for life, however, as the therapeutic effects are reversible if drug administration is stopped.

Low doses of drug administered in pulses, mimicking the release of the natural hormone, produce an elevation in gonadotrophins. If, however, the drug is given systemically in higher doses, for example as a single daily subcutaneous injection, after a transient increase the pituitary is desensitized as a result of receptor down-regulation, and the levels of FSH and LH fall sharply. As a consequence, leuprolide and other GnRH agonists are, paradoxically, being used to treat conditions where elevated GnRH levels are the cause of the condition, notably the condition where GnRH levels rise very early in life and induce sexual maturation (centrally derived precocious puberty).

Furthermore, there are some tumours that are dependent on sex steroids; some breast cancers in women are dependent on oestrogen and 80% of prostate cancers require testosterone (Vance and Smith, 1984), and these have responded to agonist therapy. A particular advantage that leuprolide therapy has for prostate cancer over the usual treatment with diethylstilboestrol, is the lack of feminizing and cardiovascular side-effects seen with the latter drug. Castration can also be used but is irreversible, and thereby unacceptable to many patients. Garnick *et al.* (1985) feel that leuprolide is at least as effective as other therapies and is a valuable addition to our present armoury of treatments for hormone-dependent cancers.

It has also been proposed that premenstrual syndrome results from a cyclic condition of elevated release of (or abnormal sensitivity to) LH and FSH (Coulson, 1986). The cyclic condition is characterized, amongst other symptoms, by weight gain, mood swings, irritability and inability to concentrate. One trial has shown that a gonadorelin analogue was effective in relieving this condition (Muse *et al.*, 1984) but further trials are required before the results can be confirmed, since the placebo effect is very marked in this condition.

References

American Diabetic Association (1979). *Diabetes*, **28**, 1027–1030.
Asch, R. H., Balmaceda, R., Borghi, M. R., Niesvisky, R., Coy, D. H. and Schally, A. V. (1983). *J. Clin. Endocrinol. Metab.*, **57**, 367–372.
Bressler, R. (1978). *Drugs*, **17**, 461–470.
Brownlee, M. and Cerami, A. (1981). *Annu. Rev. Biochem.*, **30**, 385–432.

Catt, J. K., Loumaye, E., Wynn, P. C., Iwashita, M., Hirota, K., Morgan, R. O. and Chang, J. P. (1985). *J. Steroid Biochem.*, **25**(5B), 677–689.

Conn, P. M., Staley, D., Jinnah, H. and Bates, M. (1985). *J. Steroid Biochem.*, **23**(5B), 703–710.

Coulson, C. J. (1985). *Med. Hypothes.*, **19**, 243–256.

Cutler, G. B., Hoffman, A. R., Swerdloff, R. S., Santen, R. J., Meldrum, D. R. and Comite, F. (1985). *Ann. Intern. Med.*, **102**, 643–657.

Funkenstein, B., Waterman, M. R., Masters, B. S. S. and Simpson, E. R. (1983). *J. Biol. Chem.*, **258**, 10187–10191.

Garnick, M. B., Glode, L. M., Smith, J. A. and Max, D. T. (1985). *Brit. J. Clin. Pract.*, **39**, 73–77.

Garrino, M. G., Meissner, H. P. and Henquin, J. C. (1986). *Eur. J. Pharmacol.*, **124**, 309–316.

Geisen, K., Hitzel, V., Okomonopoulos, R., Punter, J., Weyer, R. and Summ, H.-D. (1985). *Arzneim. Forsch./Drug Res.*, **35**, 707–712.

Goeddel, D. V., Kleid, D. G., Bolivar, F., Heyneker, H. L., Yansura, D. G., Crea, R., Hirose T., Kraszewski, A., Itakura, K. and Riggs, A. D. (1979). *Proc. Nat. Acad. Sci. (U.S.A.)*, **76**, 1106–1110.

Houslay, M. D., Wakelam, M. J. O. and Pyne, N. J. (1986). *Trends in Biochem. Sci.*, **11**, 393–394.

Jackson, J. E. and Bressler, R. (1981). *Drugs*, **22**, 211–245.

Johnson, I. S. (1983). *Science*, **219**, 632–637.

Kreisberg, R. A. (1985). *Ann. Intern. Med.*, **102**, 125–126.

Larner, J. (1985). In: *The Pharmacological Basis of Therapeutics*, 7th ed., Gilman, A. G., Goodman, L. S.. Rall, T. W. and Murad, F., Eds. (New York: Macmillan) Ch. 64.

Lebovitz, H. E. (1984). *Diabetes Care*, **7** (Suppl. 1), 67–71.

Lincoln, D. W., Fraser, H. M., Lincoln, G. A., Martin, G. B. and McNeilly, H. S. (1985). *Rec. Progr. Hormone Res.*, **41**, 369–411.

Lockwood, D. H. and Amatruda, J. M. (1983). *Amer. J. Med.*, **75**, Suppl. 5B, 23–31.

Malchoff, C. D., Huang, L., Gillespie, N., Palasi, C. V., Schwartz, C. F. W., Cheng, K., Hewlett, E. L. and Larner, J. (1987). *Endocrinology*, **120**, 1327–1337.

Maurer, A. C. (1979). *Amer. Sci.*, **67**, 422–431.

Mecklenburg, R. S., Benson, J. W., Becker, N. M., Brazel, P. L., Fredlund, P. N., Metz, R. J., Nielsen, R. L., Sannar, C. A. and Steenrod, W. J. (1982). *New Engl. J. Med.*, **307**, 513–518.

Montgomery, R., Dryer, R. L., Conway, T. W. and Spector, A. A. (1983). *Biochemistry: a case-orientated approach* (St. Louis: C. V. Mosby) Ch. 7.

Muse, K. N., Cetel, N. S., Futterman, L. A. and Yen, S. C. C. (1984). *New Engl. J. Med.*, **311**, 1345–1359.

Pierce, J. G. and Parsons, T. F. (1981). *Annu. Rev. Biochem.*, **50**, 465–495.

Pollet, R. J. (1983). *Amer. J. Med.*, **75**, Suppl. 5B, 15–22.

Raskin, P. and Unger, R. H. (1978). *New Engl. J. Med.*, **209**, 433–436.

Roth, J., Kahn, C. R., Lesnick, M. A., Gorden, P., De Meyts, P., Megyesi, K., Neville, D. M., Gavin, J. R., Soll, A. H., Freycher, P., Goldfine, I. D., Bar, R. S. and Archer, J. A. (1975). *Rec. Progr. Hormone Res.*, **31**, 95–128.

Schally, A. V. (1978). *Science*, **202**, 18–28.

Unger, R. H. and Orci, L. (1981a). *New Engl. J. Med.*, **304**, 1518–1524.

Unger, R. H. and Orci, L. (1981b). *New Engl. J. Med.*, **304**, 1575–1580.

Vance, M. A. and Smith, J. A. (1984). *Clin. Pharmacol. Ther.*, **36**, 350–354.

Yoon, J.-W., Austin, M., Onodera, T. and Notkins, A. L. (1979). *New Engl. J. Med.*, **300**, 1173–1179.

Appendix

Quantification of ligand/macromolecule binding

The binding of a ligand to a macromolecule, whether it be receptor, enzyme or just a binding protein, is governed by the Law of Mass Action, which states that the rate of a reaction is proportional to the concentrations of the reactants. A variety of plotting techniques have been used to elucidate the binding of ligands to macromolecules. It is useful to show how ligand binding to macromolecules, enzyme kinetics and drug/receptor binding interrelate and, in addition, to describe some other methods of plotting such as the Hill and Scatchard plots.

A.1 Enzyme kinetics: I_{50} or K_i?

Much of the study of enzyme inhibition depends on measurement of K_i (the dissociation constant of the enzyme-inhibitor complex). Another measure of inhibition is I_{50}; the concentration of inhibitor required to reduce the reaction velocity to 50% of its value in the absence of inhibitor. K_i and I_{50} are not the same, except in very unusual circumstances, and furthermore, the I_{50} values will vary if different substrate conditions are used (see Cheng and Prusoff, 1973 for a detailed discussion).

Considering the simplest case of an enzyme reaction with one substrate, S, if the concentration of substrate is very much higher than that of enzyme, the initial velocity V_0 is given by:

$$V_0 = \frac{V_{max} \times [S]}{K_m + [S]} \qquad (1)$$

where V_{max} = maximum velocity; K_m = Michaelis constant of the substrate which equals the concentration of substrate that produces half-maximal rate.

(a) In the most common form of competitive inhibition, described by the following equation, the substrate and inhibitor compete for the same active site, but with different affinities for it:

$$E + S \rightleftharpoons ES \rightleftharpoons E + P \qquad (2)$$
$$\mathbin{\updownarrow}$$
$$EI$$

Here EI = enzyme-inhibitor complex and P is product. The equation governing the reaction rate is:

$$V_i = \frac{V_{max} \times [S]}{K_m\left(1 + \dfrac{[I]}{K_i}\right) + [S]} \tag{3}$$

where V_i is the initial velocity in the presence of inhibitor at concentration i and K_i is the dissociation constant of the enzyme/inhibitor complex. If the initial velocity is one-half V_0, the inhibitor concentration i by definition equals the I_{50}, with $V_i = \frac{1}{2}V_0$ so after rearrangement:

$$I_{50} = K_i\left(1 + \frac{[S]}{K_m}\right) \tag{4}$$

Clearly, I_{50} is dependent on substrate concentration. Since different laboratories are unlikely to use the same substrate concentrations, their I_{50} values are not a sound basis for comparison of different inhibitors. Only if the substrate concentration is low compared with the K_m does I_{50} approximate to K_i. A low concentration (less than a tenth of the K_m will also render the assay more sensitive because the effect of an inhibitor will thereby be magnified, but this might lead to problems in detecting product formation (Bush, 1986).

(b) In the case of non-competitive inhibition

$$\begin{array}{ccc} E + S & \rightleftharpoons ES \rightleftharpoons & E + P \\ I\downarrow\uparrow K_{is} & I\downarrow\uparrow K_{iI} & \\ EI & ESI & \end{array} \tag{5}$$

It can be shown that:

$$I_{50} = \frac{(K_m + [S])}{\left(\dfrac{K_m}{K_{is}} + \dfrac{[S]}{K_{iI}}\right)} \tag{6}$$

If the inhibitor binds equally well to the enzyme as to the enzyme-substrate complex, $K_{is} = K_{iI} = K_i$ and the equation simplifies to:

$$I_{50} = K_i \tag{7}$$

Alternatively, if as is often the case $[S] \gg K_m$ and $K_m/[S] \ll K_{is}/K_{iI}$ the equation reduces to:

$$I_{50} = \frac{K_{is}}{K_m/[S] + K_{is}/K_{iI}} \tag{8}$$

again if $K_m/[S] \ll K_{is}/K_{iI}$, then:

$$I_{50} = K_i \qquad (9)$$

In the case of uncompetitive inhibition, provided that $[S] \gg K_m$, I_{50} is independent of substrate concentration and equal to K_i. It is not always convenient to maintain a substrate concentration that is very much higher than the K_m because this may mask the effect of weaker inhibitors. Further derivations for two-substrate reactions may be found in Cheng and Prusoff (1973).

In practical terms there is not always enough time to measure sufficient data points in order to define a K_i. If a large number of compounds have to be tested, as is normally the case, it is often expedient to establish I_{50}, and, provided that the compounds all inhibit the enzyme in a similar fashion, the I_{50}s can be used for ranking purposes. Subsequently, when a few compounds are chosen for development more work can be carried out in order to estimate their K_i values and define the type of inhibition.

A further consideration in carrying out enzyme reactions with inhibitors *in vitro* is that the enzyme concentration is set at a rate-limiting level (very much less than the substrate concentration) in order that Michaelis-Menten kinetics should be applicable. This usually requires enzyme concentrations below 10^{-7} M and sometimes as low as 10^{-10} M. This can be quite unlike the normal physiological condition where reaction rates may be regulated by substrate availability rather than by enzyme concentration. Indeed, in the glycolytic pathway of the yeast *Saccharomyces carlsbergensis*, the molarities of only two enzymes, phosphofructokinase and adenylate kinase, lie below 10^{-5} M (Hess et al., 1969). If we consider the 'concentration' of enzyme active sites, the range for all of the enzymes measured is between $2 \cdot 5 \times 10^{-5}$ and $2 \cdot 0 \times 10^{-4}$ M. The substrate concentrations vary in the same range.

Although enzymes in the glycolytic pathway may be exceptionally highly concentrated, there is no reason to suppose that other pathways may not be similar and, in addition, may show molecular organization to facilitate the passage of substrate along the pathway. It is therefore a matter of some surprise that inhibitors often do show effects *in vivo* that can be related to data obtained *in vitro*. That they do so makes life easier for the medicinal scientist.

One of the important features of enzyme inhibition is whether the inhibition is reversible or not. Irreversible inhibition is often characterized by inhibition that increases over time because it is a consequence of a chemical reaction that requires covalent bonds to be broken. Consequently, the inhibited enzyme will not recover during dialysis, unlike a reversibly-inhibited enzyme. Irreversible inhibition is frequently missed if considerable pre-incubation is not a normal part of the assay protocol, and often appears to be non-competitive inhibition because a portion of the enzyme is effectively put out of action by covalent modification (see Bush, 1986). The order of addition of substrate, inhibitor and enzyme may be crucial but is frequently unreported. In fact, a plot of log (remaining activity) against time should be a straight line for this type of inhibition. Kinetically, the reactions are normally exemplified as follows (see, for example, Coulson and Smith, 1979):

$$E + I \rightleftharpoons EI \rightarrow EI^* \tag{10}$$

EI^* represents the inactivated enzyme. Irreversible inhibitors include amino acid modifying agents such as iodoacetamide, active site directed inhibitors and suicide inactivators such as the monoamine oxidase inhibitors.

A.2 Drug binding to receptors

Just as enzymologists use a plot of substrate concentration against reaction rate to quantify an enzyme reaction, pharmacologists also need to quantify the response of an organ preparation to a drug. For drug/receptor binding:

$$D + R \rightleftharpoons DR \rightarrow Effect \tag{11}$$

where D and R represent drug and receptor respectively. The equation governing the reaction is then:

$$E = \frac{[D]E_{max}}{([D] + K_d)} \tag{12}$$

where E represents the effect produced by a given drug concentration $[D]$ and E_{max} is the maximum effect possible under the conditions, and K_d is the dissociation constant of the drug/receptor complex. This clearly bears a formal similarity to the Michaelis-Menten equation of enzyme kinetics (equation 1).

The method of plotting is different from enzyme kinetics in that log (drug concentration) is plotted against response which normally gives a sigmoid type of curve (see Figure A.1). If two drugs are compared by this approach and produce parallel lines it is likely, but not certain, that the compounds are acting in a similar way at the receptor. If they are not parallel the interpretation is more complex.

One way of analysing such data is by a Schild plot. This depends on the fact that when an antagonist inhibits the response of a given tissue to an agonist competitively, the log (dose)–response curve is shifted to the right so that one obtains a similar plot to Figure A.1, with the tissue showing the same response at a higher agonist concentration (x_A), as it did previously at the original concentration (x_A). It has been shown (Arunlakshana and Schild, 1959) that:

$$\frac{x'_A}{x_A} = 1 + \frac{x_B}{K_B} \tag{13}$$

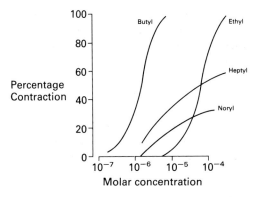

The ethyl- and butyl-trimethylammonium salts give parallel dose-response curves acting as agonists. The heptyl and nonyl analogues, however, produce curves with totally different shapes. This is interpreted as being the result of a gradual shift towards antagonism. After Stephenson, 1956.

Fig. A.1   Log dose response curve for contraction of guinea pig ileum by alkyltrimethylammomium ions.

where x_B is the concentration of antagonist and K_B its binding constant. If the ratio between these agonist concentrations giving the same response is defined as R, then:

$$R = 1 + \frac{x_B}{K_B} \tag{14}$$

and
$$\text{Log}\,(R - 1) = \text{Log}\,x_B - \text{Log}\,K_B \tag{15}$$

The Schild plot is log $(R-1)$ against $-\log x_B$ which will give a straight line with slope close to unity and a negative intercept on the x axis equal to $-\log K_B$ (see Figure A.2).

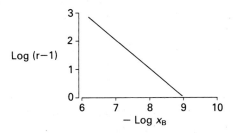

Antagonism by ligand B of a hypothetical agonist (A) at a specific receptor. The negative log of the binding constant for the antagonist is given by the intercept on the x axis (i.e. 10^{-9} M).

Fig. A.2 Schild plot.

The binding of a drug to a receptor normally depends on the dose in a linear fashion. The magnitude of the tissue response is, however, often a non-linear function of the drug concentration and reaches a maximum value, in fact a maximum tissue response can occur with only a fraction of the receptors occupied. Accordingly, it is often assumed that the tissue contains a finite number of receptors with which the drug can interact; the tissue response then depends on how many receptors are occupied by the drug which in turn depends on the affinity of the receptor for the drug.

The concept of efficacy is also useful here. Ligands are ranked on a scale from 0 to 1: an antagonist will bind to a receptor and elicit no tissue response at all, efficacy $= 0$, whereas a full agonist will be ranked at 1. Partial agonists will be ranked in between 0 and 1. These assumptions underlie much of what is discussed in this book, and have been found to hold up reasonably well in practice.

A.3 Ligand/protein binding

In order to complete this analysis, a third type of interaction must be considered — one where the binding of a ligand does not necessarily lead to any further effect. The binding of drugs to serum albumin comes into this class (Jusko and Gretch, 1976). In order to derive a mathematical formulation for the interaction a number of criteria have to be obeyed:
  (a) The binding sites must not interact with each other.
  (b) The binding must be specific and reversible.
  (c) A steady state must be reached, with equilibrium between free and bound ligand.

If we consider the simplest case of a macromolecule (M) with one binding site for one molecule of ligand L, we have the following equilibrium:

$$M + L \rightleftharpoons ML \qquad (16)$$

The dissociation constant for this equilibrium is given by:

$$K_d = \frac{[M]\,[L]}{[ML]} \qquad (17)$$

where square brackets denote concentrations and [L] is free ligand. If we consider the number of moles of ligand bound (r) to one mole of macromolecule we have:

$$r = \frac{[L]_{bound}}{[M]_{total}} = \frac{[ML]}{([M] + [ML])} \qquad (18)$$

whence
$$r = \frac{[L]}{(K_d + [L])} \qquad (19)$$

If we consider the more usual case of n identical but independent sites we have the following formula:

$$r = \frac{n\,[L]}{(K_d + [L])} \tag{20}$$

This can be rearranged in three ways to give derivations that should yield straight lines when the following graphs are drawn; (a) $1/r$ against $1/[L]$; (b) $r/[L]$ against r and (c) $[L]/r$ against $[L]$. The number of binding sites and the binding constant can be measured from the plot.

$$\text{(a)} \quad 1/r = 1/n + K_d/n\,[L] \tag{21}$$

$$\text{(b)} \quad r/[L] = n/K_d - r/K_d \tag{22}$$

$$\text{(c)} \quad n\,[L]/r = [L] + K_d \tag{23}$$

Each of these three derivations has been used to study ligand binding to macromolecules. The second is probably the best known and is attributed to Scatchard (1949). It is the best form mathematically, in that the plot does not rely heavily on measurements taken at low ligand concentrations which are the subject to the greatest error in determination. The other two plots divide by r which will magnify the effect of inaccuracies. Furthermore, as Scatchard himself noted, "double reciprocal plots tend to tempt straight lines where none exist". It is an interesting exercise to plot one set of data both by double reciprocal plot (equation 21) and by the Scatchard plot (equation 22) and compare the figures for n and K obtained.

One major use of the Scatchard plot is in studying hormone binding to receptor. The plot will also show up situations where more than one set of binding sites exist with different binding constants. The plot then gives two straight lines with different slopes connected by a boundary region where one curve blends into the other (see Figure A.3).

If we look at these equations in the light of enzyme kinetics and drug receptor binding, it is clear that there are considerable formal similarities — despite the fact that conversion to product occurs with an enzyme and pharmacological effects result from the binding of a ligand to a receptor, whereas no further change may result as a consequence of ligand binding to protein — for example the binding of drugs to serum albumin (Jusko and Gretch, 1976). The key similarity is that all of these interactions obey the Law of Mass Action.

Another method of plotting is due to Hill (1910) who derived an equation to explain the binding of oxygen to haemoglobin. This plot is of value when interaction between sites is suspected with the binding of one ligand to a macromolecule either facilitating (positive cooperativity) or hindering (negative cooperativity) the binding of a second. We take equation (18) and transform it into the Hill equation:

$$r/(n - r) = [L]/K_d \qquad (26)$$

We can call the fraction of the binding sites occupied (Z), where Z = r/n. Then the left-hand side of equation (24) becomes:

$$Z/(1 - Z) = [L]/K_d \qquad (27)$$

A plot of log $\{Z/(1-Z)\}$ against log [L] will give a straight line with a slope of 1 if there is no cooperativity. If there is cooperativity, however, the slope in the central portion of the curve will be greater than 1 for positive and less than 1 for negative cooperativity. This is because at low concentrations of drug the first binding site is being filled and at high concentrations the last; it is only in the middle that sites already filled can influence those unfilled (Figure A.4). In fact, a Scatchard plot will also indicate cooperativity by being curved.

It is interesting to note the number of double reciprocal plots used for enzyme kinetics whereas for ligand binding the Scatchard plot is usually favoured. This is not consistent but it may allow data to be made presentable. Even the Scatchard plot may produce straight lines that could be the consequence of obedience to a more complex equation as discussed by Klotz (1983).

A. One set of binding sites.

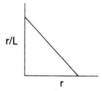

r is the number of moles of ligand bound to one mole of macromolecule, [L] is the free ligand concentration. The intercept on the *x*-axis is *n* (the number of binding sites). The slope is $- 1/K_d$.

B. Two sets of binding sites.

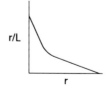

The intercept on the *x*-axis is now the sum of the number of binding sites at each of the two sets of site, $n_1 + n_2$ but the two slopes are not simply the negative reciprocals of the respective binding constants. They are rather more complex (see Feldman, 1972 for a mathematical analysis).

Fig. A.3 Scatchard plot.

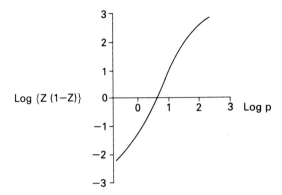

This is a Hill plot for haemoglobin in equilibrium with oxygen (*P* is the partial pressure of oxygen). The slope in the middle portion of the curve (the Hill coefficient) approximates to 2·8 whereas at both ends it is lower. The total number of binding sites for oxygen in haemoglobin is 4, but this figure is not reached because to derive the Hill equation infinite cooperativity is assumed. Since this is an ideal situation the Hill coefficient is always less than the actual number of sites and in practice indicates a minimum number, in this case 3. After Koshland (1970).

Figure A.4 Hill Plot.

References

Arunlakshana, O. and Schild, H. O. (1959). *Brit. J. Pharmacol.*, **14**, 48–58.
Bush, K. (1986). *Drugs Exptl. Clin. Res.*, **12**, 565–576.
Cheng, Y.-C. and Prusoff, W. H. (1973). *Biochem. Pharmacol.*, **22**, 3099–3108.
Coulson, C. J. and Smith, V. J. (1979). *Enzyme Microbial Technol.*, **1**, 193–196.
Feldman, H. A. (1972). *Analyt. Biochem.*, **48**, 317–338.
Hess, B., Boiteux, A. and Kruger, J. (1969). *Adv. Enzyme Regul.*, **7**, 149–167.
Hill, A. V. (1910). *J. Physiol.*, **40**, iv–vii.
Jusko, W. J. and Gretch, M. (1976). *Drug. Metab. Rev.*, **5**, 43–140.
Klotz, I. (1983). *Trends Pharmacol. Sci.*, **4**, 253–255.
Koshland, D. E. (1970). In: *The Enzymes*, Vol. 1, 3rd Ed., Boyer, P. D., Ed. (New York: Academic Press) p. 341.
Scatchard, G. S. (1949). *Ann. N.Y. Acad. Sci.*, **51**, 660–672.
Stephenson, R. P. (1956). *Brit. J. Pharmacol.*, **11**, 379–393.

Index